Iterations of Differential Operators

Iterations of Differential Operators

by

A. V. Babin
Moscow Institute of Railway Engineers,
Moscow, USSR

Translated from the Russian by
H. Zahavi

GORDON AND BREACH SCIENCE PUBLISHERS
New York • London • Paris • Montreux • Tokyo • Melbourne

Gordon and Breach Science Publishers

Post Office Box 786
Cooper Station
New York, New York 10276
United States of America

Post Office Box 161
1820 Montreux 2
Switzerland

Post Office Box 197
London WC2E 9PX
England

3-14-9, Okubo
Shinjuku-ku, Tokyo
Japan

58, rue Lhomond
75005 Paris
France

Private Bag 8
Camberwell, Victoria 3124
Australia

Library of Congress Cataloging-in-Publication Data

Babin, A. V.
Iterations of differential operators / A. V. Babin ; translated from the Russian by H. Zahavi.
p. cm.
Includes index.
ISBN 2-88124-707-5
1. Differential operators. 2. Iterative methods (Mathematics)
I. Title.
QA329.4.B33 1989
515.7'242--dc19

89-1708
CIP

Printed in Great Britain by Antony Rowe Ltd., Chippenham.

CONTENTS

ITERATIONS OF DIFFERENTIAL OPERATORS

FOREWORD

The basic theme of this book is the examination of the connections between solutions of differential equations and the iterations $A^j f$ of the differential operator A, which occurs in an equation, and which are used in the given function f, occurring in that equation. The first chapter discusses the Kotake-Narasimhan theorem, which establishes a connection between the rate of increase of the norms $\|A^f f\|$ of iterations of the elliptic operator A and the analyticity of the function u - the solution of the equation $Au=f$. Basically, however, this book describes the construction and examination of explicit formulae which express the solution u of different problems, containing the operator A, in terms of $A^j f$, $j = 0,1,\ldots$. The construction of these formulae is based on the systematic use of the methods of the theory of weighting approximations of the analytic functions of one complex variable using polynomials.

The formulae obtained are used to construct solutions of differential equations, and also to examine the properties of solutions of degenerating equations. There are also other applications to the problems of functional analysis and the theory of functions.

We shall describe the contents of the book in more detail.

The solutions of different problems for linear differential equations can be represented in the form of functions of a differential operator occurring in that equation. For example, the solution of the equation $Au-\zeta u=f$ is represented in the form $u=(A-\zeta I)^{-1}f$. The solution of the Cauchy problem $\partial_t u(t) = -Au(t)$, $u(0) = f$ can be written using the formula $u(t) = e^{-tA}f$. As is well known, we can use various methods to make sense of these formulae. For this we usually use either a spectral expansion of the operator A, or a construction of the function $g(A)$ using the resolvent of the operator A. In any case, the expression $g(A)f$ has a fully defined meaning for wide classes of differential operators A, functions $g(\lambda)$ and vectors f.

At the same time, the question of how to use $g(\lambda)$, A and f to construct $u=g(A)f$ is extremely important. This problem is solved in an elementary way if $g(A)=P(A)$ is a polynomial. In this case, to obtain $g(A)f$ it is sufficient to know how to obtain the iteration $A^j f$. If A is a differential operator with analytic coefficients, and f is an analytic function, it is not difficult, in theory, to obtain $A^j f$. Often (if the coefficients of the operator A and the function f are polynomials, for example), the corresponding calculations are quite simple. Moreover, in these cases it is also very difficult to calculate the functions $g(A)f$, for example of the form $(A-\zeta I)^{-1} f$ or $e^{-tA}f$, and we need to use finite-dimensional approximations of a different type to obtain $u=g(A)f$ (for example, finite-difference approximations or Galerkin approximations of the initial equation).

In this book we discuss another approach to calculating $g(A)f$. It consists of $g(A)f$ being represented in the form of a limit as $n\to\infty$ of the polynomials $P_n(A)f$ of the n-th degree from the operator A, applied to f. We were able to obtain similar representations for wide classes of differential operators of the first and second order with partial derivatives with analytic coefficients. The problem of constructing the polynomials $P_n(\lambda)$ turned out to be closely connected with the theory of weighting approximations of functions using polynomials on a half-line and on a line. This theory was established by Bernshtein (see his book [1]), and was then developed by a number of mathematicians (see the review articles by Akhiezer [1] and Mergelyan [1], and Mandel'broit's book [1]).

The theory of weighting approximations is developed in a new direction in the third chapter of this book. Namely, we basically consider explicitly constructing - for a specified analytic function $g(\lambda)$, a polynomial $P_n(\lambda)$, which approximates it with weight well on a straight line, and obtaining a fairly exact estimate of the error as $n \to \infty$.

In the fourth and sixth chapters we use the polynomials constructed in the third chapter to construct functions of operators in Hilbert and Banach spaces. We obtain polynomial representations of the form

$$g(A)f = \lim_{n \to \infty} P_n(A)f, \tag{1}$$

where P_n is a polynomial explicitly constructed according to the function g. The chapter contains broad classes of partial differential equations with analytic coefficients, to construct which we use (1).

A classic example of representing $g(A)f$ in the form (1) is the representation of the solution of Cauchy's hyperbolic problem $\partial_t^2 u(t) = -Au(t)$, $u(0)=f$, $\partial_t u(0)=0$, which is given in the Cauchy-Kowalewska theorem. In this case $P_n(A)$ is a finite segment of the series expansion $g(A)f = \cos(t\sqrt{A})f$ in powers of t for small $|t|$.

Using its special interpolation polynomials $P_n(\lambda) = P_n(t,\lambda)$ instead of the Taylor series of the function $g(\lambda) = g(t,\lambda) = \cos(t\sqrt{\lambda})$ enabled us explicitly to construct the solution $u(t)$ not only for small t, but also for as large t as desired. Formulae of the form (1) are also obtained for Cauchy's parabolic problem and for stationary equations.

§8 of the fourth chapter represents (1) in the form of an iteration scheme suitable for solving equations of the form $Au=f$ on a computer without recourse to finite-dimensional approximations. The chapter presents appropriate examples, and compares them with difference methods.

The explicit representations $g(A)f$ in terms of the iteration $A^j f$ of the operator using (1) enable us not only to construct $u=g(A)f$, but also to investigate the properties of the function $u=u(x)$.

The fifth chapter of the book thus analyses the properties of the function $u(x)=g(A)f(x)$ when A is a degenerating operator. In particular, estimates of the x-smoothness of the functions $u(x)$ and $u(x,t)$ of solutions of different types of degenerating differential equations are obtained. These equations have been the subject of thorough analysis (see, for example, the papers of Bony and Schapira [1-3], Moser [1], Cohn and Nirenberg [1], Oleinik [1], and Oleinik and Radkevich [1]).

Our use of the methods of the theory of functions enabled us to obtain, in a unique way, accurate estimates of the smoothness of different types of differential equations, and to establish the connection between the smoothness $u(x)=g(A)f(x)$ and the analytic properties $g(\lambda)$. In addition, we managed to include a number of cases that were not covered by previous papers, and to refine the known estimates of smoothness.

The second chapter examines the question of the existence of polynomial representations of the form (1) of solutions of the elliptic equations $Au=f$ with coefficients and right-hand sides from the Carleman classes $C(M(k))$ of infinitely differentiable functions. It turned out that this question was closely connected to the quasi-analyticity of the class $C(M(k))$. Namely, solutions u of equations with coefficients and the right-hand side from $C(M(k))$ are represented in the form (1) when, and only when, the class $C(M(k))$ is quasi-analytic.

In the seventh chapter, we discuss questions connected with expressing the solution of the nonlinear equation $F(u)=f$ in terms of $F^j(f)$, $j=0,1,\ldots$. This expression is treated in a generalised sense. Namely, it is required to obtain an expression of the value $h(u)$ of the functional h in the solution u of the equation $F(u)=f$ in terms of the value $h(F^j f)$ of this functional in iterations of the operator F, whilst it is required to obtain this expression for the complete system $\{h_i\}$ of functionals.

This problem is solved for a wide class of nonlinear elliptic second-order differential operators on the torus T^m. Its solution is based on the infinite-dimensional generalisation of Poincaré's theorem on the normal form of an analytic mapping. Namely, it is proved that we can construct in $W_p^4(T^m)$ a system of coordinates ξj, $j \in \mathbb{N}$, such that the effect of the operator F will take the form

$$F: \xi_j \to \lambda_j \xi_j, \; \xi_j \in \mathbb{R}, \; j \in \mathbb{N} \tag{2}$$

where λ_j are eigenvalues of the differential $F'(0)$ of the operator F in zero for the corresponding nonlinear differential operator F, $F(0)=0$, which acts in Sobolev's real space $W_p^4(T^m)$, $p > m$.

A mapping of the form (2) is the simplest form of mapping in an infinite-dimensional space. Thus, Poincaré's generalised theorem is, on the one hand, an analog of the theorem on the spectral expansion of linear self-adjoint differential operators to the nonlinear case. On the other, we can consider it a stationary analog of the theorem on the full integrability of evolution equations.

Chapters 2-7 basically contain a systematic discussion of the results obtained in [1-16]. This book demands of the reader a mastery of the fundamentals of analysis (the functions of a complex variable and the functions of many real variables), and also of the fundamentals of functional analysis. Those facts necessary for the discussion which exceed the minimum are formulated and provided in the references.

Chapters 3–6 form the core of the discussion, and can be read independently of the other chapters. If the reader is only interested in applications to differential equations, s/he can begin reading from the fourth chapter. The seventh chapter is completely autonomous (with the exception of §9), and the reader interested in nonlinear equations can begin reading from that chapter.

Each chapter contains formulae, theorems, lemmas and so on, which are numbered independently. If the need arises to refer to a formula or theorem from another chapter, the chapter number is added first. For example, a reference to Formula (3.5.7) is a reference to Formula (5.7) in the third chapter, the formula being in the fifth paragraph of that chapter.

The letter C with the same index can denote different constants in proofs and formulations of different theorems, lemmas and propositions, and different constants are numbered using different indices within the proof of one statement.

ITERATIONS OF DIFFERENTIAL OPERATORS

CHAPTER 1. ITERATIONS OF ELLIPTIC OPERATORS WITH ANALYTIC COEFFICIENTS

§1. Estimates of the norms of iterations of differential operators with analytic coefficients.

Suppose $K \subset \mathbb{R}^m$ is a compactum. We shall use $A(K)$ to denote a set of functions that are really analytic on K. Namely, $f \in A(K)$ if f is definite and infinitely differentiable in some neighbourhood Ω of the compactum K, whilst the constants C_0 and C_1 exist, such that $\forall x \in \Omega \subset \mathbb{R}^m$

$$|\partial^\alpha, f(x)| \leqq C_0 C_1^{|\alpha|} |\alpha|! \qquad \forall \alpha \in \mathbb{Z}_+^m \tag{1.1}$$

Here and below we will use the notation

$$\alpha = (\alpha_1, \ldots, \alpha_m), \quad \partial^\alpha = \partial_1^{\alpha_1} \ldots \partial_m^{\alpha_m}, \quad \partial_i = \partial / \partial x_i,$$
$$|\alpha| = \alpha_1 + \ldots + \alpha_m, \quad \alpha! = \alpha_1! \ldots \alpha_m!, \quad \xi^\alpha = \xi_1^{\alpha_1} \ldots \xi_m^{\alpha_m}, \qquad (\xi \in \mathbb{R}^m, \alpha \in \mathbb{Z}_+^m).$$

$\mathbb{Z}_+$ is a set of nonnegative integers, $\mathbb{Z}_+ = \mathbb{N} \cup \{0\}$.

We shall use $\mathcal{O}_\delta^C(K)$ to denote complex δ - neighbourhood of the compactum K

$$\mathcal{O}_\delta^C(K) = \{z \in \mathbb{C}^m : \exists\, x \in K : |z - x| < \delta\},$$
$$|z| = \max(|z_1|, \ldots, |z_m|). \tag{1.2}$$

We emphasise the difference $|\alpha|$ for $\alpha \in \mathbb{Z}_+^m$ from $|z|$ for $Z \in \mathbb{C}^m$ or $Z \in \mathbb{R}^m$.

Lemma 1.1 If K is a compactum and $f \in A(K)$ then the function $f(x)$ extends from K to the complex neighbourhood $\mathcal{O}_\delta^C(K)$, if $0 < \delta < C_1^{-1} m^{-1}$, C_1 is the same as in (1.1). The following estimate holds:

$$\sup_{z\in\mathcal{O}_\varepsilon^C}|f(z)|\leq\frac{C_0}{(1-mC_1\varepsilon)}\qquad 0<\varepsilon<\frac{1}{mC_1}. \tag{1.3}$$

Proof. According to (1.1), the function $f(x)$ decomposes in the neighbourhood of each point x into a Taylor series

$$f(x+y)=\sum 1/\alpha!\,\partial^\alpha f(x)y^\alpha. \tag{1.4}$$

The series converges to $f(x+y)$, since by virtue of (1.1) the residual term of Taylor's formula approaches 0 for small $|y|$. Using (1.1) and (1.4), we obtain that series (1.4) majorises using the series

$$\begin{aligned}&C_0\sum\frac{(\alpha_1+\cdots+\alpha_m)!}{\alpha_1!\cdots\alpha_m!}(C_1|y_1|)^{\alpha_1}\ldots(C_1|y_m|)^{\alpha_m}\\&=C_0(1-C_1|y_1|-\cdots-C_1|y_m|)^{-1}.\end{aligned} \tag{1.5}$$

It is obvious that series (1.5) converges, and consequently series (1.4) also converges for $|y| < 1/(mC_1)$. Estimate (1.3) follows from the estimate for the majorant (1.5). Series (1.4) also determines $f(x+y)$ for complex $y\in\mathbb{C}^m$. Therefore formula (1.4) determines the continuation $f(x)$ to $\mathcal{O}_\delta^C(K)$.

Henceforth we shall use $A(K, C_1, C_0)$ to denote the set of functions from $A(K)$, for which estimate (1.1) holds when $x\in K$ with fixed C_0 and C_1.

Lemma 1.2. If the function $f(x)$, which is defined on the compactum $K\subset\mathbb{R}^m$, continues to the function $f(z)$, which is analytic in the complex neighbourhood $\mathcal{O}_\delta^C(K)$ of the compactum K, whilst

$$|f(z)|\leq C_0(f)\qquad\text{when}\qquad z\in\mathcal{O}_\delta^C(K) \tag{1.6}$$

estimate (1.1) holds, where $C_0 = C_0(f)$, $C_1 = 2/\delta$.

Proof. When $x \in \mathcal{O}_{\delta/2}(K) = \mathcal{O}^c_{\delta/2}(K) \cap \mathbb{R}^m$ from Cauchy's integral formula we obtain

$$\frac{\partial^\alpha f(x)}{\alpha!} = \frac{1}{(2\pi i)^m} \int_{|x_1-\zeta_1|=r} \cdots \int_{|x_m-\zeta_m|=r} \frac{f(\zeta)\, d\zeta_1 \ldots d\zeta_m}{(\zeta-x)^\alpha(\zeta_1-x_1)\ldots(\zeta_m-x_m)}$$

where $r = \delta/2$. Estimating the integral on the right-hand side in absolute value using formula (1.6), we obtain that

$$|\partial^\alpha f(x)| \leqq \alpha! C_0(f) r^{-|\alpha|} \leqq |\alpha|! C_0(f)(2/\delta)^{|\alpha|}$$

whence follows (1.1).

Let us now consider, in the bounded domain $\Omega \subset \mathbb{R}^m$, the differential operators B of order p of the form

$$Bu = \sum_{|\beta| \leqq p} a_\beta \, \partial^\beta u \tag{1.7}$$

where $a_\beta \in A(\overline{\Omega})$.

Note 1.1. All the results formulated and proved in this chapter hold when a_β are matrices of order $\varkappa \times \varkappa$, $\varkappa > 1$, and u is a vector, $u = (u_1, \ldots, u_\varkappa)$ in (1.7).

Since the case $\varkappa > 1$ does not fundamentally differ from the case $\varkappa = 1$, but is more cumbersome, all the proofs will be carried out for the scalar case $\varkappa = 1$.

The following lemma estimates the increase in the norms of iterations of the operator B.

Lemma 1.3. Suppose the compactum $K_0 \subset \Omega$, $q \in \mathbb{N}$, $p \in \mathbb{N}$, $N \in \mathbb{N}$, the function $f \in C^{pN}(\Omega)$ and the derivatives f satisfy estimate

$$|\partial^\alpha f(x)| \leqq C_2 C_3^{|\alpha|} N^{|\alpha|+q} \quad \text{when} \quad |\alpha| \leqq Np, \quad x \in K_0 \tag{1.8}$$

Suppose B_j, $j = 1, \ldots, J$ are differential operators of the form (1.7) of order p_j, $p_j < p$ and the coefficients $a_{\beta j}$ of these operators belong to $A(K_0, C_1, C_0)$ where C_1 and C_0 do not depend on β and j. Suppose $0 \le j(i) \le J$ when $i = 1, \ldots, k$, $\sigma_k = p(j(1)) + \ldots + p(j(k))$. Then when $kp + |\alpha| \le Np$, $x \in K_0$ the following inequality holds:

$$|\partial^\alpha B_{j(1)} \ldots B_{j(k)} f(x)| \leqq C_2 C_5^k (C_6 N)^{|\alpha| + \sigma_k + q} \tag{1.9}$$

where we shall take $\max(2pC_1, C_3, 1)$, as the constant C_6, $C_5 = 2C_4C_0$ and C_4 is the number of terms in formula (1.7).

Proof. We shall carry out induction using k. When $k = 0$, inequality (1.9) follows from (1.8) if we assume $C_6 > C_3$. Suppose (1.9) holds when $k \le n-1$. We shall prove (1.9) when $k = n$. We will assume $x_{n-1} = B_{j(1)} \ldots B_{j(n-1)} f$. It follows from (1.9) that

$$|\partial^\alpha \partial^\beta \chi_{n-1}| \leqq C_2 C_5^{n-1} (C_6 N)^{|\alpha| + |\beta| + \sigma_{n-1} + q} \tag{1.10}$$

when $(n-1)p + |\beta| + |\alpha| < pN$. We shall now take $|\beta| < p(j(n))$ and shall consider the term $\alpha_{\beta j(n)} \partial^\beta$ from formula (1.7), which determines $B = B_j$. Since $|\beta| < p$, inequality (1.10) holds when $|\alpha| + np < pN$, whilst $|\beta| + \sigma_{n-1} < \sigma_n$. Using the fact that $a = a_{\beta j(n)}$ belongs to $A(K_0, C_1, C_0)$ and to estimate (1.10), by differentiating the product $a\partial^\beta x_{n-1}$ we obtain the estimate:

$$|\partial^\alpha (a\, \partial^\beta \chi_{n-1})| \leqq C_0 \sum_{l=0}^{|\alpha|} \frac{|\alpha|!}{l!(|\alpha| - l)!} C_1^l l! C_2 C_5^{n-1} (C_6 N)^{\sigma_n + |\alpha| - l + q}$$

$$= C_0 C_2 C_5^{n-1} (C_6 N)^{\sigma_n + |\alpha| + q} \sum_{l=0}^{|\alpha|} (C_1/(C_6 N))^l \frac{|\alpha|!}{(|\alpha| - l)!}$$

Since $|\alpha|!/(|\alpha| - l)! < |\alpha|^l$, and $|\alpha| < pN$, hence we obtain the estimate

$$|\partial^\alpha(a\,\partial^\beta\chi_{n-1})| \leqq C_0C_2C_5^{n-1}(C_6N)^{\sigma_n+|\alpha|+q}\sum_{l=0}^{|\alpha|}\left(\frac{pC_1}{C_6}\right)^l \tag{1.11}$$

Since $C_6 > 2pC_1$, the sum on the right-hand side of (1.11) does not exceed 2. Summing estimate (1.11), where $a = a_\beta$, $B = B_{j(n)}$, over β, and bearing in mind that the number of terms in (1.7) equals C_4, from (1.11) we obtain:

$$|\partial^\alpha B_{j(n)}\chi_{n-1}(x)| \leqq 2C_4C_0C_2C_5^{-1}C_5^n(C_6N)^{|\alpha|+\sigma_n+q}$$

Since $C_5 = 2C_4C_0$, hence we obtain inequality (1.9) when $k = n$, and the lemma is proved.

Theorem 1.1. Suppose $f \in A(K)$, K is a compactum, the operator B is determined using formula (1.7) and its coefficients belong to $A(K)$. Then R and C exist, such that

$$|\partial^\alpha B^k f(x)| \leqq CR^{-(pk+|\alpha|}(pk+|\alpha|!) \qquad \forall k \in \mathbb{Z}_+,\ \ x \in K \tag{1.12}$$

where $C = C_0(f)$ is the constant C_0 in estimate (1.1) for f, the number R depends on the constants C_0 and C_1 in estimate (1.1) where $f = a_\beta$, on C_4 - the number of terms in (1.7), on p and on the constant C_1 in estimate (1.1) for f.

Proof. According to Lemma 1.1 inequality (1.1) holds. Since $|\alpha|! < |\alpha|^{|\alpha|}$, the following estimate follows from (1.1) when $|\alpha| < pN$

$$|\partial^\alpha f| \leqq C_0C_1^{|\alpha|}|\alpha|! \leqq C_0C_1^{|\alpha|}(pN)^{|\alpha|} = C_0(pC_1)^{|\alpha|}N^{|\alpha|} \tag{1.13}$$

Hence it follows that inequality (1.8) holds, where $C_2 = C_0$, $q = 0$ for any $N \in \mathbb{N}$. We shall also use Lemma 1.3 when $B_j = B$. We will use $N = [k + |\alpha|/p] + 1$, [ ] is the integer part. Since $kp +$

$|\alpha| < Np$ for indicated N, inequality (1.9) holds, from which we obtain the estimate

$$\begin{aligned} |\partial^\alpha B^k f| &\leqq C_2 C_5^k C_6^{pk+|\alpha|}([k+|\alpha|/p]+1)^{pk+|\alpha|} \\ &\leqq C_2 C_5^k C_6^{pk+|\alpha|}(2[k+|\alpha|/p])^{pk+|\alpha|} \\ &\leqq C_2 C_5^k (2C_6)^{pk+|\alpha|}(pk+|\alpha|)^{pk+|\alpha|} \end{aligned}$$

According to Stirling's formula $n^n < 4^n(n!)$, therefore (1.12) holds, where $R = (8(C_5 + 1)(C_6)^{-1}$.

We shall denote the space of functions summed with the square in Ω by $H = L_2(\Omega)$, and the norm and scalar product in $L_2(\Omega)$ are specified by

$$\|u\|^2 = \int_\Omega |u(x)|^2\, dx, \quad (u, v) = \int_\Omega u(x)\bar{v}(x)\, dx \tag{1.14}$$

We shall introduce some notation to define the concept of the generalised solution $u \in L_2(\Omega)$ of the equation $Bu = f$.

The closure of the set of x for which $v(x) \neq 0$ is called the support supp v of the function $v(x)$. We shall denote the set of functions that have in Ω continuous derivatives of order not greater than k and whose supports are compact in Ω, by $C_0^k(\Omega)$. The norm in $C_0^k(\Omega)$ is specified in the usual way:

$$\|u\|_{C_0^k(\Omega)} = \max_{|\alpha| \leqq k} \sup_{x \in \Omega} |\partial^\alpha u(x)|/\alpha!$$

The set $C_0^k(\Omega)$ is everywhere dense in $L_2(\Omega)$ and since, obviously, $\|u\|_{L2(\Omega)} < \|u\|_{C_0^k(\Omega)}$ if Ω is bounded, the imbedding $C_0^k(\Omega) \subset H$ is continuous and everywhere dense.

We shall denote the spaces of linear functionals which are continuous in $C_0^k(\Omega)$ and H by $C_0^k(\Omega)'$ and H' respectively. Since $C_0^k(\Omega) \subset H$, then $H' \subset C_0^k(\Omega)'$. The linear functional $g(v)$,

$v \in H$ in the Hilbert space H can be uniquely represented in the form

$$\overline{(g_1, v)},\ g_1 \in H$$

Therefore H' is uniquely identified with H, and the inclusions

$$C_0^k(\Omega) \subset H \subset C_0^k(\Omega)'$$

hold, where the imbeddings are continuous.

Suppose B is an operator with coefficients that are infinitely differentiable in Ω, and is determined by formula (1.7). We determine the formally adjoint operator B^*,

$$B^* v = \sum_{|\beta| \leqq p} (-1)^{|\beta|} \partial^\beta (\bar{a}_\beta v) \tag{1.15}$$

Having differentiated the products, we can obviously rewrite the operator B^* in a form analogous to (1.7) (the coefficients of B^* and B naturally can differ). When $B^* = B$, the operator B is called formally self-adjoint.

Obviously, if $u \in C^p(\Omega)$, $v \in C_0^p(\Omega)$, after integrating by parts we obtain the equation

$$(Bu, v) = (u, B^* v) \tag{1.16}$$

The effect of the operator B on the function $u \in L_2(\Omega)$ is determined in the following way. If the function u is given the right-hand side of formula (1.16) determines the functional in $C_0^k(\Omega)$. The mapping $u \to Bu$ from $L_2(\Omega)$ in $C_0^k(\Omega)'$ is thus defined. Since $H = L_2(\Omega) \subset C_0^k(\Omega)'$, the functional Bu can be identical with the function f from $L_2(\Omega)$. In this case we shall say that u is a generalised solution from $L_2(\Omega)$ of the equation $Bu = f$. We shall denote the set $u \in L_2(\Omega)$ such that $B^k u \in L_2(\Omega)\ \forall\, k \in \mathbb{N}$, by $\mathcal{D}_\infty(B)$.

Proposition 1.1. If $Bu = f$, where $u \in L_2(\Omega)$, and f is infinitely differentiable in Ω, then $u \in \mathcal{D}_\infty(B)$.

Proof. Suppose $v \in C_0^{kp}(\Omega)$. Then

$$(B^k u, v) = (u, B^{*k} v) = (u, B^*(B^{*k-1} v)) = (f, B^{*k-1} v)$$

Using formula (1.16), we obtain that

$$(B^k u, v) = (B^{k-1} f, v) \tag{1.17}$$

i.e. the functional $B^k u$ continuously extends from C_0^{kp} to $L_2(\Omega)$ and therefore $B^k u \in L_2(\Omega)$.

We shall use $A_p(B)$ to denote the set of $u \in \mathcal{D}_\infty(B)$, for each of which the constants R and C exist, such that

$$\|B^k u\| \leqq C R^{-pk} (pk)!, \qquad \forall\, k \in \mathbb{Z}_+ \tag{1.18}$$

Proposition 1.2. If Ω is a bounded domain, $Bu = f$, where B is defined in (1.7), and

$$a_\beta \in \mathscr{A}(\bar{\Omega}),\ f \in \mathscr{A}(\bar{\Omega})$$

then $u \in A_p(B)$.

Proof. Since C_0^{kp} is everywhere dense in L_2, then

$$\|\varphi\| = \sup\{|(\varphi, v)| : v \in C_0^{kp},\ \|v\|_{L_2} = 1\} \qquad \varphi = B^k u$$

Assuming $\varphi = B^k u$, and using (1.17), we obtain that $\|B^k u\| < \|B^{k-1} f\|$. Hence, using Theorem 1.1, we obtain inequality (1.18).

The following paragraph proves the theorem that the analyticity of the function $u(x)$, follows from estimate (1.18), where B is an elliptic operator.

§2. Theorem on iterations of an elliptic operator.

Suppose B is determined using formula (1.7). The following function is called the principal symbol $B_{(p)}(x,\xi)$ of the operator B:

$$B_{(p)}(x,\xi)=\sum_{|\beta|=p} a_\beta(x)\xi^\beta, \qquad x\in\Omega, \quad \xi\in\mathbb{R}^m \tag{2.1}$$

The operator B is called elliptic on the compactum K, if

$$|B_{(p)}^{-1}(x,\xi)|\leqq C^{-1}|\xi|^{-p}, \qquad \forall x\in K, \quad \xi\neq 0 \tag{2.2}$$

We shall formulate the basic result of this paragraph - the theorem on iterations.

Theorem 2.1. Suppose the operator B is determined using formula (1.7), $a_\beta \in A(\bar{\Omega})$. Suppose $K\subset\Omega$, and the operator B is elliptic on K. Suppose $u\in A_p(B)$. Then u is analytic on K.

Remark 2.1. Theorem 2.1 was formulated by Kotake and Narasimhan [1]. There are many generalisations of this theorem (see, e.g., Baouendy and Goulaouic [1], Baouendy and Metivier [1], Bolley, Kamus, Mattera [1], and Lions and Magenes [1]).

The proof given here is based on Hormander's method, and is carried out in a similar way to that of the microlocal theorem on iterations given by Bolley, Kamus and Mattera [1].

We shall prove Theorem 2.1 at the end of the paragraph. We shall first prove some lemmas.

Let us first recall the properties of Fourier transforms. If $u(x)\in C_0^\infty(\mathbb{R}^m)$, its Fourier transform $\tilde{u}(\xi)$ is determined by the formula

$$\mathscr{F}u(\xi)=\tilde{u}(\xi)=\frac{1}{(2\pi)^m}\int_{\mathbb{R}^m} e^{-ix\cdot\xi}u(x)\,dx, \qquad x\cdot\xi=x_1\xi_1+\ldots+x_m\xi_m. \tag{2.3}$$

If the function $u(x)$ is finite, this integral converges for all complex ξ. Therefore when $u(x)\in C_0^\infty(\mathbb{R}^m)$, $\mathscr{F}u(\xi)$ is analytic in $\mathbb{C}^m$.

The following Parseval equality holds:

$$\int_{\mathbb{R}^m}\tilde{u}(\xi)\bar{\tilde{v}}(\xi)\,d\xi=\frac{1}{(2\pi)^m}\int_{\mathbb{R}^m}u(x)\bar{v}(x)\,dx \tag{2.4}$$

We have the following inversion formula:

$$u(x)=(\mathscr{F}^{-1}\tilde{u})(x)=\int_{\mathbb{R}^m}e^{ix\cdot\xi}\tilde{u}(\xi)\,d\xi \tag{2.5}$$

By virtue of Parseval's equality Fourier's transform continues with respect to continuity to $L_2(\mathbb{R}^m)$. Differentiating (2.5), we obtain for $u\in C_0^\infty(\mathbb{R}^m)$ that

$$(\partial^\alpha u)(x)=\mathscr{F}^{-1}((i\xi)^\alpha\mathscr{F}(u)(\xi))(x) \tag{2.6}$$

We shall formulate the sufficient condition of analyticity of the function $u\in L_2(\Omega)$.

Lemma 2.1. Suppose $u\in L_2(\Omega)$, Ω is bounded, K is a compactum, Ω_0 is a neighbourhood of K, and $K\subset\Omega_0\subset\Omega$. Suppose the sequence of function $\{u_N\}$ exists, possessing the following properties: 1) $u_N\in C_0^{pN}(\mathbb{R}^m)$; 2) $u_N=u$ at Ω_0; 3) $u_N=0$ outside Ω; 4) u_N are uniformly bounded in $L_2(\mathbb{R}^m)$; 5) the Fourier transforms of the functions u_N satisfy the estimate

$$|\tilde{u}_N(\xi)|\leqq CC_1^N N^{pN}|\xi|^{-pN}, \qquad N=1,2,\ldots, \tag{2.7}$$

where C, C_1 and p do not depend on N.

Then $u(x)$ is analytic on K : $u \in A(K)$.

Proof. Since the supports u_N are contained in Ω, we obtain, estimating the integral (2.3):

$$|\tilde{u}_N(\xi)| \leqq (2\pi)^{-m} \|u_N\| (\operatorname{mes}\Omega)^{1/2} = C_0$$

where C_0 does not depend on N by virtue of property 4. Using (2.7), from (2.5) and (2.6) we obtain

$$|\partial^\alpha u_N(x)| \leqq \int_{\mathbb{R}^m} |\xi|^{|\alpha|} |\tilde{u}_N(\xi)| \, d\xi \leqq \int_{|\xi| \leqq 1} |\tilde{u}_N(\xi)| \, d\xi$$
$$+ \int_{|\xi| \geqq 1} |\xi|^{|\alpha|} |\tilde{u}_N(\xi)| \, d\xi \leqq C' C_0 + \int_{|\xi| \geqq 1} |\xi|^{|\alpha| - pN} \, d\xi \, C C_1^N N^{pN}$$

We shall take N, such that

$$pN = |\alpha| + m + 1 + q, \; 0 \leqq q \leqq p - 1 \quad |\xi| \geqq 1$$

Then the integral with respect to $|\xi| > 1$ is finite and

$$|\partial^\alpha u_N(x)| \leqq C' C_0 + C_2 C_1^N N^{pN} \tag{2.8}$$

From the definition of N we obtain, assuming $q_1 = m + 1 + q$

$$(pN)^{pN} = (|\alpha| + q_1)^{q_1} (1 + q_1/|\alpha|)^{|\alpha|} |\alpha|^{|\alpha|} \leqq C_3^{|\alpha|} |\alpha|^{|\alpha|}$$

Using Stirling's formula, from this estimate and from (2.8) we obtain:

$$|\partial^\alpha u_N(x)| \leqq C_4 C_5^{|\alpha|} |\alpha|! \tag{2.9}$$

where C_4 and C_5 do not depend on α and N. Since $u_N(x)$ is identical with $u(x)$ when $x \in \Omega_0$, hence follows the statement of the lemma.

Lemma 2.2. The sequence of functions $\chi_n = \chi_n^1$ the class $C_0^n(\mathbb{R}^m)$ exists, satisfying the conditions:

$$\int_{\mathbb{R}^m} \chi_n^m(x)\,dx = 1, \quad \chi_n^m(x) = 0 \qquad |x| \geqq 1, \tag{2.10}$$

$$|\partial^\alpha \chi_n^m(x)| \leqq C C_0^{|\alpha|} n^{|\alpha|+m}, \qquad 0 \leqq \alpha \leqq n, \tag{2.11}$$

where C and C_0 depend only on m.

Proof. In the first stage, we shall construct $\chi_n = \chi_n^1$ in the one-dimensional case $x \in \mathbb{R}$, $m = 1$. We shall construct a Fourier transform of the functions χ_n. We will assume

$$\tilde{\chi}_n(\xi) = (2\pi)^{-1}(n+2)^{n+2} \sin^{n+2}(\xi/(n+2))\xi^{-(n+2)}. \tag{2.12}$$

Obviously, when ξ, $\eta \in \mathbb{R}$, $n \in \mathbb{N}$

$$|\chi_n(\xi + i\eta)| \leqq C e^{|\eta|}(1 + |\xi|^2 + |\eta|^2)^{-1}, \tag{2.13}$$

where $C = C(n)$. We shall use (2.5), where $u = \chi_n$, $m = 1$. Replacing the integration contour $\mathbb{R}$ by $\mathbb{R} + i\eta$ in (2.5), we obtain:

$$\chi_n(x) = e^{-\eta x} \int_{-\infty}^{\infty} e^{ix\eta} \tilde{\chi}_n(\xi + i\eta)\,d\xi. \tag{2.14}$$

Hence, using (2.13), we obtain that

$$|\chi_n(x)| \leqq C_1 e^{-\eta x} e^{|\eta|}, \ \forall \eta \in \mathbb{R}.$$

It follows from this inequality that $\chi_n(x) = 0$ when $|x| > 1$. Since

$$\tilde{\chi}_n(0)=(2\pi)^{-1},$$

by virtue of (2.3), where $\xi = 0$, $m = 1$, the integral of $\chi_n(x)$ equals 1. (2.10) thus holds.

We shall now obtain (2.11). Obivously, by virtue of (2.5), having made the substitution $\xi/(n+2) = y$, we will obtain:

$$\partial^\alpha \chi_n^1(x)=\frac{1}{2\pi}\int_{-\infty}^{\infty}(i\xi)^\alpha(n+2)^{n+2}\sin^{n+2}(\xi/(n+2))\xi^{-(n+2)}\,d\xi$$
$$=\frac{n+2}{2\pi}\int_{-\infty}^{\infty}(i(n+2)y)^\alpha\sin^{n+2}y\,y^{-(n+2)}\,dy.$$

Consequently, since $|\sin y/y| < 1$, $|\sin y| < 1$, then

$$|\partial^\alpha\chi_n^1(x)|\leqq(n+2)^{\alpha+1}\int_{-\infty}^{\infty}|\sin y|^2/y^2\,dy\leqq C(2n)^{\alpha+1}. \tag{2.15}$$

Estimate (2.11) thus holds when $m=1$ with the constant $C_0=2$.

Let us now consider the case $m>1$. We will assume

$$\chi_n^m=\chi_n^1(x_1)\times\ldots\times\chi_n^1(x_m). \tag{2.16}$$

The satisfaction of conditions (2.10) and (2.11) for $\chi_n(x)=\chi_n^m(x)$ with the constant $C_0 = 2^m$ follows directly from the satisfaction of these conditions for $\chi_n^1(x_j)$.

Lemma 2.3. The sequence χ_n of functions of the class $C_0^n(\mathbb{R}^m)$ satisfying the condition

$$\chi_n(x)=1 \quad \text{when} \quad x\in K, \qquad \chi_n(x)=0 \quad \text{when} \quad x\in\mathbb{R}^m\backslash\Omega \tag{2.17}$$

and also condition (2.11), where C and C_0 depend only on K, exists for any compactum $K\subset\Omega$.

Proof. Suppose ε is so small that 4ε - the neighbourhood - $\mathcal{O}_{4\varepsilon}(K)$ of the compactum K is contained in Ω. We shall use $\chi(x)$ to denote the characteristic function of the domain $\mathcal{O}_{\varepsilon}(K)$, $\chi(x)=1$ when $x\in \mathcal{O}_{\varepsilon}(K)$, $\chi(x)=0$ when $x\in \mathbb{R}^m\setminus\mathcal{O}_{\varepsilon}(K)$. Suppose $\chi_n^m(x)$ is a sequence of functions satisfying (2.10) and (2.11). We will assume

$$\chi_n(x)=\varepsilon^{-m}\int_{\mathbb{R}^m}\chi_n^m\left(\frac{x-y}{\varepsilon}\right)\chi(y)\,dy. \tag{2.18}$$

It is obvious that (2.17) follows from (2.10), and estimate (2.11) for χ_n follows from estimate (2.11) for χ_n^m.

The following lemma plays an important role in the proof of Theorem 2.1.

Lemma 2.4. Suppose B is determined by (1.7),

$$\alpha_\alpha\in\mathscr{A}(\bar{\Omega}),\qquad n\in\mathbb{N},\ \varphi_n(x)=\chi_{4p2n}(x),$$

where $p\in\mathbb{N}$, and the function χ_n satisfies (2.17) and (2.11). Then the functions $w_n(x,\xi)$ and $e_n(x,\xi)$ - finite in Ω - of the class $C_0^{pn}(\Omega)$ exist with respect to x, $\forall\,\xi\in\mathbb{R}^m\setminus 0$ and satisfy the equation

$$B^n(e^{-ix\cdot\xi}w_nB_{(p)}^{-n}(x,-i\xi))=e^{-ix\cdot\xi}(\varphi_n(x)-e_n(x,\xi)) \tag{2.19}$$

where $B\ (p\ )$ is determined by formula (2.1). The functions e_n and w_n satisfy the following estimates when $|\xi| > C_0 n$:

$$|e_n(x,\xi)|\leqq C^n n^{pn}|\xi|^{-pn} \tag{2.20}$$

$$|w_n(x,\xi)|\leqq C^n, \tag{2.21}$$

where C_0 and C do not depend on x, ξ, n.

Proof. Suppose w is an arbitrary smooth function of x. We shall apply the differential operator B to the function

$$e^{-ix\cdot\xi}w(x)B_{(p)}^{-1}(x,-i\xi),$$

which depends on the parameter ξ, $\xi \neq 0$. It is obvious that the result will equal the sum of expressions of the form

$$a'_\alpha(x)\,\partial^\alpha e^{-ix\cdot\xi}\,\partial^\beta(w(x)B_{(p)}^{-1}(x,-i\xi)), \qquad |\alpha|+|\beta|\leqq p,$$

where $a'_\alpha(x)$ are functions which are analytic with respect to x. It is obvious that the function

$$e^{ix\cdot\xi}\,\partial^\alpha e^{-ix\cdot\xi}=(-i\xi)^\alpha$$

is homogeneous with respect to ξ of order $|\alpha|$ and the function $B^{-1}{}_{(p)}(x,\xi)$ is homogeneous of order $-p$ with respect to ξ. Grouping the terms with identical orders of homogeneity with respect to ξ, and bearing in mind that

$$B\,e^{-ix\cdot\xi}$$

has a p-th order component, homogeneous with respect to ξ, equal to

$$B_{(p)}(x,-i\xi)\,e^{-ix\cdot\xi},$$

we obtain

$$B(e^{-ix\cdot\xi}wB_{(p)}^{-1}(x,-i\xi))=e^{-ix\cdot\xi}(w-Qw), \tag{2.22}$$

where

$$Qw = Q_1 w + \cdots + Q_p w. \tag{2.23}$$

Here Q_j is a differential operator with respect to x of an order no higher than j, and the coefficients Q_j are analytic with respect to x and homogeneous with respect to ξ of the order $-j$.

We will assume

$$w_n = B_p^n \sum_{\sigma_1 \leq pn} B_p^{-1} Q^{k_1} \ldots B_p^{-1} Q^{k_n} \varphi_n. \tag{2.24}$$

Here and below in the proof of the lemma we use the notation

$$\sigma_j = k_j + k_{j+1} + \cdots + k_n, \quad B_p = B_{(p)}(x, -i\xi). \tag{2.25}$$

We will assume

$$e_n = -((I - Q)B_p)^n B_p^{-n} w_n + \varphi_n. \tag{2.26}$$

The functions w_n and e_n satisfy (2.19). Indeed, by virtue of (2.22)

$$(I - Q)B_p v = e^{ix\cdot\xi} B\, e^{-ix\cdot\xi} v, \quad \forall v = v(x).$$

We can therefore rewrite (2.26) in the form

$$e_n = -e^{ix\cdot\xi} B^n e^{-ix\cdot\xi} B_p^{-n} w_n + \varphi_n,$$

which is equivalent to (2.19).

The definitions of (2.24) and (2.26) have the following meaning. If we assume $e_N = 0$ in (2.26), the formal solution of

(2.26), considered as an equation with respect to w, has the form

$$w = B_p^n \left(B_p^{-1} \sum_{k=0}^{\infty} Q^k \right)^n \varphi_n.$$

Obviously, formula (2.24) defines w_n as a partial sum of this series, and instead of obtaining $e_N = 0$ we will obtain e_N which satisfies estimate (2.20).

We shall now proceed to prove inequality (2.20). We shall prove that e_N can be represented in the form

$$e_n = \sum_{j=1}^{n} ((1-Q)B_p)^{n-j} \sum_{\sigma_j = pn} Q^{k_j+1} B_p^{-1} Q^{k_{j+1}} \ldots B_p^{-1} Q^{k_n} \varphi_N. \tag{2.27}$$

Indeed, according to (2.24) and (2.26)

$$-e_n + \varphi_n = ((I-Q)B_p)^n \sum_{\sigma_1 \leqq pn} B_p^{-1} Q^{k_1} \ldots B_p^{-1} Q^{k_n} \varphi_n. \tag{2.28}$$

Since

$$\sum_{\sigma_j \leqq pn} Q^{k_j+1} B_p^{-1} Q^{k_{j+1}} \ldots B_p^{-1} Q^{k_n} - \sum_{\sigma_j = pn} Q^{k_j+1} B_p^{-1} Q^{k_{j+1}} \ldots B_p^{-1} Q^{k_n}$$

$$= \sum_{\sigma_j \leqq pn} Q^{k_j} \ldots B_p^{-1} Q^{k_n} - \sum_{\sigma_{j+1} \leqq pn} Q^{k_{j+1}} \ldots B_p^{-1} Q^{k_n},$$

the following identity holds:

$$\begin{aligned} &((I-Q)B_p)^{n-j+1} \sum_{\sigma_j \leqq pn} B_p^{-1} Q^{k_j} \ldots B_p^{-1} Q^{k_n} \\ &= ((I-Q)B_p)^{n-j} \left[\sum_{\sigma_{j+1} \leqq pn} B_p^{-1} Q^{k_{j+1}} \ldots B_p^{-1} Q^{k_n} \right. \\ &\left. - \sum_{\sigma_j = pn} Q^{k_j+1} B_p^{-1} Q^{k_{j+1}} \ldots B_p^{-1} Q^{k_n} \right]. \end{aligned} \tag{2.29}$$

Transforming the right-hand side of (2.28) using (2.29) and successively taking $j=1,2,\dots,n$, we obtain (2.27).

We shall represent Q in the form of a sum using formula (2.23) and shall expand the right-hand side of (2.27) into the sum of monomials of the form

$$G=Q_{i1}B_p\cdots Q_{i_{n-j}}B_p[Q_{l_{j,1}}\cdots Q_{l_{j,k_j+1}}] \\ \times[B_p^{-1}\cdots Q_{l_{j+1},k_{j+1}}]\cdots[B_p^{-1}Q_{l_{n,1}}\cdots Q_{l_{n,k_n}}]\varphi_n. \tag{2.30}$$

We make an upper estimate of the number of such monomials. The number of different representations of the number pn in the form of the sum $\sigma_j = k_j + \dots + k_n$ does not exceed the number of combinations of $(pn+n)$, n at a time, and obviously does not exceed 2^{pn+n}. Further, each term on the right-hand side of formula (2.27) contains not more than $pn+n$ multipliers of the form Q or $(I-Q)$. Since Q has the form (2.23), each such term decomposes into the sum of not more than $(p+1)^{(p+1)n}$ monomials of the form (2.30). Bearing in mind the estimate of the number of terms in (2.27), we obtain that e_n decomposes into the sum of not more than $n(2p+2)^{(p+1)n}$ monomials G of the form (2.30).

Let us now proceed to the estimate $|G| = |G(x, \xi)|$. $G(x, \xi)$ obviously equals the result of applying the differential operators Q_j and the multiplication operators to the function B_p and B_p^{-1} to the function φ_N. These operators have coefficients that are analytic with respect to x, and these coefficients belong to $A(\overline{\Omega}, C_1, C)$ when $|\xi| = 1$ for some C_1 and C, that do not depend on ξ and j. By virtue of (2.11), where pN is represented instead of n, the function $\varphi_n = \chi_{4p^2n}$ satisfies estimate (1.8), where $N=4pn$, $q=m$.

We shall use Lemma 1.3 to estimate $|G|$, where we shall take the differential operators Q_i, and also the multiplication operators to the function B_p and B_p^{-1} as *Bj*, and we shall take φ_n as *f*.

When $|\xi|=1$, we obtain from inequality (1.9), where $\alpha = 0$, $N = 4pn$,

$$|G| \leqq C_4 C_5^k (4C_6 pn)^{\sigma+m}, \tag{2.31}$$

where k is the number of multipliers in (2.30), and $k \leq 4n$. The number σ equals the sum of orders of homogeneity with respect to $|\xi|^{-1}$ of the operators Q_j and it forms an upper bound of the sum of orders of the operators Q_j:

$$\sigma = i_1 + \cdots + i_{n-j} + [l_{j,1} + \cdots + l_{j,k_j+1}] \\ + [l_{j+1,1} + \cdots + l_{j+1,k_{j+1}}] + \cdots + [l_{n,1} + \cdots + l_{n,k_n}]. \tag{2.32}$$

Obviously, since $\sigma_j = pn$ in (2.27), and the operators $((1-Q)B_p)^{n-j}$ have an order no higher than pn, $\sigma \leqslant 2pn$. It follows from (2.27) that for each j there are $n-j$ multipliers B_p and B_p^{-1} in (2.30) - the sum order of homogeneity of which with respect to ξ equals zero. Therefore the sum order of homogeneity $G(x, \xi)$ with respect to ξ equals $-\sigma$, and from (2.31) (where $|\xi|=1$), follows the estimate for any $|\xi| \neq 0$:

$$|G(x,\xi)| \leqq C_4 C_5^{4n} (C_0 n)^{\sigma+m} |\xi|^{-\sigma}, \tag{2.33}$$

where $C_0 = 4C_6 p$. The order of homogeneity $l_{j,k}$ of the operator $Q_{l_{j,k}}$ is not less than 1, therefore $\sigma \geqslant pn$ by virtue of (2.32). When $|\xi| \geqslant C_0 n$ we obtain from (2.33):

$$|G(x,\xi)| \leqq C_4 C_5^{4n}(C_0 n)^{pn+m}|\xi|^{-pn}(C_0 n/|\xi|)^{\sigma-pn}$$
$$\leqq C_7^n (C_0 n)^{pn}|\xi|^{-pn}. \tag{2.34}$$

Using the estimate of the number of monomials $G(x, \xi)$ in the expansion e_n obtained above, from (2.34) we obtain estimate (2.20).

Let us now procced to the proof of estimate (2.21). Substituting the expression Q into the right-hand side of (2.24) using formula (2.23), we obtain the expansion w_n into the sum of monomials G_1 of a form that is analogous to (2.30). The number of such monomials does not exceed $n2^{pn}p^{pn}$. Using Lemma 1.3 in the same way as when estimating $|G|$, we will obtain the following inequality in a similar way to (2.33):

$$|G_1| \leqq C_4'(C_7')^n (C_0 n)^{\sigma+m}|\xi|^{-\sigma}.$$

Hence when $|\xi| > C_0 n$ we obtain estimate (2.21). The finiteness of e_n and w_n follows from the finiteness of φ_n and from 2.4 is proved.

Proof of Theorem 2.1. We will assume $u_n(x) = u(x)\varphi_n(x)$, where φ_n is the same as in Lemma 2.4. According to (2.3)

$$\tilde{u}_n(\xi) = (2\pi)^{-m} \int_{\mathbb{R}^m} \varphi_n(x) u(x) e^{-ix\cdot\xi}\, dx. \tag{2.35}$$

We shall use Lemma 2.4, where we shall take the operator B^* as the operator B. Since an operator that is adjoint to an elliptic operator is obviously elliptic, and the coefficients B^* are analytic, Lemma 2.4 is applicable.

Using (2.19), where we have B^* instead of B, substituting an expression for

$$e^{ix\cdot\xi}\varphi_n(x)$$

into (2.35) from (2.19) we obtain

$$\begin{aligned}\tilde{\bar{u}}_n(\xi)&=(2\pi)^{-m}\int_{\mathbb{R}^m}\bar{u}(x)B^{*n}(e^{ix\cdot\xi}w_nB^{*-n}_{(p)}(x,i\xi))\,dx\\&+(2\pi)^{-m}\int_{\mathbb{R}^m}\bar{u}(x)e_n(x,-\xi)\,e^{ix\cdot\xi}\,dx.\end{aligned}\tag{2.36}$$

Since

$$u\in\mathscr{A}_p(B),\ v=e^{ix\cdot\xi}w_nB^{*-n}_{(p)}\in C_0^{pn}(\Omega),$$

using (1.16) from (2.36) we obtain:

$$\begin{aligned}|\tilde{u}_n(\xi)|\leqq C\Big[&\|B^nu\|\sup_x|w_n(x,\xi)B^{*-n}_{(p)}(x,i\xi)|\\&+\|u\|\sup_x|e_n(x,-\xi)|\Big].\end{aligned}\tag{2.37}$$

Estimating $|w_n|$ and $|e_n|$ using inequalities (2.20) and (2.21), and

$$B^{*-n}_{(p)}(x,i\xi)$$

using (2.2), and bearing in mind that suppe_n and suppe_n are compact (i.e. e_n and w_n are finite) in Ω, and also (2.2), when $|\xi| > C_0n$ we obtain:

$$|\tilde{u}_n(\xi)|\leqq C[\|B^nu\|C^nc^{-n}|\xi|^{-pn}+\|u\|C^nn^{pn}|\xi|^{-pn}].$$

Using estimate (1.18), when $|\xi| > C_0n$ we obtain hence

$$|\tilde{u}_n(\xi)|\leqq C_1C_2^nn^{pn}|\xi|^{-pn}.\tag{2.38}$$

We shall now estimate $|\tilde{u}_n(\xi)|$ when $|\xi| < C_0n$. It follows from (2.11), where $\alpha = 0$, that $\varphi_n = \chi_{4p^2n}$ satisfies the inequality $|\varphi| <$

C_3n^m. Therefore, bearing in mind the compactness of suppφ_n in Ω, we obtain from (2.35) that

$$\|\tilde{u}_n(\xi)\| \leqq C_4 n^m \|u\|.$$

Therefore, when $|\xi| < C_0 n$

$$|\tilde{u}_n(\xi)| \leqq C_4 n^m |\xi|^{pn} |\xi|^{-pn} \leqq C_4 n^m (C_0 n)^{pn} |\xi|^{-pn}. \tag{2.39}$$

It follows from (2.38) and (2.39) that estimate (2.7) holds. By virtue of Lemma 2.1 $u(x)$ is analytic in $\overline{\Omega}_0$, and Theorem 2.1 is proved.

§3. The analyticity of the solutions of nonstationary equations.

In this paragraph we obtain theorems on the x-analyticity of solutions of some parabolic and hyperbolic equations as applications of Theorem 2.1.

We shall first briefly introduce some facts from the theory of self-adjoint operators which we shall need later. See, for example, Dunford and Schwartz [2], Yosida [1], and Kato [1].

Suppose L is a self-adjoint operator with the domain of definition $\mathcal{D}(L)$ in the Hilbert space H. Suppose L is semi-bounded from below:

$$(Lv, v) \geqq b\|v\|^2, \ Lv \in \mathcal{D}(L).$$

Then the spectrum of L is arranged on the semiaxis $\lambda > b$. We shall use E_λ to denote the family of projectors in H on to the invariant subspaces of the opeator L, which forms the resolution of unity for the operator L

$$Lf = \int_b^\infty \lambda \, dE_\lambda f, \qquad \forall f \in \mathscr{D}(L). \tag{3.1}$$

If $g(\lambda)$ is a function that is continuous on the semiaxis $\lambda > b$, the function $g(L)$ is with respect to the operator L:

$$g(L)f = \int_b^\infty g(\lambda)\, dE_\lambda f. \tag{3.2}$$

The value of the scalar product $(g(L)f, \varphi)$ equals the Stieltjes integral of the continuous function $g(\lambda)$ with respect to the function of the bounded variation $E_\lambda f,\varphi)$. The following formula holds:

$$\|g(L)f\|^2 = \int_b^\infty |g(\lambda)|^2 \, d(E_\lambda f, f), \tag{3.3}$$

which expresses the norm $g(L)f$ in terms of the integral of the function $|g(\lambda)|^2$ using the Lebesgue-Stieltjes measure produced using the monotonic bounded function $(E_\lambda f, f)$. The domain of definition of the operator $g(L)$ is given by the formula

$$\mathscr{D}(g(L)) = \{f \in H : \|g(L)f\| < \infty\}, \tag{3.4}$$

where $\|g(L)f\|$ is determined in (3.3). Henceforth we will denote $\bigcap_k \mathcal{D}(L^k)$ by $\mathcal{D}_\infty(L)$.

Consider the abstract parabolic equation

$$\partial_t u(t) = -Lu(t) + f_0 \tag{3.5}$$

with the initial condition

$$u(0) = f_1. \tag{3.6}$$

The function $u(t)$ of the numerical argument t with values in H, which is t-continuous when $t \geqslant 0$, is continuously differentiable with respect to t when $t > 0$, $u(t) \in \mathcal{D}(L)$ when $t > 0$ and Eq.(3.5) holds when $t > 0$, and the initial condition (3.6) when $t = 0$, is the solution of this equation.

Proposition 3.1. The solution (3.5), (3.6) exists and is unique when f_0, $f_1 \in H$ and is determined using the formula

$$u(t) = g_{10}(L) f_0 + g_{11}(L) f_1, \tag{3.7}$$

where

$$g_{11}(\lambda) = e^{-t\lambda}, \qquad g_{10}(\lambda) = (1 - e^{-t\lambda})/\lambda. \tag{3.8}$$

This solution is t-analytic in H when $\operatorname{Re} t > 0$.

Proof. To verify the uniqueness, consider the integral

$$\int_0^T \partial_t \|u(t)\|^2 \, dt = 2 \operatorname{Re} \int_0^T (u, \partial_t u) \, dt \quad = 2 \operatorname{Re} \int_0^T (-Lu, u) \, dt + 2 \operatorname{Re} \int_0^T (f_0, u) \, dt$$

$$\leqq T \|f_0\|^2 + \int_0^T (1 + 2|b|) \|u(t)\|^2 \, dt.$$

Hence when $t \leq t_0$ we obtain

$$\|u(t)\|^2 \leqq t_0 \|f_0\|^2 + \|f_1\|^2 + \int_0^t \|u(\tau)\|^2 \, d\tau. \tag{3.9}$$

We recall Gronwall's inequality. Suppose $x(t)$ and $v(t)$ are non-negative functions that are continuous in $[a, b]$, and $C \geq 0$ is a constant. Suppose

$$x(t) \leqq C + \int_a^t v(s) x(s) \, ds, \qquad a \leqq t \leqq b.$$

Then

$$x(t) \leqq C \exp \int_a^t v(s)\, ds.$$

Assuming

$$x(t) = \|u(t)\|^2, \quad v(t) = 1 + 2|b|, \quad C = t_0\|f_0\|^2 + \|f_1\|^2$$

by virtue of Gronwall's inequality we obtain from (3.9) such that when $f_0 = 0$ and $f_1 = 0$ the solution equals zero when $0 \leqslant t \leqslant t_0$.

We shall verify the validity of formula (3.7). For brevity, we shall assume $f_0 = 0$. Then

$$(u(t), \varphi) = \int_b^\infty e^{-t\lambda}\, d(E_\lambda f, \varphi).$$

Since $e^{-t\lambda}$ and the derivative $-\lambda e^{-t\lambda}$ of $e^{-t\lambda}$ with respect to t is bounded when $\lambda \geqslant b$, $\operatorname{Re} t \geqslant \tau_0 > 0$, and all the continuous bounded functions are integrable over $d(E_\lambda f,\varphi)$, it is possible to differentiate under the integral sign. Therefore $(\partial_t u(t), \varphi) = -\int_b^\infty \lambda e^{-t\lambda}\, d(E_\lambda f, \varphi)$ $(t>0)$. At the same time

$$(Lu(t), \varphi) = \int_b^\infty \lambda e^{-t\lambda}\, d(E_\lambda f, \varphi) \qquad (t>0).$$

Hence it follows that $(\partial_t u(t) + Lu(t), \varphi) = 0$ when $t > 0$, $\forall \varphi \in H$, i.e. Eq.(3.5) holds. In order to verify (3.6), note that by virtue of (3.3)

$$\|u(t) - u(0)\|^2 = \int_b^\infty (e^{-t\lambda} - 1)^2\, d(E_\lambda f, f).$$

Since $(e^{-t\lambda}-1)^2 \leqslant C$ when $t \geqslant b$, and $(e^{-t\lambda}-1) \to 0$ when $t \to 0$, then according to Lebesgue's theorem the integral approaches zero. Firstly, therefore, $u(t)$ is continuous with respect to t, and secondly, $u(0) = f$.

Let us now consider the abstract hyperbolic equation:

$$\partial_t^2 u(t) = -Lu(t) + f_0 \tag{3.10}$$

with the initial conditions

$$u(0) = f_1, \quad \partial_t u(0) = f_2. \tag{3.11}$$

The function $u(t)$, which is doubly continuously differentiable in H with respect to t when $t \in \mathbb{R}$, $u(t) \in \mathcal{D}(L)$ when $t \in \mathbb{R}$, and satisfies (3.10) is the solution of Eq.(3.10).

Proposition 3.2. Suppose

$$f_0 \in H, \quad f_1 \in \mathcal{D}(L), \qquad f_2 \in \mathcal{D}(\sqrt{L+bI}).$$

Then problem (3.10), (3.11) has a unique solution. This solution has the form

$$u(t) = g_{20}(A)f + g_{21}(A)f_1 + g_{22}(A)f_2, \tag{3.12}$$

where

$$g_{20}(\lambda) = (\cos(t\sqrt{\lambda}) - 1)/\lambda, \quad g_{21}(\lambda) = \cos(t\sqrt{\lambda}),$$
$$g_{22}(\lambda) = \sin(t\sqrt{\lambda})/\sqrt{\lambda}. \tag{3.13}$$

Proof. Integrating $\partial_t \|\partial_t u\|^2$ from O to T and using Gronwall's inequality, as in Proposition 3.1 we obtain that the solution (3.10), (3.11) is unique. Direct verification, analogous to that carried out in Proposition 3.1, shows that (3.12) determines the solution of problem (3.10), (3.11).

Below we shall take as the operator L a self-adjoint extension of the differential operator B.

Let us formally consider, in the bounded domain $\Omega \subset \mathbb{R}^m$ with the piecewise smooth boundary $\partial\Omega$, the self-adjoint operator B:

$$Bu = \sum_{|\alpha| \leqq \nu} (-1)^{|\alpha|} \partial^\alpha (a_\alpha \partial^\alpha u), \tag{3.14}$$

where $a_\alpha(x)$ are real functions belonging to $A(\overline{\Omega})$. It is obvious that the operator B, which is determined using formula (3.1), is symmetric in the set $C_0^\infty(\Omega)$, which, for brevity, we shall put $\mathcal{D}$, $\mathcal{D} = C_0^\infty(\Omega)$:

$$(Bu, v) = (u, Bv), \tag{3.15}$$

where (u,v) is a scalar product in $(L_2(\Omega) = H$. We shall require that the condition of lower boundedness of the operator B holds:

$$(Bu, u) \geqq b\|u\|^2, \qquad \forall u \in \mathcal{D}. \tag{3.16}$$

We shall give Hording's inequality, which gives, in particular, the sufficient condition for the satisfaction of (3.16). (See, for example, Bers, John and Schechter [1] for the proof.)

Theorem 3.1. If the operator B, determined by (3.14), is elliptic in Ω, i.e. $\forall x \in \overline{\Omega}$

$$\sum_{|\alpha| = \nu} a_\alpha(x) \xi^{2\alpha} \geqq C|\xi|^{2\nu}, (c > 0), \qquad \forall \xi \in \mathbb{R}^m, \tag{3.17}$$

then $\exists\ C_1 > 0$ such that

$$(Bu, u) \geq C_1 \sum_{|\alpha| \leq \nu} \|\partial^\alpha u\|^2 - C_2 \|u\|^2, \qquad \forall u \in C_0^\infty(\Omega). \tag{3.18}$$

Corollary 3.1. If B is elliptic in $\overline{\Omega}$, inequality (3.3) holds.

If the differential operator B satisfies conditions (3.15) and (3.16) in some set $\mathcal{D}$ of infinitely differentiable functions, Friedrichs' self-adjoint extension exists in it. We shall denote this extension by $B_{\mathcal{D}}$, and if what $\mathcal{D}$ equals is clear from the context, we shall denote it by B. We recall that the domain of definition $B_{\mathcal{D}}$ is a restriction of the operator B^*, which is adjoint to B, to the subspace H_0, which is the closure $\mathcal{D}$ using the norm $(u, u)_0 = (Bu + bu + u, u)$.

Let us now consider problems (3.5), (3.6) and (3.10), (3.11), where $L = B_{\mathcal{D}}$, B is determined using formula (3.14) and satisfies (3.16).

Remark 3.1. If $\mathcal{D} = C_0^\infty(\Omega)$, and the operator B is elliptic on $\partial\Omega$, it follows from the condition $u \in \mathcal{D}(B_\infty)$ that all the derivatives of $u(x)$ of an order from 0 to $\nu-1$ vanish on $\partial\Omega$. (See, for example, Lions and Magenes [1], vol.1, for the proof.) This indicates that the solutions of the above problems satisfy zero boundary conditions. Other boundary conditions can arise with another choice of sets $\mathcal{D}$.

Remark 3.2. If $\mathcal{D} \supset C_0^\infty(\Omega)$, the action of the operator $B_{\mathcal{D}}^k$ on $\mathcal{D}_\infty(B_{\mathcal{D}})$ agrees with that of the operator B^k. (See §1 after Formula (1.16) for a definition of the action of the operator B on functions from $L_2(\Omega)$.) Indeed, if $\varphi \in C_0^\infty(\Omega)$ and

$$u \in \mathcal{D}_\infty(B_{\mathcal{D}}) \subset L_2(\Omega),$$

then

$$(Bu, \varphi) = (u, B^*\varphi) = (u, B\varphi) = (u, L\varphi) = (Lu, \varphi).$$

Since C_0^∞ is everywhere dense in H, then $Lu = Bu$. It follows hence that

$$L^k u = B^k u, \ \forall k \in \mathbb{Z}_+.$$

Theorem 3.2. Suppose the operator B is determined using Formula (3.14) and satisfies conditions (3.15) and (3.16) in some set of functions $\mathcal{D}$, containing $C_0^\infty(\Omega)$. Suppose $u(t)$ is the solution of problem (3.5), (3.6), where

$$L = B_{\mathcal{D}}, \quad f_0 \in \mathcal{D}_\infty(L), \quad f_1 \in H = L_2(\Omega).$$

Suppose $f_0 \in A(\overline{\Omega})$ and the operator B is elliptic in the compactum $K \subset \Omega$. Then for each $t > 0$ $u(t) = u(t, x)$ is analytic in K.

Proof. Suppose $t > 0$. According to Formula (3.7) $u(t) = u_0 + u_1$. It is obvious that $u_1 = e^{-tA} f_1 \in \mathcal{D}_\infty(A)$.
Using (3.3) and Remark 3.2, we obtain:

$$\begin{aligned} \|B^k u_1\|^2 &= \|L^k u_1\|^2 = \|L^k e^{-tL} f_1\|^2 \\ &\leqq \sup_{\lambda \geqq b} |\lambda^k e^{-t\lambda}|^2 \|f_1\|^2 \leqq C_0^2 C_1^{2k} (k!)^2 \|f_1\|^2. \end{aligned} \tag{3.18}$$

We shall use Theorem 2.1, where $\Omega = \Omega_1$ (the domain containing K, and such that $\Omega_1 \subset \Omega$, B is elliptic in $\overline{\Omega}_1$, is denoted by Ω_1). Since, by virtue of (3.18), $u_1 \in A_1(B)$, according to Theorem 2.1 $u_1 \in A(K)$.

Let us now consider the function $u_0 = g_{10}(L) f_0$ where g_{10} is determined by (3.8). Obviously, since $f_0 \in \mathcal{D}_\infty(L)$, then

$$\|B^k u_0\| = \|L^k g_{10}(L) f_0\| \leqq C \|L^k f_0\| = C \|B^k f_0\|. \tag{3.19}$$

Since $f_0 \in A(\overline{\Omega})$, then by virtue of Theorem 1.1 $f_0 \in A_{2\nu}(B)$ and, according to (3.19), also $u_0 \in A_{2\nu}$. Using Theorem 2.1 we obtain that by virtue of the ellipticity of B on K the function $u_0 \in A(K)$.

Theorem 3.3. Suppose the operator B satisfies the conditions of Theorem 3.2. Suppose $u(t)$ is the solution of problem (3.10), (3.11), where $L=B_{\mathscr{D}}$, and the functions f_0, f_1 and f_2 belong to $\mathcal{D}_\infty(L)$ and simultaneously belong to $A(\bar{\Omega})$. Suppose the operator B is elliptic on the compactum $K \subset \Omega$. Then for any $t \subset \mathbb{R}$ $u(t, x)$ are analytic with respect to x in K. The function $u(t, 0)$ is analytic with respect to t in $L_2(\Omega)$.

Proof. It is obvious that it is sufficient to prove the analyticity of $u_i = g_i(A)f_i$ on K, where $g_i(A)$ is determined by (3.13), $i=0, 1, 2$. Since the functions g_i are bounded, and

$$f_i \in \mathscr{D}_\infty(A), \quad f_0 \in \mathscr{A}(\bar{\Omega}),$$

the analyticity of $u_i(x)$ on K is proved in completely the same way as that of the analyticity of u_0 in Theorem 3.2. We shall prove the analyticity of $u(t, \cdot)$ with respect to t. For simplicity we shall take $f_0 = 0$, $f_2 = 0$. Obviously, by virtue of (3.12) and (3.13) from (3.3) we obtain:

$$\|\partial_t^j u(t)\|^2 = \int_b^\infty |\lambda|^j \cos^2(t\sqrt{\lambda})d(E_\lambda f, f) \leqq \int_b^\infty |\lambda|^j d(E_\lambda f, f).$$

Therefore for even j

$$\|\partial_t^j u(t)\| \leqq \|L^{j/2} f\| = \|B^{j/2} f\|.$$

Using Theorem 1.1, where $p = 2$, $\alpha = 0$, $k = j/2$, for even j we obtain

$$\|\partial_t^j u(t)\| \leqq CR^{-j} j!.$$

For odd j we obtain the above estimate after integration with respect to t $\partial_t^{j+2}u$.

Remark 3.3. Eq.(3.10) is hyperbolic in the usual sense when $A = B$ is a second-order operator.

Remark 3.4. In Chapter 4 we shall give examples of second-order operators for which

$$\mathscr{D}_\infty(L) \cap \mathscr{A}(\bar{\Omega}) = \mathscr{A}(\bar{\Omega}).$$

CHAPTER 2. THE POLYNOMIAL SOLVABILITY OF SELF-ADJOINT DIFFERENTIAL EQUATIONS WITH INFINITELY SMOOTH COEFFICIENTS

In the previous chapter, the analyticity of $u = B^{-1}f$ was derived from estimates of the increase in the iteration norms $\|B^k f\|$.

In this chapter we examine the question of the possibility of explicitly expressing the solution of the elliptic equation $Bu=f$ in terms of the iteration $B^k f$, $k \in \mathbb{Z}_+$. It is assumed that B is a second-order operator on the torus $T^m = \mathbb{R}^m/(2\pi\mathbb{Z})^m$ with coefficients and a right-hand side from Carleman's class $C(M(k))$ of infinitely differentiable functions. The basic result of the chapter is the theorem that the explicit expression of $u = B^{-1}f$ in terms of $B^k f$ exists when, and only when, the class $C(M(k))$ is quasi-analytical (see Theorem 21).

§1. The polynomial sovability of equations in a Hilbert space.

Suppose A is a self-adjoint operator in the Hilbert space H with the domain of definition $\mathcal{D}(A)$, and suppose A is non-negative

$$(Av, v) \geqq 0, \qquad \forall v \in \mathcal{D}(A). \tag{1.1}$$

Consider the equation

$$Au + \rho^2 u = f, \tag{1.2}$$

where $\rho > 0$, $f \in \mathcal{D}_\infty(A)$. It is obvious that the solution of this equation is given by the formula

$$u = g^+(A)f, \quad g^+(\lambda) = (\lambda + \rho^2)^{-1}. \tag{1.3}$$

Definition 1.1. Eq.(1.2) is called polynomially solvable in H if

$$u = \lim_{n\to\infty} P_n(A)f, \quad P_n(A) = \sum_{j=0}^{n} C_{nj}A^j, \tag{1.4}$$

where $\{P_n(A)\}$ is some sequence of polynomials. Formula (1.4) is called a polynomial representation of u.

To obtain (1.4), it is required to estimate $\|g(A)f - P_n(A)f\|$.

Lemma 1.1. Suppose the function $\Phi(\lambda)$ is continuous when $\lambda > 0$, $\Phi(\lambda) > C_0 > 0$. Suppose $f \in \mathcal{D}(\Phi(A))$, $f \in \mathcal{D}_\infty(A)$. Suppose $g^+(\lambda)$ is a function that is bounded and continuous when $\lambda > 0$. Then

$$\|g^+(A)f - P_n(A)f\| \leqq \mu^+(P_n, g^+, \Phi)\|\Phi(A)f\|, \tag{1.5}$$

where

$$\mu^+(P_n, g^+, \Phi) = \sup_{\lambda \geqq 0} \frac{|g^+(\lambda) - P_n(\lambda)|}{\Phi(\lambda)}. \tag{1.6}$$

Proof. According to Formula (1.3.3)

$$\begin{aligned}\|g(A)f - P_n(A)f\|^2 &= \int_0^\infty \frac{|g^+(\lambda) - P_n(\lambda)|^2}{\Phi^2(\lambda)} \Phi^2(\lambda)d(E_\lambda f, f) \\ &\leqq (\mu^+(P_n, g^+, \Phi))^2 \int_0^\infty \Phi^2(\lambda)d(E_\lambda f, f) \\ &= [\mu^+(P_n, g^+, \Phi)\|\Phi(A)f\|]^2,\end{aligned}$$

whence (1.5) directly follows.

Thus to prove the existence of the polynomial representation (1.4) it is required to obtain the function Φ, such that the sequence of polynomials P_n exists, such that $\mu^+(g^+, P_n, \Phi) \to 0$ as $n\to\infty$. The problem of the conditions on Φ which guarantee the existence of these polynomials is examined in Bernshtein's theory of the weighting approximation of functions. We shall present the results of this theory that we need.

Suppose $\Phi(\lambda)$ when $\lambda > 0$ is a continuous positive function, satisfying the condition

$$\ln \Phi(e^t) \quad \text{is convex downwards} \tag{1.7}$$

We shall now formulate the Bernshtein-Mandel'broit theorem on weighting functions (see Mandel'broit [1]).

Theorem 1.1. Suppose the function $\Phi(\lambda)$ satisfies condition (1.7). Suppose, in addition,

$$\int_1^\infty \frac{\ln |\Phi(x^2)|}{x^2} dx = \infty. \tag{1.8}$$

Then for any function $\varphi(x)$ that is continuous in $\mathbb{R}$ and satisfies the condition

$$\lim_{x \to \infty} \frac{\varphi(x)}{\Phi(x^2)} = 0,$$

and for any $\varepsilon > 0$ the polynomial $P(x)$ exists, such that

$$|\varphi(x) - P(x)| < \varepsilon \Phi(x^2), \qquad \forall x \in \mathbb{R}. \tag{1.9}$$

The function $\Phi(x^2)$, which has the property formulated in the statement of Theorem 1.1, is called a weighting function on $\mathbb{R}$.

Remark 1.1. If the function $\varphi(x)$ is even, the polynomial $P(x)$ in (1.19) can also be assumed to be even. Indeed, if φ is even, then

$$\varphi(x) = \tfrac{1}{2}(\varphi(x) + \varphi(-x)).$$

By virtue of (1.9)

$$|\varphi(x)-P(x)|\leqq\varepsilon\Phi(x^2),\quad |\varphi(-x)-P(-x)|\leqq\varepsilon\Phi(x^2),$$
$$|\varphi(x)-\tfrac{1}{2}(P(x)+P(-x))|=\tfrac{1}{2}|\varphi(x)-P(x))+(\varphi(x)-P(-x))|$$
$$\leqq\tfrac{1}{2}|\varphi(x)-P(x)|+\tfrac{1}{2}|\varphi(-x)-P(-x)|\leqq\varepsilon\Phi(x^2).$$

Thus, instead of the polynomial $P(x)$ in (1.9) we can take the even polynomial $1/2(P(x) + P(-x))$ for even φ.

The basic result of this paragraph is easily derived from Theorem 1.1.

Theorem 1.2. Suppose $\Phi(\lambda)$, $\lambda > 0$ satisfies conditions (1.7), (1.8). Suppose $f \in \mathcal{D}(\Phi(A))$. Then Eq.(1.2) is polynomially solvable, i.e. the representation (1.4) holds.

Proof. We shall use Lemma 1.1, where we will assume $g^+(\lambda) = (\rho^2 + \lambda)^{-1}$. We will assume

$$\varphi(x)=(\rho^2+x^2)^{-1}=g^+(x^2).$$

According to Theorem 1.1, the sequence of even polynomials $P_n(x)$ exists, such that

$$|\varphi(x)-P^0_{2n}(x)|/\Phi(x^2)\leqq\varepsilon_n,$$

as $n\to\infty$. Assuming $x = \sqrt{\lambda}$, and using the fact that $P^0_n(x)$ are even, and

$$\varphi(\sqrt{\lambda})=g^+(\lambda),$$

we obtain that

$$\mu^+(P_n, g^+, \Phi)\to 0 \qquad \text{as } n\to\infty. \tag{1.10}$$

Using (1.10), we obtain from (1.5) that (1.4) holds.

Lemma 1.2. Suppose A_1 is a self-adjoint operator, which is semi-bounded from below. Suppose there is only a finite number of points $\lambda_1,\dots,\lambda_s$ of the spectrum of the operator A_1 on the straight line $\lambda < \rho^2$, where $\rho > 0$, and suppose zero does not belong to the spectrum A_1. Suppose u is the solution of the equation $A_1 u = f$, where $f \in \mathcal{D}_\infty(A)$. Then the polynomial $Q_0(A)$ of the S-1-th degree exist, such that

$$u = Q_0(A_1)f + u_1, \tag{1.11}$$

where u_1 is the solution of the equation

$$A_1 u_1 = f_1, \quad f_1 = (I - A_1 Q_0(A_1))f \equiv Q_1(A_1)f, \tag{1.12}$$

whilst u_1 and f_1 lie in the closed subspace $H_1 \subset H$, which is invariant with respect to A_1, whilst

$$(A_1 v, v) \geqq \rho^2(v, v), \qquad \forall v \in H_1.$$

Proof. We shall use $Q_0(\lambda)$ to denote Lagrange's interpolation polynomial of the function λ^{-1} with the interpolation nodes $\lambda_1,\dots,\lambda_s$. We will assume $u_1 = u - Q_0(A_1)f$. Obviously, u_1 satisfies Eq.(1.12). Since the projectors E_λ when $\lambda < \rho^2$ only have points of increase when $\lambda = \lambda_1, \dots, \lambda = \lambda_s$,

$$E_{\rho^2-\varepsilon} Q_1(A_1) = \int_b^{\rho^2-\varepsilon} (1 - \lambda Q_0(\lambda))\, dE_\lambda = 0.$$

We will assume $H_1 = (I - E_{\rho^2-\varepsilon})H$, where ε is fairly small. It is obvious that the spectrum of the contraction of A_1 to H_1 lies in the domain $\lambda > \rho^2$. It is also obvious that $f_1 = Q_1(A_1)f \in H_1$, and $u_1 = A_1^{-1} f_1 \in H_1$.

Lemma 1.3. Suppose the conditions of Lemma 1.2 hold and the polynomial representation (1.4), where $u = u_1$, $A = A_1$, $f = f_1$ holds for the solution of the equation $A_1u_1 = f_1$. Then the polynomial representation holds for u (with the other polynomials P_n).

Proof. If $P_n(A_1)f_1 \to u_1$, then $u = Q_0(A_1)f + u_1$ is represented in the form

$$u=\lim_{n\to\infty}[P_n(A_1)Q_1(A_1)f+Q_0(A_1)f],$$

which is obviously equivalent to Formula (1.4) with $A=A_1$.

§2. The classes $C(M(k))$ of infinitely smooth functions and the class of equations $E(M(k))$.

Suppose $M(k)$, $k = 0, 1, 2, \ldots$ is a monotonically increasing sequence of positive numbers. We shall consider the sequences, such that the number C_M exists (depending on the sequence), such that the following inequality holds:

$$\frac{M(k)}{k!}\frac{M(s)}{s!}\leqq C_M\frac{M(k+s)}{(k+s)!},\qquad (k,s\in\mathbb{N}). \tag{2.1}$$

Suppose T^m is an m-dimensional torus,

$$T^m=\mathbb{R}^m/(2\pi\mathbb{Z})^m.$$

Suppose $C^\infty(T^m)$ is a set of functions that are infinitely differentiable in $\mathbb{R}^m$, and that are periodic with respect to each variable with the period 2π.

Definition 2.1. If $M(k)$ is a sequence of numbers, then $C(M(k)) = C(M)$ is a set of functions $v(x) \in C^\infty(T^m)$ such that for each of them the numbers r and Λ exist, such that the following condition will hold:

$$\max_{x\in\Omega\,|\alpha|=k} |\partial^\alpha v(x)| \leqq \Lambda r^k M(k), \qquad k=0,1,\ldots, \tag{2.2}$$

where $\Omega = [-\pi, \pi]^m$. We shall denote the set of functions which satisfies estimate (2.2) with fixed Λ and r by $C(M(k), r, \Lambda)$.

Remark 2.1. Condition (21) is imposed in order to guarantee the closure of the class $C(M)$ with respect to multiplication (see Lemma 3.1 in §3).

Remark 2.2. As follows from (1.1.1), $C(k!)$ is identical with the set of analytic functions $A(T^m)$. It follows from condition (21) that

$$C(k!) \subset C(M(k)). \tag{2.3}$$

Indeed, assuming $s = 1$ in (2.1), we obtain:

$$M(k+1) \geqq \frac{(k+1)!}{C_M}\frac{M(1)M(k)}{k!} = \frac{M(1)}{C_M}M(k)(k+1). \tag{2.4}$$

Assuming $C = M(1)/C_M$, using induction we derive hence that

$$M(k) \geqq C^{k-1}M(1)k!. \tag{2.5}$$

Remark 2.3. The Gevrey classes of infinitely differentiable functions are the classes $C(M(k))$, for which $M(k) = k^{\mu k}$, $\mu > 1$. These classes satisfy condition (2.1). Indeed

$$\frac{k^{\mu k}}{k!}\frac{s^{\mu s}}{s!}\frac{(k+s)!}{(k+s)^{\mu(k+s)}} = \frac{(k+s)!}{k!\,s!}\left(\frac{k}{k+s}\right)^{\mu k}\left(\frac{s}{k+s}\right)^{\mu s}$$
$$\leqq \left[\left(\frac{k}{k+s}\right)^{\mu} + \left(\frac{s}{k+s}\right)^{\mu}\right]^{k+s} \leqq 1,$$

whence follows (2.1) with $C_M = 1$.

Remark 2.4. The Denjoy classes are the most well-known quasi-analytic classes of functions. For these classes

$$M(k)=(k\ln k\ln\ln k\ldots\ln\ldots\ln k)^k, \qquad (k\geqq k_0),$$

where k_0 is so great that all the logarithms are defined. When $0 < k < k_0$ $M(k) = M(k_0)$. The Denjoy classes satisfy condition (2.1). We shall verify this in the case $M(k) = (k\ln k)^k$. In this case

$$\frac{(k\ln k)^k}{k!}\frac{(s\ln s)^s}{s!}\frac{(k+s)!}{[(k+s)\ln(k+s)]^{k+s}}$$

$$=\frac{(k+s)!}{k!\,s!}\left(\frac{k\ln k}{(k+s)\ln(k+s)}\right)^k\left(\frac{s\ln s}{(k+s)\ln(k+s)}\right)^s \tag{2.6}$$

$$\leqq\left(\frac{k\ln k}{(k+s)\ln(k+s)}+\frac{s\ln s}{(k+s)\ln(k+s)}\right)^{k+s}.$$

Simple verification shows that

$$k\ln k+s\ln s\leqq(k+s)\ln(k+s) \tag{2.7}$$

when $k > 1$, $s > 1$. It follows from (2.6) and (2.7) that condition (2.1) holds with $C_M = 1$.

Consider the class of functions $C(M(k))$, where when $k \in \mathbb{N}$ $M(k))$ satisfies condition

$$M(k+1)\leqq C_0C_1^kM(k), \tag{2.8}$$

where C_0, C_1 are some constants.

Remark 2.5. If condition (2.8) holds, then $C(M(k)) = C(M(k+1))$, and the class $C(M(k))$ is closed with respect to differentiation. Indeed, if v satisfies (2.2), then $\partial_i v$ satisfies the inequality

$$|\partial^\alpha\partial_i v(x)|\leqq\Lambda r^{k+1}M(k+1)\leqq r\Lambda C_0(C_1r)^kM(k), \qquad (|\alpha|=k),$$

i.e. $\partial_j v \in C(M(k))$.

Remark 2.6. The Gevrey and Denjoy classes satisfy condition (28). For the Gevrey classes this follows from the inequality

$$M(k+1)/M(k)=(k+1)^{\mu(k+1)}k^{-\mu k}=(k+1)^{\mu}(1+1/k)^{\mu k}\leqq(k+1)^{\mu}e^{\mu}\leqq C_0 2^k$$

for some C_0.

The following inequality holds for the Denjoy class in the case $M(k) = (k\ln k)^k$ when $k\geq 2$:

$$M(k+1)/M(k)=(k+1)\ln(k+1)\left[\frac{(k+1)\ln(k+1)}{k\ln k}\right]^k$$

$$\leqq(k+1)\ln(k+1)(1+1/k)^k(1+1/\ln k)^k$$
$$\leqq 2(k+1)\ln(k+1)e2^k.$$

Inequality (28) is verified in a similar way for the case of a Denjoy class of general form.

Consider in T^m the second-order elliptic differential equation

$$Bu\equiv-\sum_{i,j=1}^{m}\partial_i(a_{ij}(x)\,\partial_j u(x))+a_{00}(x)u(x)=f(x), \tag{2.9}$$

where

$$a_{ij}(x),\quad f(x)\in C(M(k))\subset C^{\infty}(T^m),\quad a_{ij}(x)=\bar{a}_{ji}(x)$$

and the following condition of ellipticity holds:

$$\sum_{i,j=1}^{m}a_{ij}(x)\xi_i\xi_j\geqq C_0|\xi|^2,\qquad (C_0>0),\forall\xi\in\mathbb{C}^m, \tag{2.10}$$

and also the condition of invertibility

$$\ker B=0. \tag{2.11}$$

Obviously, the operator B is defined in the set $\mathscr{D}=C^{\infty}(T^m)$, whilst conditions (1.3.15) and (1.3.16) hold. Henceforth we shall also denote the Friedrichs extension $B_{\mathfrak{D}}$ of the operator B by B.

Definition 2.2. The equation $Bu=f$ of the form (2.9) belongs to the class $E(M(k))=E(M)$, if its coefficients and right-hand side f belong to $C(M)$, and conditions (2.10) and (2.11) hold and the operator B is invertible, ie. $\ker B = 0$ and the operator B^{-1} is bounded in $L_2(T^m)$.

Definition 2.3. The equation $Bu=f$ of the form (2.9) with coefficients and a right-hand side from $C^{\infty}(T^m)$ is called polynomially solvable, if its solution $u \in \mathfrak{D}(B)$ exists and is represented in the form

$$u=\lim_{n\to\infty} P_n(B)f, \tag{2.12}$$

where $P_n(B)$ is a polynomial of a degree no higher than n of the operator B, and we understand the limit using the norm $L_2(T^m)$.

Remark 2.7. If $c \in \mathbb{C}$, the representation (2.12) is equivalent to a representation of the form

$$u=\lim_{n\to\infty} P_n^0(A)f, \qquad A=B-cI, \qquad c\in\mathbb{C}, \tag{2.13}$$

where $P_n^0(\lambda)$ is a polynomial of a degree no higher than n, since with the substitution $B=A+cI$ a polynomial of the n-th degree with respect to B becomes a polynomial of the n-th degree, and vice versa.

We shall formulate the basic result of this chapter.

Theorem 2.1. Suppose $M(k)$ is a sequence which satisfies conditions (2.1) and (2.8). For all the equations from the class $E(M(k))$ to be polynomially solvable, it is necessary and sufficient that the class $C(M(k))$ is a quasi-analytic class of functions.

Remark 2.8. For simplicity, we shall confine ourselves to the case of equations of the form (2.9) in a torus. Similarly, we can consider equations in a bounded domain of the type considered in Examples 3.2, 3.3 or 4.2 of Chapter 4 (see Babin [12]) and also systems of equations.

We shall prove Theorem 2.1 in the following paragraphs.

§3. Proof of the necessity of quasi-analyticity for the polynomial solvability of equations.

Here we shall prove a rather stronger statement than that on necessity in Theorem 1.1. Namely, the following theorem holds.

Theorem 3.1. Suppose the coefficients of the operator B are fixed functions from $C^\infty(T^m)$, whilst conditions (2.10), (2.11) hold. Suppose the equation $Bu = f$ is polynomially solvable for any $f \in C(M(k))$. Then $C(M(k))$ is a quasi-analytic class of functions.

The necessity for the condition of quasi-analyticity can be easily seen from the following. The expression of the solution $u(x)$ using Formula (2.12) is local, i.e. the right-hand side of (2.12) for x lying in some neighbourhood w of the point x_0, by virtue of the locality of the differential operator B, does not depend on the values $f(x)$ for $x \notin w$. The nonquasi-analytic class $C(M(k))$ contains finite functions. Therefore we manage to change the function $f(x)$ outside w in the nonquasi-analytic case, without changing it in w. Together with the values $u(x)$, the solutions of the equation $Bu=f$ in w depend on the behaviour of f not only in w, but also in $T^m \backslash w$, and this contradicts the locality of Formula (2.12).

The proof of Theorem 3.1 is based on the following two lemmas. We shall first prove the lemma on the estimate of the product of the functions from $C(M(k))$.

Lemma 3.1. Suppose $s > 0$ is a fixed number,

$$u \in C(M(s+k), R_u, \Lambda_u),\ v \in C(M(k), r_v, \Lambda_v),$$

whilst $R_u > R_v$ (the classes $C(M(k), r, \Lambda)$ are described in Definition 2.1). Then

$$uv \in C(M(s+k), R_u, \Lambda_{uv}), \tag{3.1}$$

where

$$\Lambda_{uv} = \Lambda_u \Lambda_v C_M (1 - R_v/R_u)^{-1}. \tag{3.2}$$

Proof. By virtue of the Leibnitz formula

$$|\partial^\alpha uv| \leqq \sum_{k=0}^{|\alpha|} \frac{|\alpha|!}{(|\alpha|-k)!\,k!} \max_{|\beta|=|\alpha|-k} |\partial^\beta u| \max_{|\gamma|=k} |\partial^\gamma v|.$$

Using the definition $C(M(k), R, \Lambda)$, we derive hence and from (2.2)

$$|\partial^\alpha uv| \leqq \Lambda_u \Lambda_v \sum_{k=0}^{|\alpha|} \frac{|\alpha|!}{(|\alpha|-k)!\,k!} R_u^k M(k+s) R_v^{|\alpha|-k} M(|\alpha|-k). \tag{3.3}$$

Bearing in mind that when $k < |\alpha|$ and $s > 0$

$$|\alpha|!/k! \leqq (|\alpha|+s)!/(k+s)!,$$

we derive from (3.3) that

$$|\partial^{|\alpha|} uv| \leqq \Lambda_u \Lambda_v \sum_{k=0}^{|\alpha|} \frac{M(k+s)}{(k+s)!} \frac{M(|\alpha|-k)}{(|\alpha|-k)!} R_v^{|\alpha|-k} R_u^k (|\alpha|+s)!. \tag{3.4}$$

Substituting $k + s$ into condition (2.1) instead of k and $|\alpha| - k$ instead of s we obtain:

$$\frac{M(k+s)}{(k+s)!} \frac{M(|\alpha|-k)}{(|\alpha|-k)!} \leqq C_M \frac{M(|\alpha|+s)}{(|\alpha|+s)!}. \tag{3.5}$$

Using this inequality to estimate the terms on the right-hand side of (3.4), from (3.4) we derive:

$$|\partial^\alpha uv| \leqq C_M \Lambda_u \Lambda_v \sum_{k=0}^{|\alpha|} R_v^{|\alpha|-k} R_u^k M(|\alpha|+s)$$
$$\leqq C_M \Lambda_u \Lambda_v R_u^{|\alpha|} M(|\alpha|+s) \sum_{k=0}^{|\alpha|} \left(\frac{R_v}{R_u}\right)^{|\alpha|-k}$$
$$\leqq C_M \Lambda_u \Lambda_v R_u^{|\alpha|} (1 - R_v/R_u)^{-1} M(|\alpha|+s),$$

whence, comparing it with (2.2), we obtain (3.1) and (3.2).

Lemma 3.2. If $C(M(k))$ is not a quasi-analytic class of functions, the function $\varphi(x) \in C(M(k))$ exists, satisfying the two conditions:

$$\varphi(x) = 1 \quad \text{when} \quad |x| \leqq \tfrac{1}{2} \tag{3.6}$$

and

$$\varphi(x) = 0 \quad \text{when} \quad |x| \geqq 1. \tag{3.7}$$

Proof. According to Theorem IV of Chapter IV of Mandel'broit's book [1], if the class $C(M(k))$ is not quasi-analytic, the function $\varphi_0(x)$ of the one variable x exists, defined in $\mathbb{R}$, which satisfies the condition

$$\varphi_0(0) > 0, \quad \varphi_0(x) \geqq 0, \qquad \forall x \in \mathbb{R}, \tag{3.8}$$

the condition

$$\varphi_0(x) = 0 \quad \text{when} \quad |x| > \tfrac{1}{8}, \tag{3.9}$$

and belongs to the class $C(M(k))$ in $\mathbb{R}$. Multiplying φ_0 by the corresponding constant, we obtain the function φ_1, which satisfies

(3.9), and the integral of which equals 1 over $\mathbb{R}$. We will assume

$$\varphi_2(x_1) = \int_{-3/4}^{3/4} \varphi_1(x_1 - \xi)\, d\xi. \tag{3.10}$$

It is obvious that $\varphi_2(x_1)$ satisfies conditions (3.6) and (3.7). Since, obviously,

$$\sup |\partial^\alpha \varphi_2(x_1)| \leqq \tfrac{3}{2} \sup |\partial^\alpha \varphi_1(x_1)|, \tag{3.11}$$

then $\varphi_2 \in C(M(k))$ and the required function $\varphi(x) = \varphi_2(x_1)$ is constructed in the one-dimensional case. In the case $m > 1$ we will assume

$$\varphi(x) = \varphi_2(x_1) \ldots \varphi_2(x_m). \tag{3.12}$$

Obviously, since $|x| = \max |x_j|$ conditions (3.6) and (3.7) hold. Since $\varphi_2(x_j) \in C(M(k))$, it follows from Lemma 3.1, where $s = 0$, that $\varphi \in C(M(k))$.

Proof of Theorem 3.1. Suppose the function $u_0(x)$ identically equals unity in T^m. We will assume $f_0 = Bu_0$. Obviously,

$$f_0 = a_{00} \in C(M(k)),\ u_0 = B^{-1} f_0.$$

If the class $C(M)$ is not quasi-analytic, then, according to Lemma 3.2, the function $\varphi \in C(M)$ will be obtained, satisfying (3.6) and (3.7). We will assume $\Psi(x) = 1 - \varphi(Nx)$, where $N > 0$ is a fairly large number. It is obvious that $\Psi(x) \in C(M(k))$. We will obtain $f(x) = f_0(x)\Psi(x)$. Since f_0 and Ψ belong to $C(M(k))$, then $f \in C(M(k))$ by virtue of Lemma 3.1. Obviously

$$f - f_0 = -\varphi(Nx) f_0 \tag{3.13}$$

and, by virtue of (3.7)

$$\|f-f_0\| \leqq CN^{-m/2}. \tag{3.14}$$

Let us now consider the equation $Bu=f$. Since the operator B is elliptic and invertible in $L_2(T^m)$, then B^{-1} operates continuously from $L_2(T^m)$ to the Sobolev space $W_2^2(T^m)$ with the norm $\| \ \|_2$ (see, for example, Bers, John and Schechter [1]). (See (4.2.15) for a definition of the norm in $W_2^s(T^m)$ Therefore, by virtue of (3.14)

$$\|u-u_0\|_2 = \|B^{-1}(f-f_0)\|_2 \leqq C_1\|f-f_0\| \leqq C_1' N^{-m/2}. \tag{3.15}$$

According to Sobolev's imbedding theorem,

$$W_2^2(T^m) \subset L_p(T^m),$$

where $p > 2$ (the exact value of p is not important for us here). Therefore from (3.15) follows the estimate

$$\|u-u_0\|_{L_p} \leqq C_2 N^{-m/2}. \tag{3.16}$$

We will now assume that the equation $Bu=f$ is polynomially solvable, i.e. (2.12) holds. Then for any function χ from $L_2(T^m)$

$$(u, x) = \lim_{n\to\infty} (P_n(B)f, \chi). \tag{3.17}$$

We will assume $\delta = 1/(2N)$. Note that since $\Psi(x) = 0$ when $|x| < \delta$, then $P_n(B)f(x) = 0$ for these x. We shall take as $\chi(x)$ in (3.17) the function

$$\chi(x) = 1 \text{ when } |x| < \delta, \ \chi(x) = 0 \text{ when } |x| > \delta.$$

It is obvious that for the above choice χ the right-hand side of (3.17) equals zero, and we obtain

$$\int_{|x|<\delta} u(x)\,dx=0. \tag{3.18}$$

We shall now note that since $u_0 = 1$, then

$$\int_{|x|<\delta} u_0(x)\,dx=(2\delta)^m=N^{-m}. \tag{3.19}$$

Note that by virtue of Holder's inequality

$$\left|\int_{|x|<\delta} (u_0-u)\,dx\right|\leqq\left(\int_{|x|<\delta} 1\,dx\right)^{1-1/p}\left(\int_{|x|<\delta} |u_0-u|^p\right)^{1/p} \tag{3.20}$$
$$=(2\delta)^{m(1-1/p)}\|u_0-u\|_{L_p}.$$

Expressing the left-hand side of (3.20) using (3.18) and (3.19), and estimating the right-hand side of (3.20) using (3.16), we obtain

$$N^{-m}\leqq C_2N^{-m+m/p}N^{-m/2}. \tag{3.21}$$

Since $p > 2$, then $m/p - m/2 < 0$, and we arrive at a contradiction for fairly large N. Consequently, (3.17) cannot occur, and Theorem 3.1 is proved.

Remark 3.1. Considering the proof of Theorem 3.1, it is easy to note that the polynomial solvability of the equation $Bu=f$ is only used to establish the "locality" of the dependence of u on f, ie. the fact that $u(x_0)$ is uniquely defined using the values

$$f(x_0),\quad Bf(x_0),\ldots,B^kf(x_0),\ldots$$

At the same time, the way in which $u(x_0)$ is expressed in terms of $B^kf(x_0)$, $k = 0, 1, \ldots$ is completely unimportant.

A number of papers examine the question of which boundaries on the increase in $\|B^k f\|$ for different classes of operators B guarantee satisfaction of the property of "locality", of the dependence of the solutions of the equation $Bu=f$ on f (see, e.g, Malliavin [1], Lyubich and Tkachenko [1], Khryptun [1] and Chernyavskii [1]).

§4. Proof of the sufficiency of the quasi-analyticity of the class $C(M(k))$ for the polynomial solvability of the equations from $E(M(k))$.

It follows from Theorem 1.2 that it is sufficient to prove that $f \in \mathcal{D}(\Phi(A))$, where Φ is a function that satisfies (1.7) and (1.8), for the polynomial solvability of $Au=f$. The following theorem henceforth plays an important role.

Theorem 4.1. Suppose the sequence $M(k)$ satisfies condition (2.1). Suppose the function $\Phi_0(\xi)$, $\xi > 0$ is determined by the formula

$$\Phi_0(\xi)=\sum_{k=0}^{\infty}\frac{\xi^k}{M(k)}. \tag{4.1}$$

Suppose the equation $Bu=f$ belongs to the class $E(M(k))$, whilst the operator B is nonnegative in an invariant subspace which contains f. Then $q > 0$ exist, such that

$$f\in\mathcal{D}(\Phi_0(q\sqrt{B})). \tag{4.2}$$

The proof of Theorem 4.1 is based on $\|B^k f\|$, which we will now obtain.

Lemma 4.1. Suppose Eq.(2.9) belongs to the class $E(M)$, where $M(k)$ satisfies (2.1). Then the numbers R, R_0 and Λ_0 exist, such that when $l = 0, 1, \ldots$

$$B^l f\in C(M(2l+k), R_0, \Lambda_0 R^l) \tag{4.3}$$

(the class $C(M(k), r, \Lambda)$ is defined immediately after (22)).

Proof. Suppose $R_0 > 1$ and $\Lambda_0 > 0$ are numbers, such that

$$a_{ij} \in C(M(k), R_0/2, \Lambda_0) \qquad i,j=0,\dots,m, \quad f \in C(M(k), R_0, \Lambda_0). \tag{4.4}$$

The number R is defined later. We shall prove the lemma by induction with respect to l. When $l = 0$ the statement of the lemma is obvious. Suppose the statement holds when $l - 1$. Obviously

$$\partial^\alpha BB^{l-1}f = \partial^\alpha \sum_{i,j=0}^{m} \partial_i a_{ij} \partial_j B^{l-1} f. \tag{4.5}$$

To simplify the notation we will assume $a_{0j} = a_{j0} = 0$ when $j \neq 0$, and shall denote the unit operator by ∂_0.

It follows from the definition $C(M(k), r, \Lambda)$ that if

$$u \in C(M(k), r, \Lambda), \quad \text{then} \quad \partial^\beta u \in C(M(k+|\beta|), r, r^{|\beta|}\Lambda). \tag{4.6}$$

Using the assumption of induction and (4.6), when $j = 0, \dots, m$ we obtain:

$$\partial_j B^{l-1} f \in C(M(2(l-1)+k+1), R_0, R_0\Lambda_0 R^{l-1}) \tag{4.7}$$

(when $j = 0$ we used the monotony of $M(k)$ instead of (4.6), and the fact that $R_0 > 1$). Using Lemma 3.1, from (4.7) and (4.4) we derive:

$$a_{ij} \partial_j B^{l-1} f \in C(M(2l+k-1), R_0, \Lambda_1), \tag{4.8}$$

where by virue of (3.2)

$$\Lambda_1 = C_M R_0 \Lambda_0 R^{l-1} \Lambda_0 (1-\tfrac{1}{2})^{-1}. \tag{4.9}$$

Using (4.6) and (4.8) we obtain:

$$\partial_i a_{ij} \partial_j B^{l-1} f \in C(M(2l+k), R_0, R_0\Lambda_1).$$

Hence, summing the corresponding estimate (2.2) over i and j and using (4.5), we obtain:

$$BB^{l-1} f \in C(M(2l+k), R_0, (m^2+1)R_0\Lambda_1). \tag{4.10}$$

Using (4.9), we obtain that when

$$2C_M(m^2+1)R_0^2\Lambda_0^2 R^{l-1} \leqq R^l \tag{4.11}$$

(4.6) holds. It is obvious that inequality (4.11) holds when

$$R = 2(m^2+1)R_0^2\Lambda_0^2 C_M.$$

Lemma 4.2 is thus proved.

Proof of Theorem 4.1. Firstly, it is obvious that $C(M(k)) \subset \mathcal{D}_\infty(B)$. We shall first note that for integer $p > 0$ the following estimate holds:

$$\|B^{p/2} f\| \leqq CR_1^p M(p). \tag{4.12}$$

Indeed, it follows from (4.3) that

$$\sup_{x \in T^m} |\partial^\alpha B^l f| \leqq \Lambda_0 R^l R_0^{|\alpha|} M(|\alpha| + 2l). \tag{4.13}$$

If p is even, then assuming $l = p/2$, $|\alpha| = 0$ we will derive estimate (4.12) from (4.13), where $R_1 = \sqrt{R}$. If $p = 2l + 1$, l is integral,

$$\begin{aligned}\|B^{p/2}f\|^2 &= (B^{l+1/2}f, B^{l+1/2}f) = (BB^lf, B^lf)\\ &= \int_{T^m}\Big(-\sum_{i,j}\partial_i a_{ij}\partial_j B^lf + a_{00}B^lf\Big)B^lf\,dx\\ &= \int_{T^m}\Big(\sum_{i,j} a_{ij}\partial_j B^lf\,\partial_i B^lf + a_{00}|B^lf|^2\Big)dx\\ &\leq C_1\sum_{i,j}\sup_{x\in T^m}|\partial_j B^lf\|\partial_i B^lf| + C_0\|B^lf\|^2.\end{aligned} \tag{4.14}$$

Using inequality (4.13) where $|\alpha| = 1$ to estimate $|\partial_j B^l f|$ on the right-hand side of (4.14), we obtain (4.12) when $p = 2l + 1$. Thus (4.15) is proved for all integral $p > 0$. By virtue of (4.12) and (4.1) when $q < R_1^{-1}/2$

$$\|\Phi_0(q\sqrt{B})f\| \leq \sum_{k=0}^{\infty}\frac{q^k\|B^{k/2}f\|}{M(k)} \leq \sum_{k=0}^{\infty}\frac{C}{2^k} = 2C.$$

Using Fatu's theorem and (1.33), we obtain that

$$f \in \mathscr{D}(\Phi_0(q\sqrt{B})),$$

and Theorem 4.1 is proved.

Proof of Theorem 2.1. The necessity of the quasi-analyticity of $C(M(k))$ has already been proved. We shall prove the sufficiency. We shall first reduce the general case to the case of the positive operator B. For this we shall use Lemma 1.2. Since the operator B is elliptic, its spectrum is discrete and the conditions of Lemma 1.2, where A_1 is an extension of the Friedrichs operator B, hold. Therefore $u = Q_0(B)f + u_1$, where u_1 is the solution of the equation

$$Bu_1 = f_1, \quad f_1 = (I - BQ_0(B))f, \qquad f_1 \in H_1.$$

According to Lemma 1.3 it is sufficient to prove that the equation $Bu_1 = f_1$ is polynomially solvable.

We shall now note that $f_1 \in C(M(k))$. Indeed, according to Lemma 4.1 $B^l f \in C(M(2l+k))$ and according to Remark 2.5 and Lemma 3.1 $B^l f \in C(M(k))\ \forall\ l$. Therefore the function f_1, which also equals the sum of the functions $C_l B^l f$, also belongs to $C(M(k))$. Since $f_1 \in H_1$, and operator B is positive in H_1 by virtue of Lemma 2.1, $B > \rho^2 I$, $\rho > 0$. Theorem 4.1 is solvable. According to this theorem

$$f_1 \in \mathscr{D}(\Phi_0(q\sqrt{B})).$$

Let us now consider the function

$$T(t) = \max_{n \geqq 0} t^n / M(n), \qquad t \geqq 0. \tag{4.15}$$

It is obvious that $T(t) < \Phi_0(t)$ when $t > 0$. Therefore

$$f_1 \in \mathscr{D}(T(q\sqrt{B}))$$

by virtue of (1.3.4) and (1.3.3). If $C(M(k))$ is a quasi-analytic class of functions, then according to the Ostrovskii criterion (see Mandel'-broit [1])

$$\int_1^\infty \frac{\ln T(t)}{t^2}\, dt = \infty, \tag{4.16}$$

where $T(t)$ is determined using Formula (4.15). We shall use Theorem 1.2, where we shall take

$$\Phi(\lambda) = T(q\sqrt{\lambda}),\ A = B - \rho^2 I \geqq 0.$$

We shall verify that the conditions of Theorem 1.2 hold. Firstly, since the function $T(t)$ increases with respect to t, it follows from the condition

$$f \in \mathscr{D}(T(q\sqrt{B})) = \mathscr{D}(T(q\sqrt{A+\rho^2 I}))$$

that

$$f \in \mathscr{D}(T(q\sqrt{A})).$$

We shall further verify that

$$\Phi(\lambda) = T(q\sqrt{\lambda})$$

satisifies conditions (1.7) and (1.8). We first note that condition (1.7) - the convexity of the function

$$\ln \Phi(e^t) = \ln T(q\, e^{t/2}) = \ln T(e^{(t+2\ln q)/2})$$

is equivalent to the convexity of the function $\ln T(e^t)$. According to (4.15)

$$\varphi(t) = \ln T(e^t) = \max_{n \geq 0} \ln\left[\frac{e^{nt}}{M(n)}\right] = \max_{n \geq 0} [nt - \ln M(n)]. \tag{4.17}$$

We shall verify the convexity of the function $\varphi(t)$. Suppose t_1, $t_2 > 0$, $0 < \theta < 1$. Obviously, by virtue of (4.17)

$$\begin{aligned} \varphi(\theta t_1 + (1-\theta)t_2) &= \max[n(\theta t_1 + (1-\theta)t_2) - \ln M(n)] \\ &= n_0(\theta t_1 + (1-\theta)t_2) - \ln M(n_0) \\ &= \theta[n_0 t_1 - \ln M(n_0)] + (1-\theta)[n_0 t_2 - \ln M(n_0)] \\ &\leqq \theta\varphi(t_1) + (1-\theta)\varphi(t_2). \end{aligned} \tag{4.18}$$

Condition (1.7) thus holds for $\Phi = T$.

We shall verify condition (1.8). It will take the form

$$\int_1^\infty \frac{\ln T(qx)}{x^2}\,dx = \infty.$$

Obviously, this condition is equivalent to the condition

$$\int_q^\infty \frac{\ln T(x)}{x^2}\,dx = \infty,$$

and this condition holds by virtue of (4.16).

Therefore all the conditions of Theorem 1.2 hold. By virtue of this theorem the equation $Bu_1 \equiv Au_1 + \rho^2 u_1 = f_1$ is polynomially solvable. Hence, according to Lemma 1.3 and Remark 2.7, it follows that the equation $Bu=f$ is also polynomially solvable, and Theorem 2.1 is completely proved.

§5. Construction of the polynomials P_n(A) in explicit form.

In this paragraph we shall consider the case of the positive definite operator B, $B = A + \rho^2 I$, $A > 0$.

In Theorem 2.1, the existence of the polynomials $P_n(A)$, representing a solution of the equation $\rho^2 u + Au = f$ in the form (1.4), is stated in the case of the quasi-analyticity $C(M(k))$. However these polynomials were not constructed in explicit form. Below, using Bernshtein's method of constructing fractions which deviate from zero the least on a straight line, we shall construct the polynomials $P_n(\lambda)$ for the classes $C(M(k))$ with a fairly regular increase in $M(k)$.

We shall now describe a class of sequences $M(k)$, for which we can construct a weighting function explicitly given in the form of an infinite product, instead of the weighting function $T(q\sqrt{\lambda})$ used in the previous paragraph.

We shall set the following sequence in correspondence with the sequence $M(k)$ $k \in \mathbb{Z}_+$:

$$\beta(k) = M(k)/M(k-1), \qquad k = 1, 2, \ldots. \tag{5.0}$$

We require that $M(k)$ satisfies the condition

$$M(2k) \leqq C^k (M(k))^2, \tag{5.1}$$

where $C > 0$ is some constant, and also the conditions of regularity of increase $\beta(k)$

$$\beta(k) \leqq C_0 \beta(l) \qquad \text{when } k < l \tag{5.2}$$

$$\sum_{k=l+1}^{\infty} \frac{1}{\beta^2(k)} \leqq C_\beta \frac{l}{\beta^2(l)} \qquad \text{when } l > 1 \tag{5.3}$$

Remark 5.1. Condition (5.3) obviously holds for the analytic class of functions $\beta(k)=k$:

$$\sum_{k=l+1}^{\infty} \frac{1}{k^2} \leqq C_\beta \frac{1}{l}$$

for some C_β.

We shall now give the simple condition on $\beta(k)$ necessary for (5.3) to hold.

Proposition 5.1. Suppose $\delta>0$ and the sequence $l^{1+\delta}/\beta^2(l)$ monotonically decreases for large l. Then condition (5.3) holds.

Proof. We shall continue the function $k^{1+\delta}/\beta^2(k) = \varphi(k)$ from integer numbers k to real numbers like the step function $\varphi(k - \Theta) = \varphi(k)$ when $0 < \Theta < 1$, $k = 1, 2, \ldots$. Obviously, when $t>0$ the function $\beta^2(t) = t^{1+\delta}/\varphi(t)$ is thereby determined. We have a formula of integration by parts

$$\int_l^\infty \frac{dt}{\beta^2(t)} = -\frac{1}{\delta}\int_l^\infty \frac{t^{1+\delta}}{\beta^2(t)}\,dt^{-\delta} = \frac{1}{\delta}\frac{l}{\beta^2(l)} + \frac{1}{\delta}\int_l^\infty t^{-\delta}\,d\frac{t^{1+\delta}}{\beta^2(t)}. \tag{5.4}$$

(We learned that the estimate $l/\beta^2(l) \leqq Cl^{-\delta}$ holds by virtue of the monotony of $l^{1+\delta}/\beta^2(l)$.) The function $t^{1+\delta}/\beta^2(t)$ decreases when $t \in [l, \infty[$ for fairly large l, and therefore the last integral on the right-hand side of (5.4) is negative. Therefore

$$\int_l^\infty \frac{dt}{\beta^2(t)} \leqq \frac{1}{\delta}\frac{l}{\beta^2(l)}. \tag{5.5}$$

Since

$$\beta^2(k-\theta) = \frac{(k-\theta)^{1+\delta}}{\varphi(k-\theta)} = \frac{(k-\theta)^{1+\delta}}{\varphi(k)} = \beta^2(k)\left(\frac{k-\theta}{k}\right)^{1+\delta}, \qquad (k \in \mathbb{N}),$$

then for large l

$$\int_l^\infty \frac{dt}{\beta^2(t)} \geqq \frac{1}{2}\sum_{k=l+1}^\infty \frac{1}{\beta^2(k)}.$$

Hence and from (5.5) follows (5.3).

The following proposition shows that Denjoy's quasi-analytic classes satisfy the condition of Proposition 5.1, where $\delta=1$.

Proposition 5.2. Suppose the sequence $M(k)$ corresponds to Denjoy's quasi-analytic class

$$M(k) = (k \ln k \ln\ln k \ldots \ln \ldots \ln k)^k, \qquad (k \gg 1).$$

Then the corresponding function $\beta(k)$ is such that $\beta(k)/k$ will increase monotonically for large k.

Proof. It is obvious that it is sufficient to prove (5.3) for fairly large l. The condition of monotony $\beta(k)/k$ will have the form:

$$\frac{1}{k}\frac{M(k)}{M(k-1)} \leqq \frac{1}{k+1}\frac{M(k+1)}{M(k)},$$

which is equivalent to the inequality

$$\ln M(k+1) + \ln M(k-1) - 2\ln M(k) + \ln k - \ln(k+1) \geqq 0. \tag{5.6}$$

Bearing in mind the definition $M(k)$, we obtain that

$$\ln M(k) = k[\ln k + \varphi(k)],\ \varphi(x) = \ln\ln x + \ldots + \ln\ldots\ln x. \tag{5.7}$$

Calculating the left-hand side of (5.6) bearing in mind (5.7), we obtain:

$$\begin{aligned}
&(k+1)[\ln(k+1)+\varphi(k+1)] + (k-1)[\ln(k-1)+\varphi(k-1)] \\
&\quad -2k[\ln k + \varphi(k)] + \ln k - \ln(k+1) \\
&= k\{\ln(k+1) + \ln(k-1) - 2\ln k + \varphi(k+1) + \varphi(k-1) - 2\varphi(k)\} \\
&\quad + \ln(k+1) - \ln(k-1) + \varphi(k+1) - \varphi(k-1) + \ln k - \ln(k+1) \\
&= k\ln\frac{k^2-1}{k^2} + k\{\varphi(k+1) - 2\varphi(k) + \varphi(k-1)\} \\
&\quad + \ln k - \ln(k-1) + \varphi(k+1) - \varphi(k-1).
\end{aligned} \tag{5.8}$$

Obviously, $\varphi(x) = \psi(\ln\ln x)$, where

$$\psi(x) = x + \ln x + \ldots + \ln\ldots\ln x.$$

It is obvious that the derivatives of the function $\psi(x)$ are bounded for large x, and obviously

$$\begin{aligned}
\ln\ln(k\pm 1) &= \ln\left[\ln k + \ln\left(1 \pm \frac{1}{k}\right)\right] = \ln\left[\ln k \pm \frac{1}{k} + O\left(\frac{1}{k^2}\right)\right] \\
&= \ln\ln k \pm \frac{1}{k\ln k} + O(k^{-2}).
\end{aligned}$$

Therefore, using the Taylor expansion of the function ψ at the point $\ln\ln k$, we obtain

$$k\{\varphi(k+1)-2\varphi(k)+\varphi(k-1)\}=k\psi''(\ln\ln k)\frac{1}{k^2\ln^2 k}+O(k^{-3})$$
$$=O\left(\frac{1}{k\ln k}\right);$$
$$\varphi(k+1)-\varphi(k-1)=2\psi'(\ln\ln k)\frac{1}{k\ln k}+O(k^{-2}).$$

Since $\psi'(x) > 1/2$ for large x, hence we obtain that

$$k\{\varphi(k+1)-2\varphi(k)+\varphi(k-1)\}+\varphi(k+1)-\varphi(k-1)>0 \tag{5.9}$$

for large k. Note that

$$\begin{aligned}k\ln\left(\frac{k^2-1}{k^2}\right)+\ln k-\ln(k+1)&=k\ln\left(1-\frac{1}{k^2}\right)-\ln\left(1-\frac{1}{k}\right)\\ &=k\left(-\frac{1}{k^2}+O\left(\frac{1}{k^4}\right)-\left(-\frac{1}{k}-\frac{1}{2k^2}+O\left(\frac{1}{k^3}\right)\right)\right)=\frac{1}{2k^2}+O\left(\frac{1}{k^3}\right)\end{aligned} \tag{5.10}$$

for large k. Thus, by virtue of (5.9), (5.10) and (5.8) inequality (5.6) holds for large k and Proposition 5.2 is proved.

Proposition 5.3. The sequence $M(k)$, which corresponds to Denjoy's quasi-analytic class, satisfies condition (5.1).

Proof. Condition (5.1) is equivalent to the condition

$$\ln M(2k)\leqq C_1k+2\ln M(k).$$

For Denjoy's quasi-analytic class, this condition takes the form

$$\begin{aligned}&2k[\ln(2k)+\ln\ln(2k)+\dots+\ln\dots\ln(2k)]\\ &\quad\leqq C_1k+2k[\ln k+\ln\ln k+\dots+\ln\dots\ln k].\end{aligned}$$

This inequality is equivalent to the following:

$$\ln(2k)-\ln k+\ln\ln(2k)-\ln\ln k+\ldots+\ln\ldots\ln(2k)-\ln\ldots\ln k\leqq C_1/2.$$

The last inequality obviously holds for large k if C_1 is fairly great.

Remark 5.1. Proposition 5.1 shows that condition (5.3) can be considered the condition of regularity of the increase in $\beta(l)/l$. In the most regular case, corresponding to Denjoy's quasi-analytic classes, $\beta(l)/l$ monotonically increases for large l. Proposition 5.1 shows that condition (5.3) also allows of less regular behaviour of $\beta(l)/l$, namely an increase of $l^{\gamma}(\beta(l)/l)$, $\gamma<1/2$ is sufficient.

We shall now proceed to construct a weighting function corresponding to the sequence $\beta(k)$. We will assume

$$\psi_N(z)=\prod_{k=1}^{N}\left(1+\frac{z}{\beta^2(k)}\right),\quad \psi_\infty(z)=\prod_{k=1}^{\infty}\left(1+\frac{z}{\beta^2(k)}\right). \tag{5.11}$$

Theorem 5.1. Suppose the equation $Bu=f$ belongs to the class $E(M(k))$, and, in addition, $B=A+\rho^2 I$, $A>0$. Suppose $M(k)$ satisfies condition (2.1) and (5.1), and the sequence $\beta(k)$, determined by (5.0), satisfies conditions (5.2) and (5.3). Then $q>0$ exists, such that $f\in\mathcal{D}(\psi_\infty(qA))$.

Proof. Obviously

$$\psi_N(qA)=\sum_{l=0}^{N} b_l q^l A^l. \tag{5.12}$$

We shall use Cauchy's formula to estimate the coefficients b_l:

$$|b_l|=\left|\frac{1}{2\pi}\int_{|\zeta|=r}\frac{\psi_N(\zeta)}{\zeta^{l+1}}\,d\zeta\right|\leqq\frac{\psi_N(r)}{r^l}\leqq\frac{\psi_\infty(r)}{r^l}.$$

Therefore, using (5.11), we obtain

$$|b_l| \leqq \inf_{r>0} \frac{\psi_\infty(r)}{r^l} = \inf_{r>0} \prod_{k=1}^{l} \left(\frac{1}{r} + \frac{1}{\beta^2(k)}\right) \prod_{k=l+1}^{\infty} \left(1 + \frac{r}{\beta^2(k)}\right).$$

Taking $r = \beta^2(l)$ we will obtain that by virtue of (5.2) when $l > 1$

$$\begin{aligned} |b_l| &\leqq \prod_{k=1}^{l} \left(\frac{1}{\beta^2(l)} + \frac{1}{\beta^2(k)}\right) \prod_{k=l+1}^{\infty} \left(1 + \frac{\beta^2(l)}{\beta^2(k)}\right) \\ &\leqq (C_0^2+1)^l \prod_{k=1}^{l} \frac{1}{\beta^2(k)} \exp \sum_{k=l+1}^{\infty} \ln\left(1 + \frac{\beta^2(l)}{\beta^2(k)}\right). \end{aligned} \tag{5.13}$$

Using (5.3) we obtain:

$$\exp \sum_{k=l+1}^{\infty} \ln\left(1 + \frac{\beta^2(l)}{\beta^2(k)}\right) \leqq \exp \sum_{k=l+1}^{\infty} \frac{\beta^2(l)}{\beta^2(k)} \leqq \exp(C_\beta l) = C_1^l.$$

We therefore obtain the following estimate from (5.13):

$$|b_l| \leqq C_2^l \prod_{k=1}^{l} \frac{1}{\beta^2(k)} = C_2^l \prod_{k=1}^{l} \left(\frac{M(k-1)}{M(k)}\right)^2 = C_2^l \frac{M^2(0)}{M^2(l)}. \tag{5.14}$$

From (5.12) and (5.14) we obtain

$$\|\psi_N(qA)f\| \leqq \sum_{l=0}^{N} (qC_2)^l \|A^l f\| \frac{M^2(0)}{M^2(l)} \tag{5.15}$$

(where $l = 0$ we shall assume the product with respect to k equals unity). We shall now use Lemma 4.1, where we assume $B = A$. By virtue of (4.3), (5.15) and (5.1) we have:

$$\begin{aligned} \|\psi_N(qA)f\| &\leqq C_3 \sum_{l=0}^{\infty} (qC_2)^l \Lambda_0 R^l M(2l) M^2(0)/M^2(l) \\ &\leqq C_3 \Lambda_0 M^2(0) \sum_{l=0}^{\infty} (qC_2 CR)^l. \end{aligned}$$

If we assume $q < 1/(CC_2R)$, we obtain that $\|\psi_N(qA)f\|$ is uniformly bounded with respect to N. Using (1.3.3), we obtain that by virtue of Fatu's theorem $\|\psi_\infty(qA)f\| < C_4 < \infty$.

We shall now proceed to construct the polynomials $P_n(\lambda)$. We will assume

$$S_n(z)=\prod_{k=1}^{n}\left(1-\frac{iz}{\beta(k)}\right)^2, \tag{5.16}$$

$$T_n(z)=\tfrac{1}{2}(S_n(z)+S_n(-z)). \tag{5.17}$$

Lemma 5.1. For any $n \in \mathbb{N}$, $R \in \mathbb{R}$, an even polynomial $P_n(x^2)$ of the $2n$-th degree exists, satisfying the equation

$$1-(\rho^2+x^2)P_n(x^2)=T_{n+1}(Rx)/T_{n+1}(iR\rho). \tag{5.18}$$

When $\lambda > 0$ this polynomial satisfies the inequality

$$|(\rho^2+\lambda)^{-1}-P_n(\lambda)|\leqq(\rho^2+\lambda)^{-1}\psi_\infty(R^2\lambda)/T_{n+1}(iR\rho). \tag{5.19}$$

Proof. It is obvious that the polynomial $T_n(z)$, determined by (5.17), is even. The polynomial (5.18) is a solution of the equation

$$(\rho^2+x^2)P_n(x^2)=1-T_{n+1}(Rx)/T_{n+1}(iR\rho). \tag{5.20}$$

The right-hand side of this equation obviously equals zero when $x=i\rho$ and when $x=-i\rho$ and can therefore be divided by $x^2 + \rho^2$. The even polynomial P_n, satisfying (5.18), consequently exists. Dividing (5.18) by $(\rho^2 + x^2)$, we obtain

$$|(\rho^2+x^2)^{-1}-P_n(x^2)|\leqq(\rho^2+x^2)^{-1}|T_{n+1}(Rx)|/T_{n+1}(iR\rho). \tag{5.21}$$

Note that for real x

$$T_{n+1}(qx)=\operatorname{Re} S_{n+1}(qx).$$

Therefore

$$|T_{n+1}(Rx)| \leqq |S_{n+1}(Rx)| \leqq \prod_{k=1}^{n+1}\left(1+\frac{R^2x^2}{\beta^2(k)}\right) \leqq \psi_\infty(R^2x^2). \qquad (5.22)$$

Assuming $x^2 = \lambda$, $\lambda > 0$, from (5.21), (5.22) we obtain inequality (5.19).

Theorem 5.2. Suppose $f \in \mathcal{D}(\psi_\infty(qA))$, $q > 0$, and the polynomials $P_n(\lambda)$ are determined by Formula (5.18), where $R = \sqrt{q}$, $\lambda = x^2$. Suppose

$$\sum_{k=0}^{\infty} \frac{1}{\beta(k)} = \infty. \qquad (5.23)$$

Then Formula (1.4) holds.

Proof. We shall use Lemma 1.1, where

$$g^+(A) = (\rho^2 + A)^{-1},\ \Phi(\lambda) = \psi_\infty(q\lambda).$$

Bearing in mind (5.19), from inequality (1.5) we obtain

$$\|u - P_n(A)f\| \leqq (\rho^2)^{-1} T_{n+1}(iR\rho)^{-1} \|\psi_\infty(qA)f\|. \qquad (5.24)$$

Note that by virtue of (5.17) and (5.16)

$$T_{n+1}(iR\rho) \geqq \tfrac{1}{2} S_{n+1}(iR\rho) \geqq \prod_{k=1}^{n+1}\left(1+\frac{R\rho}{\beta(k)}\right)^2. \qquad (5.25)$$

By virtue of condition (5.23), it follows from (5.25) that $T_{n+1}(iR\rho) \to \infty$ as $n \to \infty$. Hence and from (5.24) follows (1.4).

Remark 5.2. Condition (5.23) on the numbers $\beta(k)$, determined by (5.0), is the necessary condition of the quasi-analytic class $C(M(k))$ (see Mandel'broit [1], Ch.IV, Theorem I.V).

CHAPTER 3. APPROXIMATION OF ANALYTIC FUNCTIONS ON A STRAIGHT LINE WITH THE WEIGHT ch(Rx) USING INTERPOLATION POLYNOMIALS

§1. Formulation of the problem. The polynomials $T_n(x)$.

Lemma 2.1.1, which is proved in the second chapter, shows that, to obtain formulae of the form (1.4) for $f \in \mathcal{D}(\Phi(A))$, it is required to construct a sequence of polynomials P_n, such that $\mu^+(P_n, g^+, \Phi) \to 0$ as $n \to \infty$, for the specified function g^+. As Theorems 4.1 and 5.1 show, we can compare the function Φ_0, such that if $f \in C(M(k))$, then

$$f \in \mathscr{D}(\Phi_0(q\sqrt{A})),$$

where A is a second-order operator with coefficients from $C(M(k))$, to a class of infinitely differentiable functions $C(M(k))$. According to Theorem 2.1, we can only obtain formulae of the form (2.1.4) when the class $C(M(k))$ is quasi-analytic. The most important of these quasi-analytic classes $C(M(k))$ is the class of $C(k!)$ analytic functions. According to Theorem 5.1, in this case we can take as the function $\Phi_0\sqrt{\lambda}$ the function $\psi_\infty(\lambda)$, which is determined by the formula (2.5.11), where $\beta(k)=k$, i.e.

$$\psi_\infty(\lambda) = \prod_{k=1}^{\infty}\left(1+\frac{\lambda}{k^2}\right).$$

It will be more convenient for us to take as the function $\Phi(z)$ the hyperbolic cosine

$$\Phi_0(z) = \operatorname{ch} z = \prod_{k=1}^{\infty}\left(1+\frac{4z^2}{\pi^2(2k-1)^2}\right). \tag{1.1}$$

It is obvious that the constants C_1 and C_2 exist, such that

$$\psi_\infty(\lambda) \leqq \Phi_0(C_1\sqrt{\lambda}), \quad \Phi_0(\sqrt{\lambda}) \leqq \psi_\infty(C_2\lambda).$$

Therefore we shall set the function (1.1) in correspondence with the class of $C(k!)$ analytic functions.

In this chapter we shall consider in detail the problem of approximating the functions $g^+(\lambda)$, which are analytic in

$$\mathbb{R}_+ = \{z \in \mathbb{R} : z \geqq 0\}$$

with the weight $\mathrm{ch}(R\sqrt{\lambda})$. The polynomials $P_n(\lambda)$, which approximate $g^+(\lambda)$, will be obtained like interpolation polynomials of the function $g^+(\lambda)$ using way interpolation nodes. We shall obtain the upper bounds for

$$\mu^+(P_n, g^+, \mathrm{ch}(R\sqrt{\lambda})).$$

For brevity we introduce the notation

$$\mu_R^+(P_n, g^+) = \mu^+(P_n, g^+, \mathrm{ch}(R\sqrt{\lambda})) \tag{1.2}$$

It is obvious that the problem of approximating the function $g(\lambda)$ on $\mathbb{R}_+$ with the weight $\mathrm{ch}(R\sqrt{\lambda})$ using the substitution $\lambda = x^2$ reduces to that of approximating the even function $g^+(x^2) = g(x)$ using even polynomials with the weight $\mathrm{ch}(Rx)$. The chapter will therefore essentially examine approximation on $\mathbb{R}$ with the weight $\mathrm{ch}(Rx)$, using polynomials, and shall construct these polynomials such that the approximating polynomials are even for even functions. We shall derive corollaries on approximation on $\mathbb{R}_+$ from the results on approximation on $\mathbb{R}$ at the end of the chapter (in §.6).

The deviation of P from g on $\mathbb{R}$ with the positive weight $\Phi(x)$ is determined by the formula

$$\mu(P,g,\Phi)=\sup_{x\in\mathbb{R}}\frac{|P(x)-g(x)|}{\Phi(x)}. \tag{1.3}$$

We shall assume, for brevity,

$$\mu_R(P,g)=\mu(P,g,\operatorname{ch}(Rx)). \tag{1.4}$$

Henceforth we shall construct the polynomials P_n, approximating the function g, which is analytic on $\mathbb{R}$, in the form of interpolation polynomials. The interpolation nodes will agree with the roots of the polynomial T_{n+1}. The polynomial $T_n(x)$ will be constructed in this paragraph, and in this and the following paragraphs we shall establish the important properties of these polynomials.

We shall now construct $T_n(x)$. We will assume

$$\Phi_n(z)=\prod_{k=1}^{n}\left(1+\frac{4z^2}{\pi^2(2k-1)^2}\right). \tag{1.5}$$

According to (1.1), the following inequality holds:

$$\Phi_n(x)\leqq\operatorname{ch}x,\qquad \forall x\in\mathbb{R},\qquad n\in\mathbb{N}. \tag{1.6}$$

We shall now determine the polynomials $T_n(z)$, which play the most important role in the subsequent discussion. We will assume

$$S_n(z)=\prod_{k=1}^{n}\left(1-\frac{iz}{2k-1}\right)^2,\quad (i=\sqrt{-1}), \tag{1.7}$$

$$T_n(z)=\tfrac{1}{2}(S_n(z)+S_n(-z)). \tag{1.8}$$

It is obvious that $T_n(z)$ is an even polynomial of the $2n$-th degree.

Lemma 1.1. Suppose $r = 2R/\pi$. Then

$$\mu(T_n(r,x),\ 0,\ \Phi_n(Rx))=1. \tag{1.9}$$

Proof. It follows from (1.8) and (1.7) that for real x

$$T_n(x)=\tfrac{1}{2}(S_n(x)+\bar{S}_n(x))=\operatorname{Re} S_n(x). \tag{1.10}$$

Further, comparing (1.5) and (1.7) we note that when $x \in \mathbb{R}$, $r = 2R/\pi$

$$|S_n(rx)|=\Phi_n(Rx). \tag{1.11}$$

Using (1.10) and (1.11), we obtain:

$$\mu(T_n(x),0,\Phi_n(Rx))=\sup_{x\in\mathbb{R}}\frac{|T_n(rx)|}{|\Phi_n(Rx)|}=\sup_{x\in\mathbb{R}}\frac{|\operatorname{Re} S_n(rx)|}{|S_n(rx)|}\leqq 1.$$

Bearing in mind that $T_n(0) = \Phi_n(0) = 1$, we obtain a statement of the lemma.

Lemma 1.1 shows that the deviation of the polynomial $T_n(rx)$ from zero with the weight $\Phi_n(Rx)$ equals 1. We shall prove below that the polynomial $T_n(rx)$ is (apart from the constant multiplier) a polynomial which deviates from zero the least in $\mathbb{R}$ with the weight $\Phi_n(Rx)$.

We shall use $\mathcal{N}_n(\rho)$ to denote the set of polynomials $Q(x)$ of a degree no higher than $2n$ which are real on $\mathbb{R}$, and which satisfy the normalising condition

$$Q(i\rho)=1, \quad Q(-i\rho)=1 \tag{1.12}$$

We will assume

$$Q^0(x)=T_n(rx)/T_n(ir\rho). \tag{1.13}$$

It follows from formula (1.7) that

$$S_n(\pm ir\rho)>0$$

and $T_n(ir\rho)$ is a real number.

Theorem 1.1. $Q^0 \in \mathcal{N}_n(\rho)$,

$$\mu(Q^0(x), 0, \Phi_n(Rx))=1/T_n(ir\rho). \tag{1.14}$$

The polynomial $Q^1 \in \mathcal{N}_n(\rho)$, differing from Q^0 and satisfying the inequality

$$\mu(Q^1(x), 0, \Phi_n(R_x))\leqq 1/T_n(ir\rho). \tag{1.15}$$

does not exist.

The proof of Theorem 1.1 is based on the following two lemmas.

Lemma 1.2. When changing $y \in \mathbb{R}$ from $-\infty$ to $+\infty$ arg $S_n(y)$ continuously monotonically changes from $n\pi$ to $-n\pi$.

Proof. By virtue of (1.7) we obtain:

$$\arg S_n(y)=-2\sum_{j=1}^{n} \operatorname{arctg}\left(\frac{y}{2j-1}\right). \tag{1.16}$$

The monotony of $\arg S_n(y)$ follows from this formula by virtue of the monotony of the function arctg. Since

$$\arg S_n(y) \to n\pi$$

as $y \to -\infty$, and

$$\arg S_n(y) \to -n\pi$$

as $y \to +\infty$, the statement of the lemma follows hence.

Lemma 1.3. The fraction $T_n(rx)/\Phi_n(Rx)$ specifies $2n-1$ different points and $\pm\infty$ alternating values of +1 and -1.

Proof. By virtue of (1.10) and (1.11) when $x \in \mathbb{R}$

$$T_n(rx)/\Phi_n(Rx) = \operatorname{Re} S_n(rx)/|S_n(rx)|. \tag{1.17}$$

When $\arg S_n(rx) = \pi k$

$$\operatorname{Re} S_n(rx)/|S_n(rx)| = (-1)^k, \qquad k = 0, \pm 1, \ldots, \pm(n-1).$$

When $x \to \pm\infty$

$$\operatorname{Re} S_n(rx)/|S_n(rx)| \to \pm(-1)^n$$

and the lemma is proved.

Proof of Theorem 1.1. We shall prove the theorem in the same way as the proof of Chebyshev's theorem (see Akhiezer [1]). Consider the fraction

$$\mathscr{D}(x) = [Q^0(x) - Q^1(x)]\,T_n(ir\rho)/\Phi_n(Rx). \tag{1.18}$$

To prove the theorem it is sufficient to prove that this fraction vanishes on $\mathbb{R}$ at at least $2n-1$ points (bearing in mind the multiplicity and values at ∞). Indeed, by virtue of (1.12) $Q^0(x) - Q^1(x) = 0$ when $x = i\rho$ and $x = -i\rho$. If this difference vanishes for the $2n-1$ real values of x, the polynomial $Q^0 - Q^1$, of the $2n$-th degree, must identically equal zero, which also forms the statement of Theorem 1.1.

We shall now prove that $\mathcal{D}(x)$ has $2n-1$ zeros in $\mathbb{R}$, (bearing in mind multiplicity).

According to Lemma 1.3

$$T_n(ir\rho)Q^0(x)/\Phi_n(Rx)=\pm 1$$

at $2n-1$ points

$$-\infty=x_1<x_2\leqq\ldots\leqq x_{2n}<+\infty=x_{2n+1},$$

whilst the signs + and - are taken alternately. By virtue of (1.15) and (1.14) $\mathcal{D}(x_i)$ has either the same sign as $Q^0(x_i)$ or $\mathcal{D}(x_i) = 0$. Suppose x_{i-1} and x_{i+k+1} are two points at which $\mathcal{D} \neq 0$, and at the intermediate $k+1$ points ($k > -1$), $\mathcal{D} = 0$, i.e.

$$\mathcal{D}(x_{i-1})\neq 0, \quad \mathcal{D}(x_i)=0\ldots=\mathcal{D}(x_{i+k})=0, \quad \mathcal{D}(x_{i+k+1})\neq 0. \tag{1.19}$$

Since the sign $\mathcal{D}(x_{i-1})$ is identical with the sign $Q^0(x_{i-1})$, and the sign $\mathcal{D}(x_{i+k+1})$ is identical with the sign $Q^0(x_{i+k+1})$, $\mathcal{D}(x_{i-1})$ and $\mathcal{D}(x_{i+k+1})(-1)^{k+2}$ have one and the same sign. Therefore the number of zeros $\mathcal{D}(x)$ (bearing in mind multiplicity) in the interval

$$]x_{i-1}, x_{i+k+1}[$$

has the same evenness as k. Bearing in mind by virtue of (1.19) there are at least $k+1$ zeros $\mathcal{D}$ in $]x_{i-1}, x_{i+k+1}[$, we obtain that the number of zeros in the interval $]x_{i-1}, x_{i+k+1}[$ is not less than $k+2$. We shall use x_j to denote the minimum of such numbers

x_l, that $\mathcal{D}(x_i)\neq 0$, and x_{j+m} to denote the maximum of such numbers: $-\infty \leq x_j < x_{j+m} \leq +\infty$. Dividing $[x_j, x_{j+m}]$ into the segments $]x_{l-1}, x_{l+k+1}[$, which satisfy (1.19), we obtain that there are no fewer than m zeros in $]x_j, x_{j+m}[$. If $x_j = -\infty$, $x_{j+m} = +\infty$, then $m=2n$ and $\mathcal{D}(x)$ indeed vanishes at $2n$ points. If

$$x_j > -\infty, x_{j+m} < +\infty,$$

then $\mathcal{D}(x_i) = 0$ when

$$2 \leqq i \leqq j-1, \; j+m+1 \leqq i \leqq 2n,$$

and at $\pm\infty$, i.e. it has $j - 2 + m + 2n - (j + m) = 2n - 2$ zeros and the degree of the numerator is not higher than $2n-1$.

Thus $\mathcal{D}(x)$ has $2n-1$ zeros in $\mathbb{R}$, whence follows, as shown above, the statement of Theorem 1.1.

Remark 1.1. Bernshtein (see [1]), constructed polynomials that deviate the least from zero on a straight line with the weight $Q_n(x)$, and constructed two-parametric families of such polynomials. The even polynomial Bernshtein constructed agrees with $T_n(x)$ apart from the multiplier.

In the following paragraphs we shall approximate the analytic functions $g(x)$, $x \in \mathbb{R}$ using interpolation polynomials. To approximate, in the segment $[-1, 1]$, functions that are analytic in an ellipse containing this segment, we shall take interpolation polynomials whose nodes are the roots of Chebyshev's polynomial (see e.g., Gel'fond [1]), as polynomials which approximate these functions at the maximum rate. In the same way, in the following paragraphs we shall approximate in $\mathbb{R}$ with the weight $\mathrm{ch}(R\sqrt{\lambda})$ functions that are analytic in a strip containing $\mathbb{R}$, using interpolation polynomials whose interpolation nodes agree with the roots of the polynomial T_n, determined by (1.8). The fact that the polynomials T_n deviate the least in $\mathbb{R}$ from zero with the weight Φ_n in the same way as Chebyshev's polynomials deviate the least of all from zero in a segment is the guiding consideration for this choice of nodes. Naturally, a similar choice must be justified, and we do so in the following paragraphs.

We shall now prove the lemma which justifies the possibility of taking the roots of the polynomial T_n as the interpolation nodes.

Lemma 1.4. The polynomial $T_n(x)$, determined by (1.8), is even, has $2n$ different real roots, and its positive roots x_1, ..., x_n are the solutions of the equation

$$2\sum_{j=1}^{n} \operatorname{arctg}\frac{x}{2j-1}=(2k-1)\frac{\pi}{2}, \qquad k=1,\dots,n. \tag{1.20}$$

Proof. Since Formula (1.10) holds, the real roots of the equation $T_n(x) = 0$ are the roots of the equation

$$\arg S_n(x)=(\pi/2)-\pi k.$$

Calculating $\arg S_n(x)$ (see (1.16)), we obtain hence Eq.(1.20). It is obvious that this equation has a unique positive root for each $k = 1, \dots, n$, and these roots are different for different k.

To estimate the rate of approximation of the function $g(x)$ using the interpolation polynomials $P_n(x)$ with the interpolation nodes in the roots of the polynomial T_n, we require to examine the behaviour of the polynomials T_n in a complex domain. We shall do this in the next paragraph.

§.2. Estimates of the modulus of the polynomial $T_n(z)$ on straight lines that are parallel to the real axis.

The following theorem, in which the upper limit of the polynomial modulus is given, is the basic result of this paragraph.

Theorem 2.1. Suppose the polynomial $T_n(z)$ is determined by Formula (1.8), where $n \in \mathbb{N}$. Suppose $z = x + iy$, $x, y \in \mathbb{R}$. Then

$$|T_n(x+iy)| \geqq \theta(y)|S_n(i|y|+x)|(1+x^2)^{-1}, \tag{2.1}$$

where $S_n(z)$ is determined by Formula (1.7), $\theta(y) > C$ when $|y| > 1$, $\theta(y) > C|y|$ when $|y| < 1$, and C does not depend on y, n and x.

We will obtain the lower bound for $S_n(iy + x)$ later, see Lemma 2.3.

The proof of Theorem 2.1 is based on two lemmas, which we shall formulate and prove below.

From the definition of the polynomial $T_n(z)$ (Formula (1.8), we obtain that

$$T_n(z) = \tfrac{1}{2}S_n(z)(1+Q_n(z)), \tag{2.2}$$

where

$$Q_n(z) = \prod_{k=1}^{n} q_k, \qquad q_k = \left(\frac{2k-1+iz}{2k-1-iz}\right)^2. \tag{2.3}$$

Since by virtue of (1.8) $T_n(z)$ is an even function, in this paragraph we will henceforth assume $z = x + iy$, $x \in \mathbb{R}$, $y > 0$. We shall first obtain the lower bound $|T_n(x + iy)|$ for fairly large values of y in comparison with n.

Lemma 2.1. Suppose $Q_n(z)$ is determined using Formula (2.3), $z = x + iy$, $x, y \in \mathbb{R}$. Suppose

$$|z| \geqq 4n^2. \tag{2.4}$$

Then

$$|1+Q_n(z)| \geqq 1. \tag{2.5}$$

Proof. It is sufficient to prove that under condition (2.4)

$$\operatorname{Re} Q_n(x+iy) \geqq 0. \tag{2.6}$$

According to (2.3)

$$Q_n(z) = \prod_{k=1}^{n} \left(\frac{1+(2k-1)/(iz)}{1-(2k-1)/(iz)} \right)^2. \tag{2.7}$$

It follows from (2.4) that

$$|(2k-1)/(iz)| \leqq (2k-1)/(4n^2). \tag{2.8}$$

It is obvious that

$$\arg Q_n(z) = 2 \sum_{k=1}^{n} \left[\arg\left(1+\frac{2k-1}{iz}\right) - \arg\left(1-\frac{2k-1}{iz}\right) \right]. \tag{2.9}$$

It is easy to see that when $|\xi| < 1$

$$|\arg(1+\zeta)| \leqq \arcsin|\zeta| \leqq \frac{\pi}{2}|\zeta|. \tag{2.10}$$

Bearing in mind (2.8), from (2.9) and (2.10) we obtain

$$|\arg Q_n(z)| \leqq 4 \sum_{k=1}^{n} \frac{2k-1}{4n^2} \frac{\pi}{2} = \frac{\pi}{2}.$$

Hence directly follows (2.6) and, consequently, (2.5).

The estimate for $T_n(z)$ for bounded y is based on the following lemma.

Lemma 2.2. If $x, y \in \mathbb{R}$, $y > 0$, $0 < y < 4n^2$, inequality (2.1) holds.

Proof. The modulus of the co-factor q_k from Formula (2.3) will be rewritten in the form

$$|q_k| = \frac{|2k-1-y+ix|^2}{|2k-1+y-ix|^2} = 1 - \frac{4y(2k-1)}{(2k-1+y)^2+x^2}. \tag{2.11}$$

Obviously, since $y > 0$, then $|q_j| < 1 \, \forall j$, therefore $|Q_n(z)| < q_k$, and

$$|1+Q_n(z)| \geqq 1 - |q_k(z)| \geqq 1 - |q_k|, \ \forall k. \tag{2.12}$$

Obviously,

$$\begin{aligned} 1-|q_k| &= \frac{1}{1+x^2} \frac{4y(2k-1)}{(2k-1+y)^2/(1+x^2)+x^2/(1+x^2)} \\ &\geqq \frac{4y(2k-1)}{(2k-1+y)^2+1} \frac{1}{1+x^2}. \end{aligned} \tag{2.13}$$

Let us first consider the case when $y = \operatorname{Im} z$ satisfies the condition

$$2 \leqq y \leqq 2(2n-1). \tag{2.14}$$

In this case integer k exist, $2 < k < n$, such that

$$(2k-2) \leqq y \leqq 2(2k-1).$$

Therefore

$$y(2k-1) \geqq (2k-2)(2k-1), \quad (2k-1+y)^2 \leqq 9(2k-1)^2.$$

Bearing these inequalities in mind, from (2.13) and (2.12) we obtain the estimate

$$|1+Q_n(x+iy)| \geqq C(1+x^2)^{-1}, \qquad C>0, \tag{2.15}$$

C does not depend on k.

Let us now consider the cases

$$0 \leqq y \leqq 2. \tag{2.16}$$

Assuming $k=1$ in (2.13), we obtain the inequality

$$1-|q_1| \geqq 4y/((1+y)^2+x^2) \geqq 4y/(9+x^2).$$

Hence, using (2.12) we obtain the inequality

$$|1+Q_n(z)| \geqq Cy(1+x^2)^{-1}, \qquad C>0. \tag{2.17}$$

Let us now consider the case

$$2(2n-1) \leqq y \leqq 4n^2. \tag{2.18}$$

According to (2.11)

$$|q_k| = 1 - \frac{4(2k-1)/y}{((2k-1)/y+1)^2+x^2/y^2}. \tag{2.19}$$

By virtue of (2.18), since $k < n$, $(2k-1)/y < 1/2$. Therefore

$$1/[(2k-1)/y+1)^2+x^2/y^2)] \geqq 1/[9/4+x^2/y^2]$$

$$\geqq 1/[9/4+x^2/4] = 4/9(1+x^2/9)^{-1} \geqq 4/9(1+x^2)^{-1}.$$

Hence and from (2.19) we obtain:

$$|q_k| \leqq 1 - \frac{16}{9}\frac{(2k-1)}{y}\frac{1}{(1+x^2)}. \tag{2.20}$$

From (2.3) and (2.20) follows the inequality

$$\ln|Q_n(x+iy)| \leqq \sum_{k=1}^{n} \ln\left(1 - \frac{16}{9}\frac{(2k-1)}{y}\frac{1}{(1+x^2)}\right). \tag{2.21}$$

Using the fact that $\ln(1+\alpha) < \alpha$ when $-1 < \alpha$, from (2.21) we obtain the estimate

$$\ln|Q_n(x+iy)| \leqq -\frac{16}{9}\frac{1}{y(1+x^2)}\sum_{k=1}^{n} 2k-1 = -\frac{16}{9y}\frac{n^2}{(1+x^2)}. \tag{2.22}$$

According to (2.18) $n^2/y > 1/4$, and therefore from (2.22) follows the estimate

$$\ln|Q_n(x+iy)| \leqq -4/9(1+x^2)^{-1}. \tag{2.23}$$

Therefore

$$|1+Q_n(x+iy)| \geqq 1 - e^{-4/9(1+x^2)^{-1}}.$$

Since $1 - e^{-t} > e^{-1}t$ when $0 < t < 1$, hence follows inequality (2.15). We shall now note that according to (2.2)

$$|T_n(z)| \geqq \tfrac{1}{2}|S_n(z)||1+Q_n(z)|. \tag{2.24}$$

From inequality (2.15), which holds when $2 < y < 4n^2$, and from inequality (2.17), which holds when $0 < y < 2$, we obtain inequality

(2.1) when

$$0 \leqq y \leqq 4n^2, \ y = \operatorname{Im} z,$$

and the lemma is proved.

Proof of Theorem 2.1. By virtue of the evenness of $T_n(z)$ it is sufficient to take $y > 0$. It follows from Lemma 2.1 that when $y > 4n^2$ inequality (2.5) holds. Using (2.5), from (2.24) we obtain inequality (2.1).

If $0 < y < 4n^2$, the validity of (2.1) follows from Lemma 2.2, and the theorem is proved.

In Theorem 2.1 we obtain the estimate $|T_n(x + iy)|$ through $|S_n(x + i|y|)|$. We shall now obtain the estimates $|S_n(x + iy)|$ when $y > 0$.

We will assume

$$S_n^j(z) = \sum_{k=j}^{n} \left(1 - \frac{iz}{2k-1}\right)^2, \ (S_n^1 = S_n). \tag{2.25}$$

Lemma 2.3. Suppose

$$\operatorname{Re} z \geqq 0, \ 1 \leqq j \leqq n, \ \varphi(\zeta) = \zeta \ln(1+\zeta).$$

Then

$$\begin{aligned} \ln S_n^j(iz) = {} & \varphi(2n+z) - \varphi(2j-2+z) - \varphi(2n) + \varphi(2j-2) \\ & + \theta(n, y, x, j), \qquad |\theta(n, y, x, j)| \leqq c, \end{aligned} \tag{2.26}$$

where C does not depend on

$$n,\ y,\ x,\ j;\ (x=\operatorname{Re} iz,\ y=\operatorname{Im} iz).$$

Proof. Obviously, when $y=\operatorname{Re} z=\operatorname{Im} iz\geqq 0$

$$\ln S_n^j(iz)=2\sum_{k=j-1}^{n-1}[\ln(2k+1+z)-\ln(2k+1)]. \qquad (2.27)$$

The following identity holds:

$$2\ln(2k+1+z)=(2k+2+z)\ln(2k+3+z)-(2k+z)\ln(2k+1+z)$$
$$-(2k+2+z)\ln\left(\frac{1+2k+2+z}{1+2k+z}\right). \qquad (2.28)$$

Substituting the expression $\ln(2k+1+z)$ and $\ln(2k+1),\ (z=0)$ using Formula (2.28) into (2.27) and summing over k, we obtain (2.26), where

$$\theta=\sum_{k=j-1}^{n-1}\left[(2k+2)\ln\left(\frac{1+2k+2}{1+2k}\right)-(2k+2+z)\ln\left(\frac{3+2k+z}{1+2k+z}\right)\right]. \qquad (2.29)$$

We shall now prove the boundedness of θ.

We shall introduce the notation

$$(1+2k+2+z)/(1+2k+z)=1+\alpha, \qquad \alpha=2/(1+2k+z). \qquad (2.30)$$

Since

$$|1+2k+z|^2=(1+2k+y)^2+x^2\geqq 9$$

when $y > 0$, $k > 1$, then $|\alpha| < 2/3$. For these α

$$\ln(1+\alpha)=\alpha-\alpha^2/2+\alpha^3 q(\alpha), \qquad |q(\alpha)|\leqq q_0. \qquad (2.31)$$

Using (2.31) we shall rewrite the terms of the sum (2.29), containing z, in the form

$$(2k+2+z)\left(\frac{2}{1+2k+z}-\frac{2}{(1+2k+z)^2}+\frac{8q(\alpha)}{(1+2k+z)^3}\right)$$

$$2-\frac{2}{(1+2k+z)^2}+\frac{8(2k+2+z)q(\alpha)}{(1+2k+z)^3}=2+\frac{q_1(k,z)}{(1+2k+z)^2}, \qquad |q_1(k,z)|\leqq c.$$

Substituting here $z=0$, we obtain a similar formula for the terms (2.29), that do not contain z. Substituting the expressions obtained into (2.29), and bearing in mind that Re$z > 0$, we obtain

$$|\theta|=\left|\sum_{k=j-1}^{n-1}\left[\frac{q_1(k,z)}{(1+2k+z)^2}-\frac{q_1(k,0)}{(1+2k)^2}\right]\right|\leqq C\sum_{k=j-1}^{n-1}\frac{2}{(1+2k)^2}\leqq C_1,$$

where C_1 is an absolute constant.

Remark 2.1. Using, as in Babin [15], the Euler-Maclauren formula to estimate the sum in (2.27), we can prove that when x = Rez = 0, j = 1 the residual term $\Theta(n, y)$ in (2.26) satisfies the estimate $0 < \Theta(n, y, 0, 1) < 2/g$.

We shall derive from Lemma 2.3 the lower limits, which we shall subsequently need, as corollaries.

Corollary 2.1. Suppose $y > 0$. Then

$$\begin{gathered}\ln S_n^j(iy)\geqq(2n+y)\ln(1+2n+y)-2n\ln(1+2n)\\ -(y+2j-2)\ln(1+y)-C(j),\end{gathered} \tag{2.32}$$

where $C(j)$ does not depend on n or y.

Proof. Using (2.26), where $z=y$, and bearing in mind that in (2.26)

$$\varphi(2j-2+y)=(2j-2+y)\ln(1+2j-2+y)$$
$$=(2j-2+y)\ln(1+y)$$
$$+(2j-2+y)\ln(1+(2j-2)/(1+y)),$$

we obtain (2.32).

Lemma 2.4. Suppose $y > 0$, $iz = x + iy$. Then

$$\operatorname{Re}\ln S_n(iz)\geqq n\ln(1+2n+y)^2+x^2)-2n\ln(1+2n)-c. \tag{2.33}$$

Proof. From (2.26), where $j=1$, we obtain

$$\ln S_n(iz)=(2n+z)\ln(1+2n+z)$$
$$-z\ln(1+z)-2n\ln(1+2n)+\theta$$
$$=2n(\ln(1+2n+z)-\ln(1+2n)) \qquad +z\ln(1+2n/(1+z))+\theta.$$

Hence we obtain

$$\begin{aligned}\operatorname{Re}\ln S_n(iz)&=2n(\ln|1+2n+z|-\ln(1+2n))\\&+y\ln|1+2n/(1+z)|\\&+x\arg(1+2n/(1+z))]+\operatorname{Re}\theta.\end{aligned} \tag{2.34}$$

Obviously,

$$1+2n/(1+z)=1+2n(1+y+ix)/[(1+y)^2+x^2].$$

Since $y > 0$, then $|1 + 2n/(1 + z)| > 1$. Further,

$$x\arg(1+2n/(1+z))$$
$$=x\,\mathrm{arctg}[2nx/((1+y)^2+x^2+2n(1+y))]\geqq 0$$

for all x. Therefore the right-hand side of (2.34) is not less than

$$2n(\ln|1+2n+z|-\ln(1+2n))+\operatorname{Re}\theta,$$

whence we obtain (2.33).

Lemma 2.5. Suppose

$$n \geqq j,\ j \geqq 2, \quad z = x + iy,\ x, y \in \mathbb{R}.$$

Then

$$\begin{aligned} |T_n(x+iy)| &\geqq C\theta(y)(1+x^2+y^2)^{j-2} \\ &\times \exp[(2n+y)\ln(1+2n+y) - 2n\ln(1+2n) \\ &-(2j-2+y)\ln(1+y)], \end{aligned} \tag{2.35}$$

where $\Theta(y)$ is the same as in (2.1), and $C > 0$ does not depend on n or on z.

Proof. It is sufficient to prove (2.35) when $y > 0$. By virtue of (1.7) and (2.25)

$$S_n(z) = S_n^j(z) S_{j-1}(z). \tag{2.36}$$

Since the modulus of each co-factor in (2.25) has the form $(1 + y/(2k - 1))^2 + x^2/(2k - 1)^2$ and increases as $|x|$ increases,

$$|S_n^j(x + i|y|)| \geqq |S_n^j(i|y|)|. \tag{2.37}$$

Estimating

$$\ln|S_n^j(i|y|)|$$

from above using (2.32), and $\ln|S_{j-1}(z)|$ using (2.33), where $n = j - 1$, from (2.36) we obtain

$$\begin{aligned} \ln|S_n(z)| &\geqq (2n+y)\ln(1+2n+y) - 2n\ln(1+2n) \\ &-(y+2j-2)\ln(1+y) + (j-1)\ln((1+2j-2+y)^2+x^2) - C(j). \end{aligned} \tag{2.37'}$$

Since

$$j \geqq 2,\ y \geqq 0,\ (2j-1+y)^2+x^2 \geqq 1+y^2+x^2,$$

from (2.1) and (2.37′) follows (2.35).

Corollary 2.3. Suppose $n > j$, $x, y \in \mathbb{R}$. Then

$$|T_n(x+iy)| \geqq C_0(y, j) n^{|y|} \quad (1+x^2+y^2)^{j-2}, \tag{2.38}$$

where $C_0(y, j) > 0$ when $y > 0$.

Proof. Obviously, on the right-hand side of (2.35)

$$(2n+y)\ln(1+2n+y) - 2n\ln(1+2n) \geqq y\ln(1+2n+y) \tag{2.38'}$$

and further, $y\ln(1+2n+y) \geqq y\ln n$. Using these inequalities, we obtain (2.38) from (2.35).

Henceforth we will need the upper limit $|T_n(z)|$.

Lemma 2.6. Suppose $x, y \in \mathbb{R}$. Then

$$|T_n(x+iy)| \leqq C(y)\,\mathrm{ch}(\pi x/2) n^{|y|}, \tag{2.39}$$

$$|T_n(x+iy)| \le C \exp[\pi|x|/2 + (\pi/2+1)|y| + n] \tag{2.39'}$$

Proof. According to (1.8) and (1.7) when $y > 0$

$$|T_n(x+iy)| \leqq \tfrac{1}{2}\left|\prod_{k=1}^{n}\left(1-\frac{i(x+iy)}{2k-1}\right)^2 + \prod_{k=1}^{n}\left(1+\frac{i(x+iy)}{2k-1}\right)^2\right|$$

$$\leqq \prod_{k=1}^{n}\left|1-\frac{i(x+iy)}{2k-1}\right|^2 = \prod_{k=1}^{n}\left[1+\frac{2y}{2k-1}+\frac{y^2+x^2}{(2k-1)^2}\right] \tag{2.40}$$

$$= \prod_{k=1}^{n}\left(1+\frac{x^2+y^2}{(2k-1)^2}\right)\prod_{k=1}^{n}\left(1+\frac{2y}{2k-1}\left(1+\frac{x^2+y^2}{(2k-1)^2}\right)^{-1}\right)$$

Comparing the last expression with (1.5) and (1.7) and using (1.6) and (1.1), from (2.40) we obtain the inequality:

$$|T_n(x+iy)| \leqq \mathrm{ch}(\pi/2\sqrt{x^2+y^2})[S_n(2iy)]^{1/2}. \tag{2.41}$$

Note that according to (2.26)

$$\ln S_n(2iy) \leqq (2n+2y)\ln(1+2n+2y) \quad -2n\ln(1+2n)-2y\ln(1+2y)+C_1 \tag{2.42}$$

Note that

$$2n(\ln(1+2n+2y)-\ln(1+2n)) \quad =2n\ln(1+2y/(1+2n)) \leqq 2y. \tag{2.43}$$

In a similar way,

$$2y(\ln(1+2n+2y)-\ln n) \ =2y\ln(2+1/n+2y/n) \leqq C_2(y),$$

Bearing these estimates in mind, from (2.41) and (2.42) we obtain the inequality (2.39'). Since

$$\ln S_n(2iy) \leqq 2y\ln n + C(y)+C_1(y)+C_2(y).$$

then bearing in mind (2.43), from (2.42) we obtain:

$$\ln S_n(2iy) \leqq 2y\ln n + C_3(y).$$

Hence and from (2.41) follows inequality (2.39).

§.3. Approximation with the weight ch(R Re z) of functions that are analytic in a strip.

We shall introduce the following notation. We shall use $\mathcal{J}_\beta$ to denote the strip parallel to the real axis;

$$\mathscr{I}_\beta = \{z : |\mathrm{Im}\, z| \leq \beta\}. \tag{3.1}$$

We shall use $\mathscr{A}_q(\mathscr{I}_\beta)$ to denote the set of functions that are analytic in $\mathscr{I}_\beta$ and that increase no faster than a polynomial of the q-th degree

$$|f(z + iy)| < C(1 + |z|)^q \text{ when } |\mathcal{I}_m z| < \beta$$

where $C=C(f)$. We will assume $A(\mathcal{I}_\beta) = A_0(\mathcal{I}_\beta)$.

In this paragraph we will obtain estimates of the rate of approximation of the functions from $\mathscr{A}_q(\mathscr{I}_\beta)$ using polynomials. In the following paragraph we will show that the estimates obtained cannot be improved.

Definition 3.1. We shall use $\Pi_r^{2n-1} g$ to denote, for the function g, which is continuous in $\mathbb{R}$, a polynomial of a degree no higher than $2n-1$, which is identical with g in the roots of the polynomial $T_n(rx)$, i.e. at the points

$$\pm x_k/r, \qquad k=1,\ldots,n,$$

where x_k are the roots of Eq.(1.20).

It is obvious that $\Pi_r^{2n-1} g$ is uniquely defined with respect to g and Π_r^{2n-1} is a projector, since an interpolation polynomial of the n-th degree, calculated for a polynomial of the n-th degree, is identical with the initial polynomial.

Definition 3.2. We shall call the following quantity the deviation of the polynomial $P(z)$ from the function $g(z)$, defined in $\mathscr{I}_\omega$, with the weight ch(R Rez)

$$\mu_{\omega,R}(P,g) = \sup_{|\mathrm{Im}\, z| \leq \omega} \frac{|g(z) - P(z)|}{\mathrm{ch}(R\, \mathrm{Re}\, z)}. \tag{3.2}$$

When ω=0 we shall set

$$\mu_{0,R}(P,g)=\mu_R(P,g).$$

In this case the definition of (3.2l) agrees with (1.4).

Theorem 3.1. Suppose $g \in A_q(\mathcal{J}_\beta)$, $r = 2R/\pi$, and the polynomial

$$P_{2n-1}(x)=\Pi_r^{2n-1}g(x).$$

Then when $0\leqq\omega<\beta,\ 2n\geqq q+8$ the following estimate holds:

$$\mu_{\omega,R}(P_{2n-1},g)\leqq C_\omega n^{-r(\beta-\omega)}\sup_{z\in\mathcal{J}_\beta}\frac{|g(z)|}{1+|z|^q}, \tag{3.3}$$

where C_ω does not depend on n or g.

Proof. The residual term of Lagrange's interpolation formula with interpolation nodes in the roots of the polynomial $T_n(rx)$ is represented by the Hermite formula:

$$g(z)-P_{2n-1}(z)=\frac{T_n(rz)}{2\pi i}\int_\Gamma\frac{g(\zeta)}{T_n(r\zeta)}\frac{d\zeta}{\zeta-z}, \tag{3.4}$$

where Γ is a piecewise-smooth contour surrounding the roots of the polynomial $T_n(r\zeta)$ and the point z, $|\mathrm{Im}\,z|\leqq\omega$, and lying in $\mathcal{J}_\beta$. (The proof of Hermite's formula is given in Markushevich's book [1], for example.) Since the function $g\in\mathcal{A}_q(\mathcal{J}_\beta)$ is bounded in $\mathcal{J}_\beta$, the closed contour Γ can be deformed into a contour consisting of two lines that are parallel to the real axis:

$$\Gamma=\Gamma_\rho=\{z=x\pm i\rho,\ x\in\mathbb{R}\},\qquad (\omega<\rho<\beta). \tag{3.5}$$

According to (3.2) and (3.4)

$$\mu_{\omega,R}(P_{2n-1},g)\leq\frac{1}{2\pi}\sup_{|\mathrm{Im}\,z|\leq\omega}\frac{|T_n(rz)|}{\mathrm{ch}(R\,\mathrm{Re}\,z)}\sup_{\zeta\in\Gamma_\rho}\frac{|g(\zeta)|}{(1+|\mathrm{Re}\,\zeta|)^q}$$
$$\times\sup_{x\in\mathbb{R},|y|\leq\omega}\int_{-\infty}^{\infty}\frac{2\,ds(1+|s|)^q}{|T_n(r(s+i\rho)||s+i\rho-x-iy|}. \tag{3.6}$$

According to Lemma 2.6 when $|y|\leq\omega$

$$|T_n(rx+iry)|/\mathrm{ch}(Rx)\leq e^{R|y|}\,n^{r|y|}\,C(y)\leq C_1(r,\omega)n^{r\omega}. \tag{3.7}$$

Estimating $T_n(r(x+i\rho))$ from below using (2.38), we obtain that when $|y|<\omega$

$$\int_{-\infty}^{\infty}\frac{2(1+|s|)^q\,ds}{|T_n(r(s+i\rho))||s+i\rho-x+iy|}$$
$$\leq Cn^{-r\rho}\int_{-\infty}^{\infty}\frac{(1+|s|)^q\,ds}{(1+r^2s^2)^{j-2}(\rho-\omega)}. \tag{3.8}$$

Taking $j=[q/2]+4$, where [ ] is the integer part, we obtain that the integral on the right-hand side of (3.8) converges. Estimating the right-hand side of (3.6) using (3.7) and (3.8) and allowing ρ to approach β we obtain (3.3).

Let us now consider the important case $g(z)=(i\rho_0-z)^{-1}$ separately. Obviously $(i\rho_0-z)^{-1}\in\mathscr{A}(\mathscr{J}_\beta)$ when $\beta<\rho_0$ and does not belong to $\mathscr{A}(\mathscr{J}_{\rho_0})$. However, by virtue of the singularity $(i\rho_0-z)^{-1}$ on the line $\mathrm{Im}\,z=\rho_0$ having an extremely simple form, the estimate of the rate of convergence using polynomials has the same form as for functions from $\mathscr{A}(\mathscr{J}_{\rho_0})$.

Theorem 3.2. Suppose $g(z)=(i\rho_0-z)^{-1}$ and the polynomial

$$P_{2n-1}(x)=\Pi_r^{2n-1}g,\ r=2R/\pi.$$

Then when $\omega < \rho_0$, $\omega > 0$, $n > 2$, the following estimate holds:

$$\mu_{\omega,R}(P_{2n-1},(i\rho_0-z)^{-1}) \leqq C_\omega n^{-r(\rho_0-\omega)}. \tag{3.9}$$

Proof. We shall use Formula (3.4), where $g(\zeta) = (i\rho_0 - \zeta)$. Since the integrand expression on the right-hand side of (3.4) vanishes at ∞, the integral over the contour surrounding the roots $T_n(r\zeta)$, the point z and the point $i\rho_0$, equals zero. Therefore the integral over Γ equals the integral (3.4), where Γ is replaced by the contour γ, which surrounds the point $i\rho_0$ and is traversed in a negative direction. Calculating the latter integral using the theorem on the residue, we obtain:

$$\frac{1}{(i\rho_0-z)} - P_{2n-1}(z) = \frac{1}{i\rho_0 - z}\,\frac{T_n(rz)}{T_n(ir\rho_0)}. \tag{3.10}$$

Therefore, according to (3.2) and (3.7)

$$\mu_{\omega,R}(P_{2n-1},(i\rho_0-z)^{-1}) \leqq C_1(r,y)n^{r\omega}(\rho_0-\omega)^{-1}(T_n(ir\rho_0))^{-1}. \tag{3.11}$$

Using (2.38), where $j=2$, we obtain (3.9) to estimate $T_n(ir\rho_0)$.

Theorem 3.3. Suppose $g(z) = (\rho_0^2 + z^2)^{-1}$ and the polynomial

$$P^0_{2n-2}(z) = \Pi_r^{2n-1} g(z),\ r = 2R/\pi.$$

Then when $\rho_0 > 0$, $0 \leqq \omega < \rho_0$ the following inequality holds:

$$\mu_{\omega,R}(P^0_{2n-2},(\rho_0^2+z^2)^{-1}) \leqq C_\omega n^{-r(\rho_0-\omega)}. \tag{3.12}$$

Proof. Obviously,

$$(\rho_0^2+z^2)^{-1} = [(i\rho_0+z)^{-1} + (i\rho_0-z)^{-1}]/(-2i\rho_0). \tag{3.13}$$

Taking Formula (3.10), where $-z$ is substituted instead of z, and adding it to (3.10), bearing in mind the evenness of $T_n(z)$ we obtain:

$$\left(\frac{1}{i\rho_0 - z} + \frac{1}{i\rho_0 + z}\right) - (P_{2n-1}(z) + P_{2n-1}(-z)) \\ = \frac{T_n(rz)}{T_n(ir\rho_0)}\left(\frac{1}{i\rho_0 - z} + \frac{1}{i\rho_0 + z}\right). \tag{3.14}$$

Using (3.13) and putting

$$P^0_{2n-2} = (P_{2n-1}(z) + P_{2n-1}(-z))/(-2i\rho_0),$$

(this polynomial is even, and is of the $2n-2$ degree), from (3.14) we obtain:

$$\frac{1}{\rho_0^2 + z^2} - P^0_{2n-2}(z) = \frac{T_n(rz)}{T_n(ir\rho_0)} \frac{1}{\rho_0^2 + z^2}. \tag{3.15}$$

Since the right-hand side vanishes in the roots of the polynomial $T_n(rz)$, $P^0_{2n-2}(z)$ is indeed an interpolation polynomial of the function $(\rho_0^2 + z^2)^{-1}$. Estimating $T_n(rz)$ on $\mathcal{I}_\omega$ from above, and $T_n(ir\rho_0)$ from below using (2.39) and (2.38), we obtain (3.12).

Theorem 3.4. The polynomial $P^0_{2n-2}(x)$ constructed in Theorem 3.3, is an even polynomial of the $2n-2$ degree, and its deviation on $\mathbb{R}$ from $(\rho_0^2 + x^2)^{-1}$ with the weight $\Phi_n(Rx)/(\rho_0^2 + x^2)$ where Φ_n is determined by (1.5), is the least.

Proof. Suppose P_{2n-2} is some polynomial of the $2n-2$ degree. Obviously,

$$\sup_x \frac{|(\rho^2 + x^2)^{-1} - P_{2n-2}(x)|}{\Phi_n(R_x)/(\rho_0^2 + x^2)} = \sup_x \frac{|1 - P_{2n-2}(x)(\rho_0^2 + x^2)|}{\Phi_n(Rx)}. \tag{3.16}$$

We shall put

$$Q_{2n}(x)=1-P_{2n-2}(x)(\rho_0^2+x^2).$$

It is obvious that $Q_{2n}\in\Upsilon\Upsilon_n(\rho_0)$ (see (1.12)). According to Theorem 1.1, the polynomial which deviates the least from zero with the weight $\Phi_n(Rx)$ is identical with $T_n(rx)/T_n(ir\rho_0)$. Since it follows from (3.15) that

$$1-(\rho_0^2+x^2)P_{2n-2}^0(x)=T_n(rx)/T_n(ir\rho_0),$$

we obtain that the left-hand side of (3.16) is minimal when

$$P_{2n-2}(x)=P_{2n-2}^0(x),$$

Q.E.D.

§ 4. The order of error of the best approximation of the function $(\rho^2+x^2)^{-1}$ with the weight $\operatorname{ch}(Rx)$.

In the previous paragraph we constructed the polynomials P_{2n-2}^0, such that

$$\mu_R(P_{2n-2}^0,(x^2+\rho^2)^{-1})\leqq Cn^{-2R\rho/\pi} \tag{4.1}$$

(see (3.12) where $\omega=0$). According to Theorem 3.4 the polynomials P_{2n-2}^0 gave the best approximation on $\mathbb{R}$ with the weight $\Phi_n(Rx)/(x^2+\rho^2)$. Since $\operatorname{ch}(Rx)>\Phi_n(Rx)$, the best approximation with the weight $\operatorname{ch}(Rx)$ using the polynomial $P_{2n-1}(x)$ gives the smallest deviation,

$$\mu_R(P_{2n-1}^*,(x^2+\rho^2)^{-1})<\mu_R(P_{2n-2}^0,(x^2+\rho^2)^{-1}). \tag{4.2}$$

However, as can be seen from the following theorem, the order of error of the best approximation is the same as in (4.1).

Theorem 4.1. For any R, $\rho > 0$ the constant $C_0 > 0$ exists, depending only on R and ρ, such that for any polynomial $P_{2n-1}(x)$ of the $2n-1$ degree

$$\mu_R(P_{2n-1},(x^2+\rho^2)^{-1}) \geqq C_0 n^{-2R\rho/\pi}. \tag{4.3}$$

Thus, the estimates of the rate of approximation using polynomials with the weight ch (Rx) on a straight line of the function $(x^2 + \rho^2)^{-1}$, obtained in the previous paragraph, cannot be improved. Since

$$(x^2+\rho^2)^{-1} \in \mathscr{A}(\mathscr{I}_\beta), \ \forall \beta < \rho,$$

the estimates of the rate of approximation of the functions from $\mathscr{I}_\beta$ on $\mathbb{R}$ cannot be improved either.

The proof of Theorem 4.1 will be based on two important results of the theory of functions. To simplify the discussion we shall give their proof.

Lemma 4.1. (Poisson's formula). Suppose $Q(t)$ is a polynomial, $\rho > 0$. Then

$$\frac{\rho}{\pi}\int_{-\infty}^{\infty}\frac{\ln|Q(t)|}{\rho^2+t^2}dt \geqq \ln|Q(i\rho)|. \tag{4.4}$$

Proof. Suppose, first, $Q(z) = z - z_0$. Note that $\ln|Q(t)| = \operatorname{Re}\ln Q(t)$. We shall deform the straight line, along which integration is carried out in (4.4), into a circle with center at the point ρi, if $\operatorname{Im} z_0 < 0$ or with center at the point $-\rho i$, if $\operatorname{Im} z_0 < 0$. Calculating the integral, we obtain that the left-hand side of (4.4) equals

$$\operatorname{Re}\ln(\rho i - z_0) = \ln|\rho i - z_0|,$$

if $\mathrm{Im} z_0 < 0$ and equals $\ln|-\rho i - z_0|$ if $\mathrm{Im} z_0 > 0$. Therefore when $\mathrm{Im} z_0 < 0$ in (4.4) we have an equality for $Q(z) = z - z_0$. When

$$\mathrm{Im}\, z_0 > 0 \qquad \ln|-\rho i - z_0| > \ln|\rho i - z_0|$$

there is an inequality in (4.4).

If $Q(z)$ is a polynomial of arbitrary degree, we obtain (4.4), expanding Q into multipliers of the first degree.

Lemma 4.2. (Markov's inequality). Suppose the polynomial $Q\nu$ is of a degree no higher than ν. Then when $x \in \mathbb{R}$, $|x| > 1$

$$|Q_\nu(x)| \leqq M T_\nu^*(x), \qquad M = \max_{-1 \leqq x \leqq 1} |Q_\nu(x)|, \tag{4.5}$$

where $T_\nu^*(x)$ is a Chebyshev polynomial of the ν-th degree,

$$T_\nu^*(x) = \tfrac{1}{2}\{(x + \sqrt{x^2 - 1})^\nu + (x - \sqrt{x^2 - 1})^\nu\}. \tag{4.6}$$

Proof. It is obvious that it is sufficient to prove the lemma when M=1. Note that Chebyshev's polynomial, determined by (4.6), has in [-1, +1] a deviation from zero that equals 1. It takes the values

$$(-1)^{\nu - k}, k = 0, 1, \ldots, \nu$$

at the points

$$x_k = -\cos(k\pi/\nu)$$

Consider a subsidiary problem: to obtain the polynomial $Q_\nu(x)$ of the ν-th degree, which satisfies the following condition at some point x_0:

$$Q_\nu^0(x_0) = T_\nu^*(x_0), \qquad (|x_0| > 1) \tag{4.7}$$

and which deviates the least from zero at [-1, 1].

Since $T_\nu^*(x)$ satisfies condition (4.7), the deviation of the required polynomial must be not more than 1, i.e.

$$\sup_{|x| \leqq 1} |Q_\nu^0(x)| \leqq 1. \tag{4.8}$$

Using the fact that $T_\nu^*(x)$ takes the values ± 1 alternately at the $\nu + 1$ points, as in the proof of Theorem 1.1 we will obtain that the polynomial

$$\mathscr{D}(x) = T_\nu^*(x) - Q_\nu^0(x)$$

vanishes at [-1, 1] at at least the ν points. It follows from (4.7) that $\mathcal{D}(x)$ has not less than $\nu+1$ zeros in $\mathbb{R}$, i.e. $\mathcal{D}(x) \equiv 0$ and $Q_\nu^0 = T_\nu^*$. Thus, for any polynomial $Q_\nu^0(x)$, satisfying (4.7), the following estimate holds:

$$\sup_{|x| \leqq 1} |Q_\nu^0(x)| \geqq 1. \tag{4.9}$$

Let us now suppose $Q_\nu(x)$ is an arbitrary polynomial of the degree ν, $Q_\nu(x_0) \neq 0$. We will assume

$$Q_\nu^0(x) = Q_\nu(x) T_\nu^*(x_0)/Q_\nu(x_0).$$

Obviously, Q_ν^0 satisfies (4.7). Substituting Q_ν^0 into (4.9), we obtain (4.5).

Proof of Theorem 4.1. Suppose $P_{2n-1}(x)$ is a polynomial of a degree no higher than $2n-1$. We will assume $\varepsilon = \mu(P_{2n-1}, (x^2 + \rho^2)^{-1}$. Obviously,

$$|(x^2+\rho^2)^{-1}-P_{2n-1}(x)|\leqq\varepsilon\,\mathrm{ch}(Rx).$$

Hence we obtain

$$|1-(x^2+\rho^2)P_{2n-1}(x)|\leqq\varepsilon\,\mathrm{ch}(Rx)(\rho^2+x^2). \tag{4.10}$$

We will assume

$$V(x)=\varepsilon^{-1}(1-(x^2+\rho^2)P_{2n-1}(x)). \tag{4.11}$$

By virtue of (4.10)

$$|V(x)|\leqq\mathrm{ch}(Rx)(\rho^2+x^2), \qquad x\in\mathbb{R}. \tag{4.12}$$

We shall use Formula (4.4), where we will assume $Q(t)=V(t)$. Since by virtue of (4.10) $V(\rho i) = \varepsilon^{-1}$, from (4.4) we obtain:

$$\frac{\rho}{\pi}\int_{-\infty}^{\infty}\frac{\ln|V(t)|}{\rho^2+t^2}\,dt\geqq\ln(\varepsilon^{-1}). \tag{4.13}$$

We shall now use inequality (4.12) to make an upper estimate of the left-hand side of (4.13). In this estimate we wish to reflect the dependence on the number ν - the degree of $V(t)$ - which by virtue of (4.11) equals $2n+1$. Making the substitution $x = y/\xi$, $\xi > 0$ in Formula (4.5), we obtain:

$$\left|Q_\nu\left(\frac{y}{\xi}\right)\right|\leqq\left[\max_{-\xi\leqq y\leqq\xi}\left|Q_\nu\left(\frac{y}{\xi}\right)\right|\right]T_\nu\left(\frac{y}{\xi}\right) \quad (|y|\geqq\xi). \tag{4.14}$$

Note that when

$$|x|\geqq 1,\ |T_\nu^*(x)|\leqq|x|^\nu 2^\nu$$

(this is immediately obvious from (4.6)). Therefore, formulating $Q_n(y/\xi) = Q_n^1(y)$ for any polynomial $Q_\nu^1(y)$ of a degree no higher than ν we obtain:

$$|Q_\nu^1(y)| \leqq \left|\frac{2y}{\xi}\right|^\nu \max_{-\xi \leqq t \leqq \xi} |Q_\nu^1(t)|, \qquad (|y| \geqq \xi > 0). \tag{4.15}$$

Using (4.12) and (4.15), where $Q_\nu^1 = V$, $\nu = 2n + 1$, we obtain

$$|V(y)| \leqq |2y/\xi|^{2n+1} \operatorname{ch}(R\xi)(\rho^2+\xi^2), \qquad (|y| \geqq \xi > 0). \tag{4.16}$$

Using estimates (4.16) and (4.12), we shall make an upper estimate of the right-hand side of (4.13):

$$\begin{aligned} \frac{\rho}{\pi}\int_{-\infty}^{\infty} \frac{\ln|V(t)|\,dt}{\rho^2+t^2} &\leqq \frac{2\rho}{\pi}\int_0^\xi \frac{\ln \operatorname{ch}(Rt)+\ln(\rho^2+t^2)}{\rho^2+t^2}\,dt \\ &+ \frac{2\rho}{\pi}\int_\xi^\infty [\ln(2t/\xi)^{2n+1} + \ln \operatorname{ch}(R\xi) + \ln(\rho^2+\xi^2)](\rho^2+t^2)^{-1}\,dt. \end{aligned} \tag{4.17}$$

Obviously

$$\begin{aligned} \int_\xi^\infty \frac{\ln(2t/\xi)^{2n+1}}{\rho^2+t^2}\,dt &= \frac{(2n+1)\xi}{2}\int_2^\infty \frac{\ln z}{\rho^2+z^2\xi^2/4}\,dz \\ &\leqq (n+1)\xi \int_2^\infty \frac{\ln z}{z^2\xi^2/4}\,dz \leqq 4(n+1)/\xi. \end{aligned} \tag{4.18}$$

Further,

$$\int_0^\xi \frac{\ln \operatorname{ch}(Rt)}{\rho^2+t^2}\,dt \leqq \int_0^\xi \frac{Rt}{\rho^2+t^2}\,dz = \frac{R}{2}\ln\left(\frac{\rho^2+\xi^2}{\rho^2}\right),$$

$$\begin{aligned}
&\int_0^\xi \frac{\ln(\rho^2+t^2)}{\rho^2+t^2}\,dt \leqq C_0(\rho),\\
&\int_\xi^\infty \frac{\ln(\rho^2+\xi^2)}{\rho^2+t^2}\,dt = \frac{\ln(\rho^2+\xi^2)}{\rho}\operatorname{arctg}\frac{\rho}{\xi} \leqq C_1(\rho),\\
&\int_\xi^\infty \frac{\ln \operatorname{ch}(R\xi)}{\rho^2+t^2}\,dt \leqq R\xi\int_\xi^\infty \frac{dt}{\rho^2+t^2} = \frac{R\xi}{\rho}\operatorname{arctg}\frac{\rho}{\xi} \leqq R\cdot C_2,
\end{aligned} \tag{4.19}$$

From (4.17), (4.18) and (4.19) we obtain:

$$\begin{aligned}
&\frac{\rho}{\pi}\int_{-\infty}^{\infty} \frac{\ln|V(t)|}{\rho^2+t^2}\,dt\\
&\leqq \frac{2\rho}{\pi}\left\{\frac{R}{2}\ln\left(\frac{\rho^2+\xi^2}{\rho^2}\right) + C_0(\rho) + \frac{4(n+1)}{\xi} + RC_2 + C_1(\rho)\right\}.
\end{aligned} \tag{4.20}$$

Assuming $\xi = n+1$ in (4.20), from (4.13) and (4.20) we obtain:

$$\begin{aligned}
\ln(\varepsilon^{-1}) &\leqq \frac{R\rho}{\pi}\ln(\rho^2+(n+1)^2) + C_3(\rho) \leqq \frac{2R\rho}{\pi}\ln n\\
&+\frac{R\rho}{\pi}\ln\left(\frac{\rho^2}{n^2}+\frac{(n+1)^2}{n^2}\right) + C_3(\rho) \leqq \frac{2R\rho}{\pi}\ln n + C_4(\rho).
\end{aligned}$$

Therefore

$$\varepsilon \geqq e^{C_4} n^{-2R\rho/\pi},$$

and since $\varepsilon = \mu_R(P_{2n-1}, (x^2+\rho^2)^{-1})$, hence directly follows (4.3).

§ 5. Estimates of the rate of convergence of functions from $A(\mathcal{J}_\infty)$.

We shall use $A_q(\mathcal{J}_\infty)$ to denote the set of entire functions g, belonging to $A_q(\mathcal{J}_\beta)$ for any $\beta > 0$.

We will assume for $g \in A_q(\mathcal{J}_\infty)$

$$g_q^*(\beta) = \sup_{|\mathcal{T}m\, z| \leqq \beta} |g(z)|/(1+|z|)^q, \qquad g_0^*(\beta) = g^*(\beta). \tag{5.0}$$

Theorem 5.1. Suppose

$$g\in\mathscr{A}_q(\mathscr{J}_\infty),\ q\geqq 0,\ R>0,\ r=2R/\pi,$$

and the polynomial $P_{2n-1}\Pi_r^{2n-1}g$. Then when $\omega\geqq 0$ for all fairly large n

$$\mu_{\omega,R}(P_{2n-1},g)\leqq C(\omega,r,q)n^{r\omega}\sigma_n(r,\omega,g), \tag{5.1}$$

where

$$\begin{aligned}\sigma_n(r,\omega,g)&\\ =\inf_{y\geqq(\omega+1)r}&\{g_q^*(y/r)\exp[-(2n+y)\ln(1+2n+y)\\ &+2n\ln(1+2n)+(q+8+y)\ln(1+y)]\}.\end{aligned} \tag{5.2}$$

Making a lower estimate of $|T_n(r(s+i\rho)|$ on the right-hand side of (3.6) using inequality (2.35), where $j=[q/2]+4$, and using (3.7), from (3.6) we obtain when $\rho\geqq\omega+1$ in the same way as in (3.3):

$$\begin{aligned}\mu_{\omega,R}(P_{2n-1},g)&\leqq C(r,\omega)n^{r\omega}g_q^*(\rho)\\ &\times\exp[-(2n+r\rho)\ln(1+2n+r\rho)+2n\ln(1+2n)\\ &+(q+8+r\rho)\ln(1+r\rho)].\end{aligned} \tag{5.3}$$

Making the substitution $r\rho=y$, and bearing in mind that the left-hand side of (5.3) does not depend on ρ, from (5.3) we obtain inequality (5.1).

Remark 5.1. Since

$$\mathscr{A}_q(\mathscr{J}_\infty)\subset\mathscr{A}_q(\mathscr{J}_\beta),\quad\forall\beta,$$

by virtue of Theorem 3.1 $\mu_{\omega,R}(P_{2n-1}, g) \to 0$ as $n \to \infty$ faster than $n^{-\beta+\omega}$ for any $\beta > 0$.

We shall now obtain estimates of the rate of approximation for the functions $g(Z)$, which satisfy the specific conditions of increase as $\operatorname{Im} z \to \infty$. We shall consider functions of two types: the type $e^{-t\bar{z}^2}$ and the type e^{itz}. These functions will be encountered when examining parabolic and hyperbolic equations in the following chapters.

We shall use $A_{p,t}(\mathcal{J}_\infty)$ to denote a set of entire functions $g(z) \in A_0(\mathcal{J}_\infty)$ for each of which the constant C exists, such that

$$\sup_{|\mathcal{J}m z| \leq \rho} |g(z)| \leq C e^{t\rho^p}. \tag{5.4}$$

Theorem 5.2. Suppose $g \in A_{p,t}(\mathcal{J}_\infty)$, $p > 1$. Suppose $R > 0$, $r = 2R/\pi$, $\omega \geq 0$. Suppose $P_{2n-1} = \Pi_r^{2n-1} g$ is an interpolation polynomial of the function g. Then for fairly large n

$$\sigma_n(r, \omega, g) \leq \exp(-(p-1-\varepsilon)(r \ln n/p)^{p/(p-1)} t^{-1/(p-1)}), \tag{5.5}$$

where $\varepsilon > 0$ is as small as desired.

Proof. According to the definition of σ_n in (5.2), where $q=0$, and (5.4), we have:

$$\sigma_n(r, \omega, g) \leq C_1 \inf_{y \geq (\omega+1)r} e^{\varphi(y)}, \tag{5.6}$$

where

$$\varphi(y) = \tau y^p - (2n+y)\ln(1+2n+y)$$
$$+ 2n \ln(1+2n) + (y+8)\ln(1+y), \qquad (\tau = t/r^p). \tag{5.7}$$

In order to estimate the lower bound on the right-hand side of (5.6), it is sufficient to substitute the specific value y into $\varphi(y)$. We shall take the solution of the equation $d\varphi/dy = 0$ as this y. This equation has the form

$$\varphi'(y)=\tau p y^{p-1}-\ln(1+2n+y)+\ln(1+y)-\frac{2n+y}{1+2n+y}+\frac{y+8}{1+y}=0. \tag{5.8}$$

We transform it to the form

$$\frac{\tau p y^{p-1}}{\ln(2n)}=\frac{\ln(1+2n+y)}{\ln 2n}-\frac{\ln(1+y)}{\ln 2n}+\left(\frac{2n+y}{1+2n+y}\right)\frac{1}{\ln 2n}-\left(\frac{y+8}{1+y}\right)\frac{\Delta}{\ln 2n}. \tag{5.9}$$

Since we are interested in the behaviour of σ_n as $n\to\infty$, we shall seek not an exact, but an asymptotic solution $y_0 = y_0(n)$ of Eq.(5.9). We seek the solution $y_0(n)$, which satisfies the conditions

$$y_0(n) \to \infty \text{ as } n \to \infty,\ y_0(n)/n \to 0 \text{ as } n \to \infty. \tag{5.10}$$

If $y = y_0(n)$ satisfies (5.10), the right-hand side of (5.9) obviously approaches 1 as $n\to\infty$. We shall take as the approximate solution

$$y_0(n)=(\ln(2n)/(\tau p))^{1/(p-1)}. \tag{5.11}$$

Obviously, $y_0(n)$ satisfies (5.10). At the same time Eq.(5.9) is satisfied apart from terms of the order $\ln\ln n/\ln n$.

We shall substitute $y = y_0(n)$, determined using (5.11), into $\varphi(y)$. We shall isolate the terms that have the greatest increase with respect to n. Obviously, by virtue of (5.11)

$$\begin{aligned}\tau(y_0(n))^p-y_0(n)\ln(1+2n+y_0(n))&=y_0(n)(\tau(y_0(n))^{p-1}-\ln(2n))\\ -y_0(n)\ln(1+(1+y_0(n))/(2n))&=y_0(n)\tau(1-p)(y_0(n))^{p-1}+o(y_0(n)).\end{aligned} \tag{5.12}$$

Since $p>1$, the remaining terms in (5.7), where $y = y_0(n)$, have the smallest order of increase as $n\to\infty$:

$$\begin{aligned}&-2n(\ln(1+2n+y_0(n))-\ln(1+2n))=-2n(\ln(1+y_0(n)/(1+2n))\\ &=o(y_0(n)),\ (y_0(n)+8)\ln(1+y_0(n))=o((y_0(n))^{1+\varepsilon_1})),\quad \forall\varepsilon_1>0.\end{aligned}$$

Therefore, according to (5.12),

$$\varphi(y_0(n))=\tau(1-p)(y_0(n))^p+o((y_0(n))^p),$$

and, consequently, by virtue of (5.11)

$$\varphi(y_0(n))\leqq-(p-1-\varepsilon_2)t^{-1/(p-1)}(r\ln n/p)^{p/(p-1)} \tag{5.13}$$

for fairly large n, where $\varepsilon_2>0$ is as small as desired. Since $y_0(n) > (\omega+1)r$ for large n, (5.5) follows from (5.6) and (5.13).

We shall consider the limiting value $p=1$ separately.

Theorem 5.3. Suppose the assumptions of Theorem 5.2 hold when $p=1$. Then for any $\varepsilon>0$ for fairly large n

$$\sigma_n(r,\omega,g)\leqq[(1+\varepsilon)(1-e^{-t/r})]^{2n}. \tag{5.14}$$

Proof. We shall seek the asymptotic solution $y_0(u)$ of Eq.(5.8), satisfying the condition

$$y_0(n)\to\infty,\ y_0(n)/n\to\gamma\neq 0 \text{ as } n\to\infty.$$

We rewrite (5.8), where $p=1$, in the form

$$\ln\frac{1+2n+y}{1+y}=\tau+\frac{1}{1+2n+y}+\frac{7}{1+y}. \tag{5.15}$$

The right-hand side of (5.15) approaches τ as $n\to\infty$. We take as $y_0(n)$ the solution (5.15) with a changed right-hand side, which equals τ. We obtain:

$$y_1(\tau)=(1+2n-e^{\tau})/(e^{\tau}-1).$$

Isolating, in this expression, the principal part as $n \to \infty$, we shall put

$$y_0(n) = 2n/\lambda, \qquad \lambda = e^{\tau} - 1. \tag{5.16}$$

Substituting $y = y_0(n)$ into (5.7), where $p=1$, we obtain:

$$\varphi(y_0(n)) = 2n\tau/\lambda - 2n\ln((1+2n+2n/\lambda)/(1+2n))$$

$$-2n/\lambda \ln((1+2n+2n/\lambda)/(1+2n/\lambda)) + 8\ln(1+2n/\lambda).$$

Isolating the principal part here, for large n we have::

$$\begin{aligned} &\varphi(y_0(n)) = 2n(\tau/\lambda - \ln(1+1/\lambda) - 1/\lambda \ln(\lambda+1) + \varphi_1(n), \\ &|\varphi_1(n)| \leqq C|\ln n|. \end{aligned} \tag{5.17}$$

Bearing in mind that

$$\ln(1+\lambda) = \tau, \qquad \ln(1+1/\lambda) = -\ln(1-e^{-\tau}),$$

from (5.17) we obtain that for any $\varepsilon > 0$ for fairly large n

$$\varphi(y_0(n)) \leqq 2n(\ln(1+\varepsilon) + \ln(1-e^{-\tau})), \tag{5.18}$$

whence follows (5.14).

§ 6. Approximation of functions that are analytic on a semiaxis.

We shall use $\mathcal{J}_\beta^+$ to denote the domain, bounded by a parabola, that surrounds the real semiaxis

$$\mathcal{J}_\beta^+ = \{\lambda = \xi + i\eta : \xi \geqq \eta^2/(4\beta^2) - \beta^2\}. \tag{6.1}$$

It is easy to see that under the mapping $\lambda = z^2$ the strip $\mathcal{J}_\beta$, determined by (3.1), becomes $\mathcal{J}_\beta^+$.

We shall use $A(\mathcal{J}^+_\beta)$ to denote a set of functions that are analytic in $\mathcal{J}^+_\beta$ and bounded in $\mathcal{J}^+_\beta$.

If $P(\lambda)$ is a polynomial, $g^+ \in A(\mathcal{J}^+_\beta)$, we shall use $\mu^+_{\omega,R}$, $\omega < \beta$, to denote the deviation of P from g in $\mathcal{J}^+_\omega$ with the weight $\mathrm{ch}(R\,\mathrm{Re}\sqrt{\lambda})$:

$$\mu^+_{\omega,R}(P,g^+) = \sup_{\lambda\in\mathcal{J}^+_\omega} \frac{|g^+(\lambda)-P(\lambda)|}{\mathrm{ch}(R\,\mathrm{Re}\sqrt{\lambda})}. \tag{6.2}$$

For brevity, we shall denote $\mu^+_{0,R}$ by μ^+_R.

We shall use $T_n^+(\lambda)$ to denote a polynomial of the n-th degree, which is determined when $\lambda > 0$ by the equation

$$T_n^+(\lambda) = T_n(\sqrt{\lambda}), \tag{6.3}$$

where $T_n(z)$ is determined by Formula (1.1). Since $T_n(z)$ is an even polynomial, $T_n^+(\lambda)$ is indeed a polynomial.

According to Lemma 1.4 the polynomial $T_{n+1}(z)$ has $n+1$ positive roots $x_1, \ldots, x_{n+1}$, which are solutions of Eq.(1.20). It is obvious that $\lambda_j = x_j^2$, $j = 1, \ldots, n+1$ are roots of the polynomial $T^+_{n+1}(\lambda)$. Therefore for any function $g^+ \in A(\mathcal{J}^+_\beta)$ the interpolation polynomial $P_n(\lambda)$ with interpolation roots $\lambda j/r^2$, $j = 1, \ldots, n+1$ is defined, which we shall formulate as

$$P_n(\lambda) = \Pi_r^{+n} g^+(\lambda). \tag{6.4}$$

We shall now derive from the theorems on approximation of functions from $A(\mathcal{J}_\beta)$, proved in §§ 3 and 5, theorems on the approximation of functions from $A(\mathcal{J}^+_\beta)$. For this we shall use the following lemma.

Lemma 6.1. Suppose $P_{2n+1} = \Pi_r^{2n+1} g$, where g is an even function. Then P_{2n+1} is an even polynomial of the $2n$-th degree.

If the functions $g(z)$ and $g^+(\lambda)$ are connected by the relations $g^+(\lambda) = g(\sqrt{\lambda})$ when $\lambda > 0$, then

$$\Pi_r^{+n} g^+(\lambda) = P_{2n+1}(\sqrt{\lambda}). \tag{6.5}$$

At the same time the following equality holds for the polynomial $P_n = \Pi_r^{+n} g^+$

$$\mu^+_{\omega,R}(P_n, g^+) = \mu_{\omega,R}(P_{2n+1}, g). \tag{6.6}$$

Proof. Since the polynomial $T_{n+1}(rx)$ is even, its roots x_1, ..., x_{2n+2} are symmetrically arranged with respect to zero. Therefore the interpolation polynomial P_{2n+1} with interpolation nodes in these roots of the even function $g^+(x^2)$ is also even and, consequently, its degree is not higher than $2n$. By virtue of the definition P_{2n+1} at the points x_j $P_{2n+1}(x_j) = g(x_j)$. Therefore

$$P_{2n+1}(\sqrt{x_j^2}) = g(\sqrt{x_j^2}).$$

Since the numbers x_j^2 are roots of the polynomial

$$T^+_{n+1}(r^2\lambda) = T_{n+1}(r\sqrt{\lambda}),$$

(6.5) holds.

Since according to (6.5)

$$P_n(\lambda) = \Pi_r^{+n} g^+(\lambda) = P_{2n+1}(\sqrt{\lambda}),$$

then

$$\sup_{\lambda \in \mathcal{I}^+_\beta} \frac{|P_n(\lambda) - g^+(\lambda)|}{\mathrm{ch}(R\,\mathrm{Re}\sqrt{\lambda})} = \sup_{\lambda \in \mathcal{I}^+_\beta} \frac{|P_{2n+1}(\sqrt{\lambda}) - g(\sqrt{\lambda})|}{\mathrm{ch}(R\,\mathrm{Re}\sqrt{\lambda})} = \sup_{z \in \mathcal{I}_\beta} \frac{|P_{2n+1}(z) - g(z)|}{\mathrm{ch}(R\,\mathrm{Re}\, z)},$$

whence follows (6.6).

Theorem 6.1. If

$$g^+\in\mathscr{A}(\mathscr{J}_\beta^+),\ r=2R/\pi,$$

and the polynomial $P_n = \Pi_r^{+\,n-1}g^+$ then when $0 < \omega < \beta$

$$\mu_{\omega,R}^+(P_{n-1},g^+)\leqq C_\omega n^{-r(\beta-\omega)} \times\sup\{M: M=|g^+(\lambda)|,\lambda\in\mathscr{J}_\beta^+\}. \tag{6.7}$$

Theorem 6.2. If

$$g^+(\lambda)=(\lambda+\rho_0^2)^{-1},\ r=2R/\pi, \qquad P_{n-1}=\Pi_r^{+\,n-1}g^+,$$

then when $0 < \omega < \rho_0$

$$\mu_{\omega,R}^+(P_{n-1},(\lambda+\rho_0^2)^{-1})\leqq C_\omega n^{-r(\rho_0-\omega)}. \tag{6.8}$$

Theorem 6.3. If

$$g^+(z^2)=g(z)\in\mathscr{A}(\mathscr{J}_\infty),\ r=2R/\pi, \qquad P_{n-1}=\prod\nolimits_r^{+\,n-1}g^+,$$

then when $0 < \omega < \infty$

$$\mu_{\omega,R}^+(P_{n-1},g^+)\leqq C_\omega n^{r\omega}\sigma_n(r,\omega,g), \tag{6.9}$$

where σ_n is determined by Formula (5.2), in which $q=0$.

Theorem 6.4. If

$$g^+(z^2)=g(z)\in\mathscr{A}_{p,t}(\mathscr{J}_\infty),\ p>1, \qquad P_{n-1}=\Pi_r^{+\,n-1}g^+,\ r=2R/\pi,$$

then when $0 < \omega < \infty$ inequality (6.9) holds, where $\sigma_n(r, \omega, g)$ satisfies inequality (5.5) for large n.

Theorem 6.5. If

$$g^+(z^2)=g(z)\in\mathscr{A}_{1,t}(\mathscr{I}_\infty), \qquad P_{n-1}=\Pi_r^{+n-1}g^+,\ r=2R/\pi,$$

Formula (6.9) holds, in which $\sigma_n(r, \omega, g)$ satisfies inequaity (5.14) for large n.

Proof of Theorems 6.1 - 6.5. If

$$g^+(\lambda)\in\mathscr{A}(\mathscr{I}_\beta^+),$$

then $g^+(z^2) = g(z) \in A(\mathcal{J}_\beta)$. We use Theorem 3.1, where q=0, and Lemma 6.1, where instead of n we substitute n-1, and from (3.3) and (6.6) we obtain inequality (6.7).

We obtain inequality (6.8) in a similar way, if we use inequality (3.12) instead of (3.3).

We obtain inequality (6.9) from (6.6) and (5.1), and the number σ_n satisfies (5.2) by virtue of Theorem 5.1.

Theorem 6.4 follows from Theorems 6.3 and 5.2.

Theorem 6.5 follows from Theorems 6.3 and 5.3.

Remark 6.1. Similar theorems also hold for functions from $A_q(\mathcal{J}_\beta^+)$ which have a power increase in $\mathcal{J}_\beta^+$.

CHAPTER 4. CONSTRUCTION OF FUNCTIONS OF SELF-ADJOINT OPERATORS WITH ANALYTIC COEFFICIENTS

In this chapter we will obtain formulae expressing solutions of stationary equations, and also Cauchy problems for nonstationary equations in terms of polynomials of the differential operator occurring in these equations. The polynomial representations obtained are used in the next chapter to investigate the smoothness of solutions of degenerating equations.

§ 1. Polynomial representations of functions of a self-adjoint operator in a Hilbert space.

Suppose H is a Hilbert space, and the self-adjoint operator A acts in H with the domain of definition $\mathcal{D}(A)$, whilst A is semi-bounded from below. If $g(\lambda)$ is a function that is continuous on the spectrum, the function $g(\lambda)$ of the operator A is defined by Formula (1.3.2).

Note that when A is a differential operator, in rare cases we are able to construct explicitly the projectors E_λ, which occur in this formula. Therefore in this chapter we will construct $g(A)f$ in constructive form

$$g(A)f = \lim_{n\to\infty} P_n(A)f, \tag{1.1}$$

where P_n is some polynomial of the n-th degree, which depends on the function $g(\lambda)$.

Remark 1.1. If $A = A_0 + bI$, $b \in \mathbb{C}$, Formula (1.1) takes the form

$$g(A_0 + bI)f = \lim P_n(A_0 + bI)\, f.$$

Since $P_n(A_0 + bI)$ is a polynomial of the variable A_0 of the same degree as A, and $g(A_0 + bI) = g_1(A_0)$, to obtain the representation $g(A)f$ using Formula (1.1) it is sufficient to obtain the representation $g_1(A_0)f$ using this formula.

Since we can represent the operator A, which is semi-bounded from below, in the form $A = A_0 + bI$, we will henceforth assume, for simplicity, that the operator A is nonnegative, i.e. Formula (2.1.1) holds.

Remark 1.2. Henceforth we shall direct most of our attention to the functions $g(A)$, which arise in connection with the solution of differential equations. For example $(A + \rho^2 I)^{-1}f$, which is the solution of the equation $Au + \rho^2 u = f$, or $e^{-tA}f$, which is the solution of Cauchy's problem $\partial_t u = -Au$, $u(0) = f$.

In this chapter we shall consider the case when A is a second-order differential operator. As already pointed out at the beginning of the third chapter, it follows from the analyticity f and the coefficients A that in many cases

$$f \in \mathscr{D}(\mathrm{ch}(R\sqrt{A})).$$

Below we will obtain representations $g(A)f$ of the form (1.1) for the vectors f which satisfy this condition.

We will assume for $f \in H$

$$R_0(A, f) = \sup\{R \in \mathbb{R} : f \in \mathscr{D}(\mathrm{ch}\, R\sqrt{A})\}. \tag{1.2}$$

It is obvious that $R_0(A, f) \geq 0$. The condition $R_0(A, f) > 0$ indicates that

$$f \in \mathscr{D}(\mathrm{ch}\, R\sqrt{A})$$

for some $R > 0$.

Suppose $\| \ \|_E$ is some norm, determined in

$$\mathscr{D}_\infty(A) = \bigcap_{k=0}^{\infty} \mathscr{D}(A^k).$$

Then we will assume:

$$M^E(A, f, R) = \sum_{j=0}^{\infty} R^{2j} \| A^j f \|_E / (2j)!. \tag{1.3}$$

(If $E=H$ and $\| \ \|_E = \| \ \|_H = \| \ \|$, for brevity we shall write $M(A, f, R)$ instead of $M^H(A, f, R)$). We will assume

$$R^E(A, f) = \sup\{R \in \mathbb{R} : M^E(A, f, R) < \infty\}. \tag{1.4}$$

Proposition 1.1. If $f \in \mathfrak{D}_\infty(A)$, then

$$\| \operatorname{ch} R\sqrt{A} f \| \leqq M(A, f, R) \tag{1.5}$$

and

$$R_0(A, f) = R^H(A, f). \tag{1.6}$$

Proof. We take the partial sum of the series expansion of the function $\operatorname{ch}(R\sqrt{\lambda})$

$$\Phi_{0n}(\lambda) = \sum_{j=0}^{n} R^{2j} \lambda^j / (2j)!.$$

By virtue of (1.3)

$$\| \Phi_{0n}(A) f \| \leqq M = M(A, f, R)$$

Since

$$\Phi_{0n}^2(\lambda)\to \mathrm{ch}^2(R\sqrt{\lambda})$$

monotonically, it follows from Formula (1.33) and Fatu's theorem that

$$\|\mathrm{ch}(R\sqrt{A})f\|^2 \leqq M,$$

and Formula (1.5) is proved.

It follows from (1.5) that the left-hand side of (1.6) is not smaller than the right-hand side. To prove (1.6), it remained to prove that the left-hand side of (1.6) is not larger than the right-hand side. Note that if

$$f\in \mathscr{D}(\mathrm{ch}\, R\sqrt{A}),\ R>0,$$

then

$$\begin{aligned}\|A^j f\|^2 &= \int_0^\infty \lambda^{2j} e^{-2R\sqrt{\lambda}} e^{2R\sqrt{\lambda}}\, d(E_\lambda f,f)\\ &\leqq 4\sup_{\lambda\geqq 0}(\lambda^{2j} e^{-2R\sqrt{\lambda}})\|\mathrm{ch}\, R\sqrt{A} f\|^2.\end{aligned} \tag{1.7}$$

When $t \geq 0$

$$\sup_{\lambda\geqq 0} \lambda^j e^{-2t\sqrt{\lambda}} = \sup_{x\geqq 0} x^{2j} e^{-tx^2} = t^{2j} e^{-2j}(2j)^{2j}. \tag{1.8}$$

According to Stirling's formula

$$e^{-2j}(2j)^{2j} \leqq (2j)!\,. \tag{1.9}$$

Therefore, assuming $t=R$ in (1.8), from (1.7) we obtain

$$\|A^j f\| \leqq 4R^{-2j}(2j)!\|\mathrm{ch}\, R\sqrt{A} f\|. \tag{1.10}$$

Replacing R by $R-\varepsilon$, from (1.10) we obtain that when $\varepsilon > 0$

$$M(A, f, R-\varepsilon) < C_\varepsilon \| \operatorname{ch} R\sqrt{A} f \|.$$

Since ε is as small as desired, then $R_0(A, f) \leqslant R^H(A, f)$, and (1.6) therefore holds.

Proposition 1.2. The conditions

$$f \in \mathscr{D}(\operatorname{ch} R\sqrt{A})$$

and

$$f \in \mathscr{D}(\operatorname{ch} R\sqrt{A+bI}),$$

where $b \in \mathbb{R}$, are equivalent.

Proof. The following inequality is obvious:

$$C_1(b) \operatorname{ch}(R\sqrt{\lambda+b}) \leqq \operatorname{ch}(R\sqrt{\lambda}) \leqq C_2(b) \operatorname{ch}(R\sqrt{\lambda+b}), \tag{1.10'}$$

where C_1 and C_2 do not depend on λ when $\lambda > 0$. Using this inequality and Formula (1.3.4), we obtain the required statement.

Proposition 1.3. For any $j \in \mathbb{N}$

$$R_0(A, f) = R_0(A, A^j f). \tag{1.11}$$

Proof. Suppose $0 < R < R_1 < R_0(A, f)$. Then, according to (1.3.3)

$$\| \operatorname{ch}(R\sqrt{A}) A^j f \|^2 \leqq \sup_{\lambda \geqq 0} \frac{(\operatorname{ch}(R\sqrt{\lambda})\lambda^j)^2}{(\operatorname{ch} R_1\sqrt{\lambda})^2} \| \operatorname{ch}(R_1\sqrt{A}) f \|^2. \tag{1.12}$$

Obviously

$$\mathrm{ch}(R\sqrt{\lambda})/\mathrm{ch}(R_1\sqrt{\lambda}) \leqq 2\,e^{(R-R_1)\sqrt{\lambda}}.$$

Using (1.8), where $t = R_1 - R$, for $\lambda > 0$ we obtain:

$$\mathrm{ch}(R\sqrt{\lambda})\lambda^j/\mathrm{ch}(R_1\sqrt{\lambda}) \leqq (R_1-R)^{2j}e^{-2j}(2j)^{2j}.$$

Hence and from (1.12) follows inequality

$$\|\mathrm{ch}\,(R\sqrt{A})A^jf\|^2 \leqq (R_1-R)^{2j}e^{-2j}(2j)^{2j}\|\mathrm{ch}\,R_1\sqrt{A}f\|. \tag{1.13}$$

Therefore (1.11) holds.

Lemma 1.1. Suppose $R_0(A, f) > 0$. Suppose $0 < R < R_1 < R_0(A, f)$. Suppose $g^+(\lambda)$ is a function that is continuous in $\mathbb{R}_+$, whilst

$$g^+(\lambda)/\mathrm{ch}(R\sqrt{\lambda})$$

is bounded on $\mathbb{R}_+$. Suppose $P_n(\lambda)$ is some polynomial. Then when $j \in \mathbb{Z}_+$

$$\|A^jg^+(A)f - A^jP_n(A)f\| \leqq \mu_R^+(P_n, g^+)(2ej/\varepsilon)^{2j}\|\mathrm{ch}\,R_1\sqrt{A}f\|, \tag{1.14}$$

where $\varepsilon = R_1 - R$, μ_R^+ is determined using Formula (3.1.2).

Proof. Obviously,

$$\|A^jg^+(A)f - A^jP_n(A)f\| = \|[g^+(A) - P_n(A)]A^jf\|. \tag{1.15}$$

Using inequality (2.1.5), where

$$\Phi = \mathrm{ch}(R\sqrt{\lambda}),$$

and $A^j f$ is substituted instead of f, we obtain:

$$\|g^+(A)A^jf - P_n(A)A^jf\| \leqq \mu_R^+(P_n, g^+)\|\operatorname{ch}(R\sqrt{A})A^jf\|. \tag{1.16}$$

Using inequality (1.13), from (1.16) we obtain inequality (1.14).

Theorem 1.1. Suppose $0<R<R_0(A,f)$. Suppose $g^+ \in A(\mathcal{J}_\beta^+)$ (see §6, Chapter 3). We will assume $r = 2R/\pi$, and suppose $P_n(\lambda)$ is an interpolation polynomial of the function

$$g,\ P_n = \Pi_r^{+n} g^+$$

(see 3.6.4). Then Formula (1.1) holds. Moreover, when $j \in \mathbb{Z}$

$$A^j g^+(A) f = \lim_{n\to\infty} A^j P_n(A) f, \qquad j=0,1,\ldots, \tag{1.17}$$

whilst estimate (1.14), in which $R<R_1<R_0(A,f)$, holds.

Proof. According to Theorem 3.6.1, $\mu_R^+(P_n, g) \to 0$ as $n \to \infty$. We obtain (1.17) from estimate (1.14). We obtain Formula (1.1) from (1.17) when $j=0$.

§ 2. Functions of differential operators and generalized solutions of differential equations.

Suppose Ω is a domain in

$$\mathbb{R}^m, \bar{\Omega} = \Omega \cup \partial\Omega.$$

Suppose the integral $\varkappa > 1$ is specified, and $G(x)(x \in \Omega)$ is a Hermitian matrix of dimensions $\varkappa \times \varkappa$, which depends infinitely smoothly on $x \in \bar{\Omega}$. It is assumed that G is positive definite when $x \in \Omega$ and negative definite when $x \in \partial\Omega$. We shall use $L_2(\Omega, G)$ to denote the Hilbert space of vector-functions defined in Ω with values in $\mathbb{C}^\varkappa$ with the scalar product

$$(u,v)_G=(u,v)=\int_\Omega \langle Gu(x), v(x)\rangle\, dx, \tag{2.1}$$

where $\langle u, v\rangle$ is the standard scalar product in

$$\mathbb{C}^\kappa,\ \langle u,v\rangle=u_1\bar v_1+\ldots+u_\kappa\bar v_\kappa.$$

Suppose $\mathcal{D}_0$ is some linear subspace of the space $L_2(\Omega, G)$, consisting of functions that are analytic in $\bar\Omega$. We shall denote the closure $\mathcal{D}_0$ in $L_2(\Omega, G) = H$ by H_0. Suppose B is a linear differential operator with coefficients that are infinitely smooth on $\bar\Omega$ and which possess the following properties:

1) B maps $\mathcal{D}_0$ into $\mathcal{D}_0$:

$$Bv\in\mathcal{D}_0,\qquad \forall v\in\mathcal{D}_0; \tag{2.2}$$

2) B is symmetric on $\mathcal{D}_0$:

$$(Bu,v)=(u,Bv),\qquad \forall u,v\in\mathcal{D}_0; \tag{2.3}$$

3) B is semi-bounded from below on $\mathcal{D}_0$:

$$(Bu,u)\geqq b(u,u),\qquad \forall u\in\mathcal{D}_0; \tag{2.4}$$

4)

$$R^{H_G}(B,f)>0,\qquad \forall f\in\mathcal{D}_0, \tag{2.5}$$

where R^{HG} is determined by Formula (1.4), in which $E = H_G = H$.

Specific examples of the operators B and sets $\mathcal{D}_0$ will be given in the following paragraphs.

Theorem 2.1. Suppose the operator B has the properties (2.2)-(2.5). Suppose B_1 is some lower semi-bounded self-adjoint extension of the operator B with the domain of definition $\mathcal{D}(B_1) \subset H_0$, $\mathcal{D}(B_1) \supset \mathcal{D}_0$. Then:

1) the following inclusion holds:

$$\mathscr{D}_0 \subset \mathscr{D}_\infty(B_1) = \bigcap \{\mathscr{D}(B_1^j), \tag{2.6}$$

2) If $0 < R < R^{HG}(B, f)$, then

$$f \in \mathscr{D}(\mathrm{ch}(R\sqrt{B_1})), \quad \|\mathrm{ch}(R\sqrt{B_1})f\| \leq M(B, f, R) < \infty,$$

where M is determined using Formula (1.3).

3) If B_0 is a self-adjoint Friedrichs expansion of the operator B, $g(\lambda)$ is a continuous bounded function on the half-line $\lambda > b_0$, which contains the spectra of the operators B_1 and B_0, and the function $f \in \mathcal{D}_0$, then $g(B_1)f = g(B_0)f$.

Proof. The insertion (2.6) holds, since by virtue of (2.2)

$$B^k \mathscr{D}_0 \subset \mathscr{D}_0, \quad \forall k \in \mathbb{N}.$$

Statement 2) is a corollary of (2.5) and Proposition 1.1, which is applicable by virtue of (2.6). The Friedrichs expansion exists by virtue of conditions (2.3) and (2.4), which guarantee the symmetry and semi-boundedness of B. Note that since we can always proceed from the function $g(\lambda)$ to the function $g(\lambda + b_0)$ and from the operators B_1 and B_0 to the operators $B_1 - b_0 I$ and $B_0 - b_0 I$, we can assume, from the very beginning, that the operators B_1 and B_0 are nonnegative and $b_0 = 0$. By virtue of section 2) of Theorem 2.1 and Proposition 1.2

$$f \in \mathscr{D}(\mathrm{ch}(R\sqrt{B_1}))$$

and

$$f \in \mathscr{D}(\mathrm{ch}(R\sqrt{B_0})).$$

Note that since the function

$$\Phi(x) = \Phi_0(x^2) = \mathrm{ch}(Rx)$$

is a weighting function (see Theorem 2.1.1), according to Theorem 2.1.1 and Lemma 3.6.1 the sequence of polynomials $P_n(\lambda)$ exists, such that

$$\mu^+(P_n, g, \mathrm{ch}(R\sqrt{\lambda}))$$

as $n \to \infty$. From Lemma 2.1.1, where $A = B_1$, we therefore obtain that

$$P_n(B_1)f \to g(B_1)f, \quad P_n(B_0)f \to g(B_0)f$$

as $n \to \infty$. Since when

$$f \in \mathscr{D}_0 \quad P_n(B_1)f = P_n(B_0)f = P_n(B)f,$$

hence follows Statement 3).

Corollary 2.1. If the operator B satisfies conditions (2.2)-(2.5), any semi-bounded self-adjoint expansion B_1 into H_0 of the operator B is identical with the Friedrichs expansion.

Proof. We shall take as the function g from section 3) of Theorem 2.1 the function

$$(\lambda + b_0 + \rho^2)^{-1}, \ \rho > 0.$$

Obviously

$$\mathscr{D}(B_1)=(B_1-b_0+\rho^2)^{-1}H_0,\ \mathscr{D}(B_0)=(B_0-b_0+\rho^2)^{-1}H_0,$$

and the operators $(B_1 - b_0 + \rho^2)^{-1}$ and $(B_0-b_0+\rho^2)^{-1}$ are continuous in H_0. According to the statement of section 3) the operators $(B_1 - b_0 + \rho^2)^{-1}$ and $(B - b_0 + \rho^2)^{-1}$ agree in all $f \in \mathcal{D}_0$, which satisfy (2.6). Since these f are everywhere dense in $\mathcal{D}_0$ and consequently in H_0, these operators are identical, and $\mathcal{D}(B_0) = \mathcal{D}(B_1)$. Q.E.D.

For the vectors $f \in \mathcal{D}_0$ we shall use $g(B)f$ to denote $g(B_0)f$, where B_0 is the Friedrichs expansion of the operator B.

We shall also denote the Friedrichs expansion of the operator B by B. Since all considerations will henceforth be conducted on $\mathcal{D}_0$, this will not lead to confusion. We shall use A to denote the nonnegative operator $A = B - bI$, and also its self-adjoint Friedrichs operator.

The following theorem shows that functions of the operator A permit polynomial representations of the form (1.1) in $\mathcal{D}_0$.

Theorem 2.2. Suppose $f \in \mathcal{D}_0$. Then $R > 0$ will be obtained, such that $0 < R < R_0(A, f)$. Suppose the function g^+ and the polynomials $P_n = \Pi_r^{+n} g^+$ are the same as in Theorem 1.1. Then Formula (1.17) holds.

Proof. it follows from section 2) of Theorem 2.1 that $R_0(A, f) > 0$. Using Theorem 1.1, we obtain the statement of Theorem 2.2.

We shall now formulate a theorem on the polynomial representation of solutions of Cauchy's problems (1.3.5), (1.3.6) and (1.3.10), (1.3.11), and also the stationary equation $Au + \rho^2 u = f$.

Theorem 2.3. Suppose the differential operator B satisfies conditions (2.2)-(2.5), $A = B - bI$. Suppose

$$f \in \mathscr{D}_0, \qquad R_0 = R^{H_G}(A, f).$$

Suppose $0 < R < R_1 < R_0$. Suppose the functions

$$g_{ij},\ i=1,2, \quad j=0,1,2,$$

are defined using Formulae (1.3.8) and (1.3.13), $g_{00} = \lambda^{-1}$. Suppose the polynomials P_n^{ij} are interpolation polynomials of the functions $g_{ij}(\lambda + b)$:

$$P_n^{ij}(\lambda)=\Pi_r^{+n} g_{ij}(\lambda+b), \tag{2.7}$$

where

$$i=0,1,2,\ j=0,\ \ldots,\ i,\ \Pi_r^{+n}$$

is determined by (3.6.4), $r = 2R/\pi$. Suppose $M_0 = M(A, f, R_1)$.

The following statements then hold:

1) If $b = \rho^2 > 0$, then $\forall\, n \in \mathbb{N}$, $l \in \mathbb{Z}_+$

$$\|A^l B^{-1} f - A^l P_n^{00}(B-bI) f\| \leq C_{0l} n^{-2R\rho/\pi} M_0. \tag{2.8}$$

2) If $\theta > 0$, $t \in \mathbb{C}$

$$\operatorname{Re} t > 0, \quad |t|^2 \operatorname{Re} t \leq \theta, \tag{2.9}$$

then for any $n \in \mathbb{N}$, $l \in \mathbb{Z}_+$, $j = 0, 1$

$$\|A^l g_{1j}(t, B) f - A^l P_n^{1j}(B-bI) f\| \leq C_{1l}(\varepsilon)\, e^{-\sigma_1 \ln^2 n} M_0, \tag{2.10}$$

where

$$\sigma_1 = (1-\varepsilon) r^2/(4\theta), \tag{2.11}$$

and $\varepsilon > 0$ is as small as desired.

3) If $t \in \mathbb{R}$, $|t| < \theta$, then when $j = 0, 1, 2$, $n \in \mathbb{N}$, $l \in \mathbb{Z}_+$

$$\|A^l g_{2j}(t, B)f - A^l P_n^{2j}(B-bI)f\| \leqq C_{2l}(\varepsilon)(1+\varepsilon)^{2n}(1-e^{-\theta/r})^{2n} M_0, \tag{2.12}$$

where $\varepsilon > 0$ is as small as desired.

Proof. We shall use Theorem 1.1, where $g(\lambda) = g_{ij}(\lambda + b)$. By virtue of (1.14) and (1.5), and bearing in mind that $A = B - bI$, we obtain:

$$\|A^l g_{ij}(B) - A^l P_n(B-bI)\| \leqq C_l \mu_R^+(P_n, g_{ij}(\lambda+b)) M(B, f, R_1). \tag{2.13}$$

When

$$i=0, j=0, b=\rho^2, g_{00}(\lambda+b)=(\lambda+\rho^2)^{-1},$$

using Estimate (3.6.8), where $\omega = 0$, from (2.13) we obtain Estimate (2.8).

When $i=1$, $j=0$ or $j=1$, we shall use Theorem 3.6.4 to estimate μ^+_R on the right-hand side of (2.13). Consider the function

$$g_{11}(z^2) = e^{-tz^2}.$$

Suppose $t = \xi + i\eta$, $z = x + iy$. Then

$$\mathrm{Re}(-tz^2) = -\xi(x^2-y^2) + 2\eta xy = -\xi(x-\eta y/\xi)^2 + (\xi + \eta^2/\xi)y^2.$$

Therefore when

$$\xi > 0, e^{-tz^2} \in \mathscr{A}_{2,\theta}(\mathscr{I}_\infty),$$

where $\theta = (\xi^2 + \eta^2)/\xi$. It is also obvious that

$$g_{10}(t, z^2) = (e^{-tz^2} - 1)/z^2 \in \mathscr{A}_{2,\theta}(\mathscr{I}_\infty).$$

It is obvious that

$$g_{ij}(t, z^2 + b) \in \mathscr{A}_{2,\theta}(\mathscr{I}_\infty)$$

for any $b \in \mathbb{R}$. Using Theorem 3.6.4, and inequality (3.5.5), where $\omega = 0$, we obtain the estimate

$$\mu_R^+(P_{n-1}(\lambda), g_{1j}(\lambda + b)) \leqq C \exp[-(1-\varepsilon)(r \ln n/2)^2 \theta^{-1}]. \tag{2.14}$$

From (2.13) and (2.14) we obtain (2.10).

Let us now suppose i=2. Consider the function

$$G_{21}(b + z^2) = \cos(t\sqrt{b + z^2}),$$

where t, $b \in \mathbb{R}$. Since

$$\operatorname{Im} t\sqrt{b + z^2} = \operatorname{Im}(zt\sqrt{1 + bz^{-2}})$$

as $\operatorname{Im} z \to \infty$ does not exceed $t \operatorname{Im} z + C$, then

$$\cos(t\sqrt{b + z^2}) \in \mathscr{A}_{1,t}(\mathscr{I}_\infty).$$

In a similar way $g_{20}(b + z^2)$ and $g_{22}(b + z^2)$ belong to $\mathscr{A}_{1,t}(\mathscr{I}_\infty)$. Theorem 6.5 is therefore applicable. Using Formulae (3.6.9) and (3.5.14), we obtain the estimate

$$\mu_R^+(P_{n-1}(\lambda), g_{2j}(\lambda + b)) \leqq C[(1+\varepsilon)(1 - e^{-\theta/r})]^{2n}. \tag{2.15}$$

Using (2.15) to estimate the right-hand side of (2.13), we obtain (2.12).

Remark 2.1. If the p-th order operator B is elliptic in the subdomain $\omega \subset \Omega$, the norm of the function v in the Sobolev space $W_2^{pl}(\omega)$ is estimated from above by $\|A^l v\|$. The convergence of $P_n(B - bI)f$ to $g(B)f$ in $W_2^{pl}(\omega)$ therefore follows from Formulae (2.8), (2.10) and (2.12).

Corollary 2.2. Suppose the operator B and the function f satisfy the conditions of Theorem 2.3. Suppose $u_0 = (A + \rho^2)^{-1}f$, the function $u_1(t)$ is a solution of a non-strict parabolic Cauchy problem

$$\partial_t u_1(t) = -Bu(t), \quad u(0) = f_1, \tag{2.16}$$

and the function $u_2(t)$ is the solution of the non-strict hyperbolic Cauchy problem

$$\partial_t u_2(t) = -Bu_2(t) + f_0, \quad u_2(0) = f_1, \quad \partial_t u_2(0) = f_2. \tag{2.17}$$

The following polynomial representations then hold for solutions of these problems:

$$A^l u_i = \lim_{n\to\infty} A^l \sum_{j=0}^{i} P_n^{ij}(B - bI)f_j, \qquad l = 0, 1, \ldots, \tag{2.18}$$

where P_n^{ij} are the same as in Theorem 2.3, when i=1, 2 and $t > 0$ is fixed.

Proof. According to Propositions 1.3.1 and 1.3.2, where L=B, when i=1, 2

$$u_i(t) = \sum_{j=0}^{i} g_{ij}(B)f_j. \tag{2.19}$$

When i=1, 2 from Estimates (2.10), (2.12) and from (2.19) we obtain (2.18). When i=0 Formula (2.18) follows from (2.8).

Remark 2.2. Since the polynomial $P(B - bI)$ is obviously some polynomial $P_0(B)$ of B of the same degree, we immediately obtain polynomial representations of generalized solutions of the problems considered

$$u_i = \lim_{n\to\infty} \sum_{j=0}^{i} P_{0n}^{ij}(B) f_j, \tag{2.20}$$

where $i = 0, 1, 2$, and the number $t \in \mathbb{R}$ is fixed when $i=1, 2$. Estimates of the rate of convergence of (2.20) directly follow from estimates of Theorem 2.3.

Remark 2.3. If

$$P_n^{ij}(\lambda) = \Pi_r^{+n} g_{ij}(\lambda + b)$$

are interpolation polynomials of the functions $g_{ij}(\lambda + b)$ with interpolation nodes - the roots of the polynomial $T_n^{+}(r^2\lambda)$ - then

$$P_{0n}^{ij}(B) = P_n^{ij}(B - bI)$$

is an interpolation polynomial of the function $g_{ij}(\lambda)$ with the nodes $b, b + \lambda_1, \ldots, b + \lambda_n$.

Remark 2.4. We can also obtain representations similar to (2.16), when f_j depends on t,

$$t, f_j = f_j(t), \|\mathrm{ch}(R\sqrt{B}) f\| \leq M,$$

where M does not depend on t, $0 \leqslant t \leqslant T$. For this purpose we use a representation of the solutions of the equations using the variation of the constants. For example, the particular solution of the equation $\partial u = -Au + f(t)$ is written in the form

$$u(t) = \int_{t_0}^{t} e^{-A(t-\tau)} f(\tau)\, d\tau.$$

We shall use the previously obtained polynomial representation of the operator $e^{-A}(t - \tau)$ applied to $f(\tau)$ to represent the solutions of this equation polynomially. When

$$f(t) = f_0 e^{\gamma t}, \quad u(t) = G(A) f_0 e^{\gamma t},$$

where

$$G(\lambda) = (1 - e^{-(\lambda+\gamma)t})/(\lambda + \gamma),$$

we can also obtain the polynomials $P_n(\lambda)$ like interpolation polynomials of the function $G(\lambda)$.

In the following paragraph we shall give examples of second-order differential operators with analytic coefficients, for which we shall take as $\mathcal{D}_0$ all the functions that are analytic on $\overline{\Omega}$.

§ 3. Examples of differential operators in bounded domains, for which the set $\mathcal{D}_0$ is a set of analytic functions.

Example 3.1. (Differential operator in a torus).

We shall use

$$\mathscr{A}^{\kappa} = (\mathscr{A}(T^m))^{\kappa}, \ \kappa \geqq 1,$$

to denote the vector-functions

$$v(x) = (v^1(x), \ldots, v^{\kappa}(x))$$

which are really-analytic in $\mathbb{R}^m$ and periodic with the period 2π with respect to each variable x_j, $j = 1, \ldots, m$ (we can obviously assume these functions are defined on the m-dimensional torus T^m).

Consider in $A^{\varkappa}(T^m)$ the matrix differential operator

$$Bv = -\sum_{i,j=1}^{m} \partial_i(a_{ij}(x)\,\partial_j v) + a_{00}(x)v, \qquad \partial_i = \partial/\partial x_i. \tag{3.1}$$

Here $a_{ij}(x)$ for each i, j is a κ-th order square matrix, and its elements

$$a_{ijkl} \in \mathscr{A}(T^m),\ k, l, \ldots, \kappa.$$

The following conditions of symmetry are assumed to hold:

$$a_{ij} = a_{ji}^*, \qquad i, j = 0, \ldots, m. \tag{3.2}$$

Here $*$ is the sign of Hermitian conjugation with reference to the standard scalar product $\langle , \rangle$ in $\mathbb{C}^{\varkappa}$.

The condition of nonnegativity of the smooth part of the operator (3.1) is assumed to hold. Namely, for any

$$\xi = (\xi_1, \ldots, \xi_m) \in (\mathbb{C}^\kappa)^m (\xi_i \in \mathbb{C}^\kappa)$$

the condition of nonnegativeness of the quadratic form $a(x, \xi)$, determined in the following formula, holds:

$$a(\xi) \equiv \sum_{i,j=1}^{m} \langle a_{ij}(x)\xi_i, \xi_j \rangle \geqq 0, \qquad x \in \bar{\Omega}. \tag{3.3}$$

The standard scalar product (u, v) is determined in the space $(L_2(T^m))^{\varkappa} = H$ using Formula (2.1), where $G = I$, $\Omega = [0, 2\pi]^m$.

Theorem 3.1. Suppose the operator B is determined by Formula (2.1) in $\mathcal{D}_0 = (A(T^m))^{\varkappa}$. Then conditions (2.2)-(2.5) hold.

Proof. The inclusion (2.2) is obvious, since the differential operator with analytic coefficients transforms the analytic functions into analytic functions. We obtain Eq.(2.3) by using the formula of integration by parts and condition (3.2) (there are no boundary terms due to the periodicity). We shall verify (2.4), using the formula of integration by parts and condition (3.3):

$$
\begin{aligned}
(Bv, v) &= \int_\Omega \langle -(\partial_i a_{ij} \partial_j v) + a_{00} v, v \rangle \, dx \\
&= \int_\Omega \sum \langle a_{ij}(x)\, \partial_j v, \partial_i v \rangle \, dx + \int_\Omega \langle a_{00}(x) v, v \rangle \, dx \\
&\geqq \int_\Omega \langle a_{00} v, v \rangle \, dx \geqq b \|v\|^2.
\end{aligned} \tag{3.4}
$$

Inequality (2.5) for second-order operators B with analytic coefficients and for analytic functions f in the bounded domain Ω is proved in Ch.1, Theorem 1.1.1. The theorem is proved.

By virtue of Theorem 3.1, all the conditions of Theorem 2.3 are satisfied. Therefore, if the operator B on T^m is determined by Formula (3.1), the representations of the form (2.20), constructed in Theorem 2.3, hold for generalized solutions of the equation $Au + \rho^2 u = f$ and problems (3.5), (3.6) and (3.10), (3.11).

Remark 3.1. We can consider a second-order self-adjoint operator on the analytic manifold Ω without a boundary (the scalar product will be taken in $L_2(\Omega, d\omega)$, where $d\omega$ is some smooth measure in Ω), in a similar way to the operator B, determined in T^m by Formula 3.1.

Example 3.2. (Differential operator degenerating on the boundary of a domain.)

Suppose Ω is a bounded domain in $\mathbb{R}^m$ with the smooth boundary $\partial\Omega$. We shall use $(A(\Omega))^\varkappa$ to denote the set of vector-functions that are really-analytic on the closure Ω of the domain $\overline{\Omega}$.

Consider the differential operator B, determined by Formula (3.1). It is assumed that the coefficients of this operator belong to $A(\Omega)$ and that conditions (3.2) and (3.3) hold. In addition the condition of degeneracy on the boundary is imposed on the operator B:

$$\sum_{i=1}^{m} n_i(x)a_{ij}(x)=0, \qquad j=1,\dots,m, \quad \forall x\in\partial\Omega, \tag{3.5}$$

where $(n_1(x), \dots, n_m(x))$ is an external normal to $\partial\Omega$ at point x.

According to The Gauss-Ostrogradskii formula

$$\begin{aligned}(Bu, v) &= \int_\Omega \Big\langle -\sum_{i,j} \partial_i(a_{ij}\,\partial_j u) + a_{00}u, v\Big\rangle dx \\ &= -\int_{\partial\Omega} \sum_{i,j} \langle a_{ij}\,\partial_j u, v\rangle n_i\, dx + \int_\Omega \sum_{i,j} \langle a_{ij}\,\partial_j u, \partial_i v\rangle\, dx \\ &\quad + \int_\Omega \langle a_{00}u, v\rangle\, dx.\end{aligned} \tag{3.6}$$

By virtue of condition (3.5) the integral over $\partial\Omega$ equals zero. Once again integrating by parts, we obtain that $(Bu, v) = (u, Bv)$, i.e. the operator B is symmetric on $(\mathcal{A}(\Omega))^\varkappa$. As in (3.4), the upper semi-boundedness of the operator B in $(\mathcal{A}(\Omega))^\varkappa$ is verified. The following theorem thus holds:

Theorem 3.2. Suppose the operator B is determined by (3.1) in $\mathcal{D}_0 = (\mathcal{A}(\Omega))^\varkappa$ and satisfies the conditions imposed in Example 3.2. Conditions (2.2)–(2.5) then hold.

Example 3.3. (A differential operator which is self-adjoint with a weight and degenerates on the boundary of a domain.)

Let us now consider in $\Omega \subset \mathbb{R}^m$ an operator that is more common than in Example 3.2:

$$B_G u \equiv BG_u \equiv -\sum_{i,j=1}^{m} \partial_i(a_{ij}(x)\,\partial_j(G(x)u)) + a_0(x)u. \tag{3.7}$$

Here $G(x)$ is a Hermitian matrix which is positive when $x \in \Omega$ and nonnegative when $x \in \bar{\Omega}$:

$$\langle G(x)\xi, \xi\rangle \geq 0, \qquad \forall \xi\in\mathbb{C}^\varkappa, \quad x\in\bar{\Omega}. \tag{3.8}$$

We shall require that the conditions of degeneracy on the boundary hold:

$$G(x) \sum_{i,j=1}^{m} n_i(x) a_{ij}(x)\, \partial_j G(x) = 0, \tag{3.9}$$

$$G(x) \sum_{i=1}^{m} n_i(x) a_{ij}(x) G(x) = 0, \qquad j = 1, \ldots, m, \tag{3.10}$$

where $x \in \partial\Omega$. We shall also require that the following conditions hold:

$$a_0^*(x) G(x) = G(x) a_0(x), \qquad x \in \bar{\Omega}. \tag{3.11}$$

The last condition obviously holds if $a_0 = a_0^*$ and G commutes, and also if $a_0(x) = a_{00}(x) G(x)$, where $a_{00}(x) = a_{00}^*(x)$. It is assumed that the elements of the matrices a_{ij}, G belong to $A(\bar{\Omega})$ and that the conditions of symmetry and positiveness of (3.2) and (3.3) hold.

On the above assumptions the operator B_G, which is determined by (3.7), is symmetric and positive definite in $(A(\Omega))^{\varkappa}$ with reference to the scalar product $\langle\ ,\ \rangle$ in $L_2(\Omega, G)$, which is determined by Formula (2.12).

Indeed,

$$\begin{aligned} -\int_\Omega \left\langle G \sum_{i,j=1}^{m} \partial_i(a_{ij}\, \partial_j(Gu)), v \right\rangle dx + \int_\Omega \langle G a_0 u, v\rangle\, dx \\ = -\int_{\partial\Omega} \sum_{i,j=1}^{m} n_i \langle a_{ij}\, \partial_j(Gu), Gv\rangle\, dx \\ + \int_\Omega \sum_{i,j=1}^{m} \langle a_{ij}\, \partial_j Gu, \partial_i Gv\rangle\, dx + \int_\Omega \langle a_0 u, Gv\rangle\, dx. \end{aligned} \tag{3.12}$$

The integral with respect to $\partial\Omega$ obviously equals

$$-\int_{\partial\Omega} \left[G \sum_{i,j=1}^{m} n_i a_{ij} (\partial_j G) u, v \;+\; G \sum_{i,j=1}^{m} n_i a_{ij} G\, \partial_j u, v \right] dx.$$

This integral equals zero by virtue of conditions (3.9) and (3.10). Carrying out integration by parts in Formula (3.12) once again, we obtain Eq.(2.3) in $\mathscr{D}_0=(\mathscr{A}(\Omega))^{\varkappa}$, i.e. B_G is symmetric. (2.4) follows from (3.12) and (3.3) and from the boundedness of a_0 and G, i.e. B_G is semi-bounded from below. According to Theorem 1.1.1, (2.5) holds. Assuming $\mathcal{D}_0 = (\mathcal{A}(\Omega))^{\varkappa}$, we obtain the following theorem.

Theorem 3.3. Suppose the operator B_G, which is determined by (3.7) in $\mathcal{D}_0 = (\mathcal{A}(T^m))^{\varkappa}$, satisfies all the conditions imposed in Example 3.3.. Then B_G satisfies conditions (2.2)-(2.5).

According to Theorems 3.1, 3.2, 3.3, when B is an operator from 3.1, 3.2, 3.3 $\forall f \in \mathcal{A}(\bar{\Omega})$ the statements of Theorem 2.3 hold.

Remark 3.2. If $b > 0$ in inequality (3.4) and, accordingly, in (2.4), we can represent the operator B in the form $B = A + \rho^2 I$, where $\rho^2 = b$, $A > 0$, I is a unit operator and the solution of the equation $Bu_0 = f$ is represented in the form (2.18). The polynomial $P_n^{00}(\lambda) = P_n(\lambda)$ is determined by Formula (2.7), where $r = 2R/\pi$. Thus, we need to specify two numbers to obtain $P_n^{00}(\lambda)$: ρ and R; $0 < \rho^2 < b$, $0 < R < R^H(A, f)$. To obtain these numbers it is required to have the lower limits of the number b in (3.4) and of the number $R^H(A,f)$. We have the simplest upper limit when the matrix $a_{00}(x)$ is positive definite:

$$a_{00}(x) \geqq b_0 I, \quad \forall x \in \bar{\Omega},$$

where $b_0 > 0$ does not depend on x. In this case, it is immediately obvious from the derivation of (3.4) that $b > b_0$. The lower bounds of the number $R^H(A, f)$ will be obtained in the fifth paragraph.

§ 4. Examples of problems which reduce to those considered earilier.

In this paragraph we will consider Dirichlet problems for an elliptic equation, first-order systems which degenerate on a boundary, and the Dirichlet problem for a parabolic equation.

Example 4.1. A first-order system that degenerates on the boundary of a domain.

Suppose the following system is given in the bounded domain $\Omega \subset \mathbb{R}^m$:

$$Fu \equiv \sum_{i=1}^{m} a_i \partial_i u + a_0 u = f, \tag{4.1}$$

where $a_i(x)$ is a matrix of order $\varkappa \times \varkappa$, whose coefficients belong to $\mathcal{A}(\bar{\Omega})$.

We will assume that the operator F satisfies condition

$$(Fu, Fu) \geqq C_0 \|u\|^2,\ C_0 > 0,\ \forall u \in (\mathcal{A}(\bar{\Omega}))^{\varkappa}, \tag{4.2}$$

where $(\ ,\ )$ is a scalar product in $(L_2(\Omega))^{\varkappa}$.

We will also assume that $\forall\, u,\ v \in \mathcal{A}(\bar{\Omega})^{\varkappa}$

$$(F^*Fu, v) = (Fu, Fv), \tag{4.3}$$

where

$$F^*u = -\sum_{i=1}^{m} \partial_i(a_i^* u) + a_0^* u. \tag{4.4}$$

Remark 4.1. The degeneracy of F on $\partial\Omega$ is sufficient for (4.3) to hold. Indeed, by virtue of Green's formula the boundary terms have the form:

$$(F^*Fu, v) - (Fu, Fv) = -\int_{\partial\Omega} \left\langle Fu, \sum_{j=1}^{m} n_j a_j v \right\rangle dx. \tag{4.5}$$

We will assume

$$b_0(x)=\sum_{j=1}^{m} n_j(x)a_j(x). \tag{4.6}$$

If when $x \in \partial\Omega$ $b_0(x)$ is such that

$$a_i^* b_0=0, \qquad i=0,\dots,m, \tag{4.7}$$

it follows from the determination Fu in (4.1) and from (4.7) that the right-hand side of (4.6) equals zero, and Formula (4.3) holds.

Consider the equation

$$Fu=f, \tag{4.8}$$

where $f \in \mathcal{A}(\overline{\Omega})$. It is obvious that if u is a generalized solution from $(L_2(\Omega))^{\varkappa}$ of Eq,(4.8), u is a generalized solution of the equation

$$F^*Fu=F^*f. \tag{4.9}$$

We will assume $\mathfrak{D}_0 = (\mathcal{A}(\overline{\Omega}))^{\varkappa}$. It is obvious that the second-order operator $B=F^*F$ also satisfies (2.3) by virtue of condition (4.3). Condition (2.2) obviously holds, and condition (2.3) holds by virtue of Theorem 1.1.1.. Condition (2.4) holds by virtue of (4.2) and (4.3), whilst in (2.4) $b = c_0 > 0$.

Note that the operator B is a special case of the operators considered in Example 3.2, and Remark 3.2 holds.

Example 4.2. (An operator which degenerates on a boundary and is produced by an operator with zero boundary conditions.)

Consider the equation $Bu=f$, where B is an operator of the form (3.1) with coefficients from $\mathcal{A}(\Omega)$, $f \in (\mathcal{A}(\Omega))^{\varkappa}$. It is assumed that conditions (3.2) hold, and besides (3.3) the condition of ellipticity is imposed on $\partial\Omega$:

$$a(\xi) \equiv \sum_{i,j=1}^{m} \langle a_{ij}(x)\xi_i, \xi_j \rangle \geqq C_0 \sum_{k=1}^{m} |\xi_k|^2, \qquad C_0 > 0,$$
$$\forall (\xi_1, \ , \xi_m) \in \mathbb{C}^{\kappa m}, \quad \forall x \in \partial\Omega. \tag{4.10}$$

It is assumed that the bounded domain Ω is specified by the inequality

$$\Omega = \{x \in \mathbb{R}^m : g(x) > 0\}, \partial\Omega = \{x : g(x) = 0\}, \tag{4.11}$$

$$\operatorname{grad} g|_{\partial\Omega} \neq 0, \quad g \in \mathscr{A}(\bar{\Omega}). \tag{4.12}$$

We seek the solution of the equation $Bu=f$, which satisfies the zero boundary conditions:

$$u|_{\partial\Omega} = 0. \tag{4.13}$$

Remark 4.2. The operator B is symmetric and bounded from below in a set of smooth functions which equal zero on $\partial\Omega$. However, if we take as $\mathcal{D}_0$ a set of functions with a finite order of zero in $\partial\Omega$, condition (2.2) will not hold, since after differentiation the order of zero decreases. If we require an infinite-order vanishing on $\partial\Omega$, for example if we take $\mathcal{D}_0 = C_0^{\infty}(\Omega)$, condition (2.2) will hold. However in this case $\mathcal{D}_0$ will not contain analytic functions (besides 0), and condition (2.5) will not hold.

Thus, we cannot explicitly describe the set $\mathcal{D}_0$, which satisfies (2.2)-(2.5), and simultaneously (4.13). Therefore, instead of considering the equation $Bu=f$ with zero boundary conditions, we shall consider an equivalent equation that degenerates on $\partial\Omega$.

We will assume $u(x) = g(x)v(x)$ and will consider an equation, obtained from $Bu=f$, where B is determined by (3.1):

$$B_G v \equiv BGv = f, \qquad G = gI. \tag{4.14}$$

Obviously, the operator B_G has the form (3.7). Since $u(x)$ is the solution of the equation $Bu=f$, which is elliptic on $\partial\Omega$, where $f \in (\mathcal{A}(\Omega))^{\varkappa}$, then $u(x)$ is a function which is smooth in the neighbourhood of the boundary, and which equals zero on $\partial\Omega$. Therefore $v(x) = u(x)/g(x)$ will obviously be a smooth solution of (4.14) and will be identical with the generalized solution of (4.14) from $L_2(\Omega, G)$.

It is obvious that by virtue of (4.11) the operator B_G, where $G=gI$, satisfies (3.9) and (3.10), and since $g(x)I$ commutes with a_{00}, condition (3.11) holds. Thus, Theorem 2.3 is applicable by virtue of Theorem 3.3. To be able to apply this theorem to solve the equation $B_G v = f$ in the same way as in Remark 3.2, it is required that the operator B_G is positive definite in $L_2(\Omega, G)$ on $\mathcal{D}_0 = (\mathcal{A}(\Omega))^{\varkappa}$. The following theorem shows that the positive definiteness of the matrix a_{00} in (3.1) is sufficient for this.

Theorem 4.1. Suppose the operator B satisfies all the requirements of Example 4.2. Suppose, in addition, the matrix a_{00} in (3.1) satisfies the condition

$$a_{00}(x) \geqq q_0 I, \qquad q_0 > 0. \tag{4.15}$$

Then the operator B_G, determined by (4.14), where g is the same as in (4.11), satisfies condition (2.4), where $\mathcal{D}_0 = (\mathcal{A})\Omega))^{\varkappa}$, $b > 0$.

Proof. We shall use Ω_ε to denote the set

$$\Omega_\varepsilon = \{x \in \Omega : 0 \leqq g(x) \leqq \varepsilon\}. \tag{4.16}$$

It is obvious that by virtue of (4.12) for small ε Ω_ε will be contained in the small neighbourhood $\partial\Omega$. Therefore for small ε by virtue of (4.10) the operator B is elliptic in Ω_ε, i.e. the inequality (4.10) holds when $x \in \Omega_\varepsilon$ (with some constant $C_0 > 0$ it is twice as small as the original one).

By virtue of (3.12)

$$(B_G v, v)_G = \int_\Omega [a(\nabla(gv)) + \langle a_{00} gv, gv\rangle]\, dx.$$

Using (4.15) to make a lower estimate of the terms on the right-hand side, we obtain for small $\varepsilon > 0$:

$$(B_G v, v)_G \geqq \int_{\Omega_\varepsilon} a(\nabla(gv))(x)\, dx + \int_\Omega q_0 |gv|^2(x)\, dx. \tag{4.17}$$

Since $(v, v)_G = (gv, gv/g)$, to prove

$$(B_G v, v) \geqq b(v, v)_G,\ b > 0,$$

it is sufficient to prove that

$$\int_\Omega |gv|^2/g\, dx \leqq \frac{1}{b}\left[\int_{\Omega_\varepsilon} \frac{C_0}{2} \sum_{i=1}^m |\partial_i(gv)|^2\, dx + \int_\Omega q_0 |gv|^2\, dx\right] \tag{4.18}$$

for some $b > 0$. The right-hand side of this inequality is estimated from below through the squares of the norms of the function gv in the Sobolev space $W_2^1(\Omega_\varepsilon)$ and in $L_2(\Omega)$. The norm in $L_2(\Omega)$ of the function $(gv)/\sqrt{g}$ is on the left-hand side. The operator of multiplication by $1/\sqrt{g}$ of functions which equal zero on $\partial\Omega$ is continuous from $W_2^1(\Omega_\varepsilon)$ to $L_2(\Omega_\varepsilon)$ by virtue of the condition grad $g \neq 0$ on $\partial\Omega$. Hence follows estimate (4.18), and consequently also (2.4) for B_G with $b > 0$.

We shall now carry out a detailed proof of (4.18) and shall indicate the dependence of b on C_0 and on q_0.

We shall make an upper estimate of the left-hand side of (4.18). We will assume $gv(x) = u(x)$. By virtue of (4.16)

$$\int_\Omega |u|^2(x)/g(x)\, dx \leqq \delta^{-1} \int_{\Omega\setminus\Omega_\delta} |u|^2(x)\, dx + \int_{\Omega_\delta} |u|^2(x)/g(x)\, dx. \tag{4.19}$$

It is obvious that to obtain an estimate of the form (4.18) it is sufficient to make an upper estimate of the second term on the right-hand side of (4.19). To estimate the integral over to Ω_δ, we shall make a change of variable in Ω_δ, which we shall now describe.

Suppose δ is so small that $g' = \text{grad } g \neq 0$ on Ω_δ. Consider in Ω_δ a differential equation with reference to $x=x(t)$:

$$dx/dt = g'(x(t))/\langle g', g'\rangle, \qquad x(0)=\omega, \quad \omega\in\partial\Omega. \tag{4.20}$$

We shall denote the solution of (4.20) by $x(\omega,t)$. Obviously

$$dg(x(t))/dt = 1, \qquad g(x(\omega,t)) = t. \tag{4.21}$$

We shall use the formula

$$F:(\omega,t)\to x(\omega,t)$$

to determine the mapping $\partial\Omega \times [0, \delta]$ to Ω_δ. It is easy to see that F is a diffeomorphism of these sets. With this diffeomorphism the volume elements are connected by the relation

$$dx = k(\omega,t)\,dt\,d\omega, \quad 0<k_0\leqq k(\omega,t)\leqq k_1. \tag{4.22}$$

The point

$$(\omega(y), t_0(y)) = F^{-1}y$$

corresponds to each point $y \in \Omega_\delta$ by virtue of the invertibility of F. Obviously $t_0 = g(y)$. Therefore if $y \in \Omega_\delta$, then $y = x(\omega, t_0)$, where $x(\omega,t)$ is the solution of (4.20), $\omega = \omega(y)$, $t_0 = g(y)$. Consequently, if $u=0$ on $\partial\Omega$, then

$$|u(y)| = \left| \int_0^{g(y)} \frac{d}{dt} u(x(\omega, t))\, dt \right| = \left| \int_0^{g(y)} \langle \nabla_x u, dx/dt \rangle_{\mathbb{R}^m} dt \right|$$
$$= \left| \int_0^{g(y)} \langle \nabla_x u, g' \rangle_{\mathbb{R}^m} / \langle g', g' \rangle_{\mathbb{R}^m} dt \right|$$
$$\leqq C \int_0^{g(y)} |\nabla_x u|(x(\omega(y), t))\, dt.$$

Hence, bearing in mind that by virtue of (4.10)

$$C_0^{1/2} |\nabla u| \leqq a(\nabla u)^{1/2},$$

we obtain:

$$\int_{\Omega_\delta} g^{-1}(y) |u|^2(y)\, dy$$
$$\leqq \frac{C_1}{\sqrt{C_0}} \int_{\Omega_\delta} \left[g^{-1}(y) |u|(y) \int_0^{g(y)} a(\nabla u)^{1/2}(x(\omega(y), t))\, dt \right] dy. \tag{4.23}$$

Making the change $g^{-1}(y)t = \tau$, we obtain that the right-hand side of (4.28) is not greater than

$$\frac{C_1}{\sqrt{C_0}} \int_{\Omega_\delta} \left[\int_0^1 |a(\nabla u)|^{1/2}(x(\omega(y), \tau g(y)))\, d\tau |u(y)| \right] dy$$
$$\leqq \frac{C_1}{\sqrt{C_0}} \int_0^1 \left\{ \left[\int_{\Omega_\delta} a(\nabla u)(x(\omega(y), \tau g(y)))\, dy \right]^{1/2} \left[\int_{\Omega_\delta} |u(y)|^2\, dy \right]^{1/2} \right\} d\tau. \tag{4.24}$$

Making the change $y = F(\omega, t)$ and bearing in mind (4.22), we obtain:

$$\int_{\Omega_\delta} a(\nabla u)(x(\omega(y), \tau g(y)))\, dy = \int_{\partial\Omega \times [0, \delta]} a(\nabla u)(x(\omega, \tau t) k(\omega, t)\, d\omega\, dt.$$

Bearing in mind (4.22) and making the change $\tau t = \theta$, we obtain that the last integral is not larger than

$$K_1 \int_{\partial\Omega} \frac{1}{\tau} \int_0^{\tau\delta} a(\nabla u)(x(\omega, \theta))\, d\theta\, d\omega.$$

Changing variable (ω, Θ) to $x = F(\omega, \Theta)$, bearing in mind (4,.22) we obtain that this integral is not larger than

$$\frac{k_1}{k_0} \frac{1}{\tau} \int_{\Omega_{\tau\delta}} a(\nabla u)(x)\, dx \leqq \frac{K_1}{K_0} \frac{1}{\tau} (B_G v, v)$$

(we used (4.17) and the fact that $\tau\delta < \delta < \varepsilon$). Therefore the right-hand side of (4.24), and consequently also (4.23), is not larger than

$$\frac{C_2}{\sqrt{C_0}} \int_0^1 \tau^{-1/2}\, d\tau (B_G v, v)^{1/2} \|u\|.$$

Since by virtue of (4.17)

$$\|u\| \leqq [(B_G v, v)/q_0]^{1/2},$$

from (4.23) and (4.19) we obtain:

$$\int_\Omega |u|^2/g\, dx \leqq \delta^{-1} q_0^{-1} (B_G v, v) + 2C_2 (B_G v, v)/\sqrt{C_0 q_0}.$$

Hence we obtain:

$$\begin{aligned}(B_G v, v)_G &\geqq C_3 (v, v)_G [q_0^{-1} + C_0^{-1/2} q_0^{-1/2}]^{-1} \\ &= C_3 q_0 \sqrt{C_0} (\sqrt{q_0} + \sqrt{C_0})^{-1} (v, v)_G.\end{aligned}$$

An estimate for b is thus obtained in (2.4):

$$b \geqq C_3 q_0 \sqrt{C_0} (\sqrt{q_0} + \sqrt{C_0})^{-1}. \tag{4.26}$$

Remark 4.3. It is obvious from the proof of Theorem 4.1

that we can weaken condition (4.10) in Theorem 4.1. It is sufficient to require that when $x_2 \in \Omega_\delta$

$$C_0|\langle \xi, \nabla g\rangle|^2 \leqq \sum \langle a_{ij}(x)\xi_i, \xi_j\rangle = a(\xi). \tag{4.27}$$

Example 4.3. (A system which degenerates on a boundary and is produced by Dirichlet's problem for a parabolic equation.)

Consider the parabolic equation

$$\partial_t u - \partial_x(a_{11}(x, t)\, \partial_x u(x, t)) + a_{00}(x, t)u(x, t) = f(x, t), \tag{4.28}$$

where the functions

$$a_{11}(x, t),\ a_{00}(x, t),\ f(x, t)$$

are periodic with respect to t with the period 2π and belong to

$$\mathscr{A}(\bar{\Omega} \times T^1),\ \Omega =]-1, 1[\subset \mathbb{R}^1.$$

It is assumed that

$$a_{11}(x, t) \geqq q_1 > 0, \quad a_{00}(x, t) \geqq q_0 > 0. \tag{4.29}$$

We seek the solution of (4.28), which is periodic with respect to t with the period 2π and satisfies the boundary condition

$$u(x, t) = 0 \text{ when } x \in \partial\Omega \tag{4.30}$$

Note that, for simplicity, we consider the case of a scalar equation and one spatial variable. We can consider the case $x > 1$

and $m > 1$ in a similar way, if the domain Ω is defined by a condition of the form (4.11).

We shall reduce problem (4.28), (4.30) to a first-order system that degenerates on $\partial\Omega$. We shall take the function $g(x)$, which equals zero on $\partial\Omega$, $g > 0$ on Ω, and $g' \neq 0$ on $\partial\Omega$. For example, $g(x) = 1 - x^2$. We will assume

$$v_1 = u/g, \quad v_2 = \partial_x u, \quad G = \begin{pmatrix} g & 0 \\ 0 & 1 \end{pmatrix}. \tag{4.31}$$

It is easy to verify that if u is a smooth solution of (4.28), (4.30), the pair of functions $V = (v_1, v_2)$ is a smooth solution of the system:

$$A_1 V \equiv \left[\begin{pmatrix} 1 & 0 \\ 0 & 0 \end{pmatrix} \partial_t - \partial_x \begin{pmatrix} 0 & a_{11} \\ 1 & 0 \end{pmatrix} + \begin{pmatrix} a_{00} & 0 \\ 0 & 1 \end{pmatrix} \right] GV = \begin{pmatrix} f \\ 0 \end{pmatrix} \equiv F. \tag{4.32}$$

The operator A_1^*, which is adjoint to A_1 in $L_2(\Omega \times T^1, G)$, is given by the formula

$$A_1^* V \equiv \left[-\begin{pmatrix} 1 & 0 \\ 0 & 0 \end{pmatrix} \partial_t + \begin{pmatrix} 0 & 1 \\ a_{11} & 0 \end{pmatrix} \partial_x + \begin{pmatrix} a_{00} & 0 \\ 0 & 1 \end{pmatrix} \right] GV \tag{4.33}$$

(there are no boundary terms, since g equals zero on $\partial\Omega$). We will assume $B = A_1^* A_1$. Obviously, the operator B satisfies conditions (2.2), (2.3) on $\mathcal{D}_0 = (A(\Omega \times T^1))^2$ and also satisfies condition (2.4), since

$$(BV, V)_G = (A_1^* A_1 V, V)_G = (A_1 V, A_1 V)_G \geqq 0.$$

A more detailed investigation shows that in (2.4) $b > 0$.

Bearing in mind that condition (2.5) holds for any analytic f by virtue of Theorem 1.1.1., we obtain that Remark 3.2 is applicable to the equation $BV = F$, and $B = A + \rho^2 I$ and V can be represented in polynomial form (2.20). Since the equation $A_1 V = F$ is equivalent to the equation $BV = A_1^*F$, the expression V is thereby obtained in terms of the iteration A_1 and A_1^*. Bearing in mind that $u = gv_1$, we obtain the corresponding expression of the solution (4.28), (4.30).

§ 5. Lower estimates $R_0(B, f)$ for the differential operator B.

The estimates of the rate of convergence obtained in Theorem 2.3 essentially depend on the number R, which satisfies the condition $0 < R < R_0(B, f)$. It is useful to consider the following problem to estimate the number $R_0(B, f)$:

$$\partial_t^2 u = Bu, \quad u(0) = f, \quad \partial_t u(0) = 0. \tag{5.1}$$

Lemma 6.1. Suppose B is a self-adjoint operator in the Hilbert space H bounded from below. Suppose $u(t)$, $0 < t < t_0$ is a function that is doubly continuously differentiable with respect to t with values in H, $u(t) \in \mathcal{D}(A)$ and $Bu(t)$ is continuous with respect to t. Suppose $u(t)$ is a solution of problem (5.1) when $0 < t < t_0$. Then

$$f \in \mathscr{D}(\mathrm{ch}(t_0\sqrt{B}), \quad u(t) = \mathrm{ch}(t\sqrt{B})f$$

when $0 < t < t_0$.

Proof. Suppose E_λ is a projector belonging to resolution of unity of the operator B. Since E_λ commutes with B, $u_\lambda = E_\lambda u(t)$ satisfies the Cauchy problem

$$\partial_t^2 u_\lambda = E_\lambda B u_\lambda, \quad u_\lambda(0) = E_\lambda f, \quad \partial_t u_\lambda(0) = 0.$$

Since $E_\lambda B$ is a bounded operator, the solution of this Cauchy problem is unique, and is given by the formula

$$E_\lambda u(t)=u_\lambda(t)=\mathrm{ch}(t\sqrt{E_\lambda B})E_\lambda f=E_\lambda\,\mathrm{ch}(t\sqrt{E_\lambda B})E_\lambda f$$
$$=\mathrm{ch}(t\sqrt{B})E_\lambda f.$$

Therefore when $0 < t < t_0$, $\forall N \in \mathbb{R}$

$$E_N u(t)=\int_b^N dE_\lambda u(t)=\int_b^N \mathrm{ch}(t\sqrt{\lambda})\,dE_\lambda f. \tag{5.2}$$

Obviously,

$$\int_b^N \mathrm{ch}^2(t\sqrt{\lambda})\,d(E_\lambda f,f)=\|E_N u(t)\|^2\leqq\|u(t)\|^2.$$

Letting N tend to ∞, we obtain that by virtue of (1.3.4)

$$f\in\mathscr{D}(\mathrm{ch}\,t\sqrt{A})$$

when $0 < t < t_0$. Passing to the limit in (5.2) as $N\to+\infty$, and bearing in mind that $E_N u(ty) \to u(t)$, we obtain that

$$u(t)=\mathrm{ch}(t\sqrt{R})f$$

when $0 < t < t_0$.

Suppose $\Omega \subset \mathbb{R}^m$, and the operator B is determined by the formula

$$Bu=-\sum_{i,j=1}^m a_{ij}(x)\,\partial_i\partial_j u+\sum_{i=1}^m a_{0i}(x)\,\partial_i u+a_{00}(x)u, \tag{5.3}$$

where a_{ij} are square matrices of order $\varkappa$. It is assumed that the elements of these matrices are analytic in $\overline{\Omega}$.

In this paragraph it is assumed that Ω is a bounded domain.

Consider problem (5.1), where B is determined by (5.3). It is obvious that in this case (5.1) is a problem of the Cauchy-

Kovalevskoi type. According to the Cauchy-Kovalevskoi theorem the solution of this problem exists when $0 < t < \varepsilon$, where $\varepsilon > 0$. This solution is represented in the form of the series

$$u(t, x) = \sum_{k=0}^{\infty} B^k f(x) t^{2/k}/(2k)!\,. \tag{5.4}$$

We shall use $R_0^K(B, f)$ to denote the radius of the convergence of this series on $\bar{\Omega}$.

Theorem 5.1. Suppose the operator B is determined by Formula (5.3), suppose

$$\mathcal{D}_0 = \mathcal{A}^\kappa(\bar{\Omega}),$$

and B satisfies conditions (2.2)-(2.4). Suppose $f \in \mathcal{D}_0$. Then

$$R_0(B, f) \geqq R_0^K(B, f). \tag{5.5}$$

Proof. The function $v(t)$, which is determined by Formula (5.4), belongs to $\mathcal{D}_0$ when $t < R_0^K(B, f)$. Obviously, $u(t)$ and $Bu(t)$ are analytic with respect to t when

$$0 \leqq t \leqq R_0^K(B, f).$$

Using Lemma 5.1, we obtain that

$$f \in \mathcal{D}(\mathrm{ch}(R\sqrt{B}))$$

when $0 < R < R_0^K(B, f)$. Hence follows (5.5).

We will assume when $\omega \subset \Omega$

$$M^C(B, f, \omega, R) = \sum_{k=0}^{\infty} \sup_{x \in \omega} |B^k f(x)| R^{2k}/(2k)!\,. \tag{5.6}$$

Here the modulus of the vector $u \in \mathbb{C}^{\varkappa}$ is determined by the formula

$$|u| = \max\{|u_i|, i = 1, \ldots, \kappa\}. \tag{5.7}$$

We shall introduce the norm of the matrix $a = \{a_{ij}\}$ using the formula

$$|a| = \max_k \sum_l |a_{kl}|. \tag{5.7'}$$

By virtue of (5.7) and (5.7') $|av| \leq |a|\ |v|$.

We shall use $R_0{}^C(B, f, \omega)$ to denote the number

$$R_0^C(B, f, \omega) = \sup\{R \in \mathbb{R} : M^C(B, f, \omega, R) < \infty\}. \tag{5.8}$$

Obviously, from Formula (5.4) and (5.6) follows inequality

$$R_0^K(B, f) \geqq R_0^C(B, f, \Omega). \tag{5.9}$$

Theorem 5.1 enables us to obtain the exact estimates $R_0(B, f)$ in a number of cases. However it is often difficult to calculate $R_0{}^K(B, f)$ accurately. We can use the estimate (5.9) to estimate $R_0{}^K(B, f)$. Below we will obtain the lower limits $R_0{}^C(B, f, \omega)$, based on using majorants.

For a vector-funtion $v \in A(\overline{\omega})$ we shall introduce the majorant

$$\hat{v}(y) = \sum_{k=0}^{\infty} \frac{y^k}{k!} \max_{|\alpha|=k} \sup_{x \in \omega} |\partial^\alpha v(x)|, \qquad y \in \mathbb{R}. \tag{5.10}$$

Here the modulus of the vector $\partial^\alpha v$ is determined by (5.7). If $v = \{v_{ij}\}$ is a matrix $\varkappa \times \varkappa$, its majorant is also introduced using

Formula (5.10), where the norm of the matrix $a = \partial^{\alpha} v$ is determined using Formula (5.7'). As usual,

$$\partial^{\alpha} = \partial_1^{\alpha_1} \ldots \partial_m^{\alpha_m}, \quad |\alpha| = \alpha_1 + \ldots + \alpha_m.$$

Obviously

$$\partial_y^k \hat{v}(y)|_{y=0} = \sup_{x \in \omega} \max_{|\alpha|=k} |\partial^{\alpha} v(x)|. \tag{5.11}$$

We shall use $\mathcal{A}^+$ to denote the set of functions of the variable $y \in \mathbb{C}$ which are analytic in the neighbourhood of zero and expand in a series in powers of y with nonnegative coefficients. It is obvious that the majorants $\hat{v}$, determined by (5.10), belong to $\mathcal{A}^+$.

We shall now consider the operator B, determined by (5.3) in $\hat{B}$, and shall assume

$$a_2(y) = \sum_{i,j=1}^{m} \hat{a}_{ij}(\omega, y), \quad a_1(y) = \sum_{i=1}^{m} \hat{a}_{0i}(\omega, y), \quad a_0(y) = \hat{a}_{00}(\omega, y). \tag{5.12}$$

We shall use $\hat{B}$ to denote the operator determined in the neighbourhood of zero in $\mathbb{R}$ by the formula

$$\hat{B}v(y) = a_2(y)\, \partial_y^2 v + a_1(y)\, \partial_y v + a_0(y) v. \tag{5.13}$$

The following theorem shows that the operator $\hat{B}$ which is restricted to the set $\{0\}$ consisting of the one point $y=0$, is a majorant for B on ω.

Theorem 5.2. Suppose B is determined by (5.3), and $\hat{B}$ is determined by (5 .13), where a_i, i = 0, 1, 2 is determined by (5.12). Then when $f \in (\mathcal{A}(\omega))^{\varkappa}$

$$M^C(B, f, \omega, R) \leqq M^C(\hat{B}, \hat{f}, \{0\}, R), \tag{5.14}$$

$$R_0^C(B, f, \omega) \geqq R_0^C(\hat{B}, \hat{f}, \{0\}). \tag{5.15}$$

The proof of the theorem is based on the lemma we shall now formulate.

If the two functions $g_1, g_2 \in A^+$ and each Taylor coefficient in zero of the function g_1 is no greater than the corresponding coefficient of the function g_2, we will write $g \lessdot g_2$.

Lemma 5.2. Suppose $a(x)$ is a $\varkappa \times \varkappa$-dimensional matrix, v is a vector with $\varkappa$ components, whilst the elements $a(x)$ and $v(x)$ belong to $A(\omega)$. Then

$$\partial^{\hat{\alpha}}\widehat{(av)} \ll \partial_y^{|\alpha|}(\hat{a}\hat{v}). \tag{5.16}$$

Proof. We shall first prove that

$$\hat{a}_1 \ll \hat{a}_2, \quad v_1 \ll \hat{v}_2, \qquad \widehat{a_1 v_1} \ll \hat{a}_2 \hat{v}_2. \tag{5.17}$$

It is obvious that by virtue of Leibnitz's formula $\partial^\alpha(a_1 v_1)$ decomposes into the sum of $2^{|\alpha|}$ monomials of the form

$$\partial^\beta a_1\, \partial^\gamma v_1, \qquad |\beta| + |\gamma| = |\alpha|.$$

By virtue of (5.11) the modulus of each monomial is estimated from above by

$$\partial_y^{|\beta|} \hat{a}_1\, \partial_y^{|\gamma|} \hat{v}_1 |_{y=0}$$

and, consequently, through

$$\partial_y^{|\beta|}\hat{a}_2\,\partial_y^{|\gamma|}\hat{v}_2|_{y=0}.$$

by virtue of the condition in (5.17). Summing these estimates over all the monomials, we will obtain the estimate

$$|\partial^\alpha(a_1v_1)|\,|_\omega \leqq \partial_y^{|\alpha|}(\hat{a}_1\hat{v}_1)|_{y=0} \leqq \partial_y^{|\alpha|}(\hat{a}_2\hat{v}_2)|_{y=0}.$$

Hence directly follows (5.17) and (5.16) when $|\alpha| = 0$. Noting that it follows from $\hat{a} \prec g$ that

$$\widehat{\partial^\alpha a} \ll \partial_y^{|\alpha|} g,$$

we obtain (5.16) for arbitrary α.

Proof of Theorem 5.2. We shall prove, first of all, that when $x \in \omega$

$$|B^k f(x)| \leq \hat{B}^k \hat{f}(y)|_{y=0} \qquad (k=0,1,\ldots). \tag{5.18}$$

We shall write the coefficients of the operator $\hat{B}$ in the form of the sum over Formulae (5.12). We shall expand $B^k f$ and $\hat{B}^k \hat{f}$ into sums of monomials, consisting of the differentiation operators with respect to x and y and multiplication operators to

$$a_{ij}(x), \hat{a}_{ij}(y)$$

respectively. It is obvious that a monomial in majorants in the expansion $\hat{B}^k \hat{f}$ corresponds to each monomial in the expansion $B^k f$, whilst the majorant of the monomial does not, in accordance with (5.16), exceed the monomial in the majorants. Summing these inequalities over all the monomials, we will obtain that $(\widehat{B^k f}) \prec$ than sum of monomials in majorants. This sum equals $(\hat{B})^k(\hat{f})$, whence

$$(\widehat{B^k f}) \ll (\hat{B})^k \hat{f}.$$

Hence directly follows (5.14). Inequality (5.15) follows from (5.14).

Theorem 4.3 enables us to reduce the estimate R_0^c to the one-dimensional case. Let us now consider the one-dimensional case. We shall find the following lemma useful in this case.

Lemma 5.3. Suppose the operators $\hat{B}_1$ and $\hat{B}_2$ with the coefficients $a_{1i}(y)$, $a_{2i}(y)$ respectively, $i = 0, 1, 2$, are determined by Formula (5.13). Suppose

$$a_{1i},\ a_{2i},\ f_1,\ f_2 \in \mathscr{A}^+$$

Suppose

$$a_{1i} \ll a_{2i}, \qquad i = 0, 1, 2,\ \ f_1 \ll f_2.$$

Then for any $R > 0$

$$M^C(\hat{B}_1, f_1, \{0\}, R) \leqq M^C(\hat{B}_2, f_2, \{0\}, R).$$

The proof of the lemma is obvious.

It is useful to consider the following Cauchy problem to estimate $R_0^C(\hat{B}, \hat{f}, \{0\})$

$$\partial_t^2 v = a_2(y)\,\partial_y^2 v + a_1(y)\,\partial_y v + a_0(y) v, \qquad v|_{t=0} = \hat{f},\ \ \partial_t v|_{t=0} = 0. \tag{5.19}$$

Lemma 5.4. Suppose the solution $v(t, y)$ of problem (5.19) is analytic when $y=0$ with respect to t when $0 < t < t_0$. Then

$$R_0^C(\hat{B}, \hat{f}, \{0\}) \geqq t_0.$$

Proof. The function $v(t, y)$ expands in a series (5.4), where

$$x=y,\ B=\hat{B},$$

$\hat{B}$ is determined by (5.13). Since a_0, a_1, $a_2 \in A^+$, the coefficients of this series when $y = 0$ are nonnegative numbers which we denote by γ_j. When $y = 0$

$$v(t,0)=\sum \gamma_j t^j, \qquad \gamma_j \geqq 0, \tag{5.20}$$

we use t_1 to denote the radius of the convergence of this series.

We shall prove that $t_1 > t_0$. We will assume the opposite, i.e. we will assume that $t_1 < t_0$. Then $v(t, 0)$ is analytic at the point t_1. We shall use ρ_1 to denote the radius of the convergence of the series expansion with respect to $(t - t_1)$ of the function $v(t, 0)$. We will assume $\rho_2 = \rho_1/2$. If ε is a fairly small number, at the point t_2, $t_1 - \varepsilon < t_2 < t_1$ the function $v(t, 0)$ expands in a series with respect to $(t_2 - t)$, whilst the radius of convergence of this series is not less than $3/2\rho_2$. Therefore for the derivatives $v^{(k)}(t, 0)$ of the function $v(t, 0)$ at the point t_2 the following estimate holds:

$$0 \leqq v^{(k)}(t_2, 0) \leqq C\rho_2^{-k}k!\,. \tag{5.21}$$

Note that $v^{(k)}(t_2, 0) > 0$ since (5.20) converges when $t < t_2 + \varepsilon$, $\varepsilon > 0$. From (5.21) and (5.20) we obtain:

$$\sum_{j=k}^{\infty} \gamma_j t_2^{j-k} j!/(j-k)! \leqq C\rho_2^{-k}k!\,. \tag{5.22}$$

Assuming $\rho_3 = \rho_2/2$, hence we obtain

$$\sum_{k=0}^{\infty}\sum_{j=k}^{\infty} \gamma_j \rho_3^k t_2^{j-k} j!/((j-k)!) \leqq C\sum_{k=0}^{\infty} 2^{-k}. \tag{5.23}$$

Changing the order of summation on the left-hand side, we obtain

$$\sum_{j=0}^{\infty}\gamma_j\sum_{k=0}^{j}t_2^{j-k}\rho_3^k j!/((k!)(j-k)!)=\sum_{j=0}^{\infty}\gamma_j(\rho_3+t_2)^j\leqq 2C.$$

Since when $\varepsilon < \rho_3, \rho_3 + t_2 > t_1$,
we arrive at a contradiction with the fact that t_1 is the radius of convergence of series (5.20). Therefore $t_1 > t_0$, and the lemma is proved.

If the function $g(y) \in A^+$, we shall use $r(g)$ to denote the radius of the Taylor series convergence of this function at zero.

Theorem 5.3. Suppose the operator $\hat{B}$ is determined by Formula (5.13), where $a_i \in A^+$, $i = 0, 1, 2$.
Suppose $\hat{f} \in A^+$. We assume

$$r_0=\min(r(a_2), r(a_1), r(a_0), r(\hat{f})).$$

Then

$$R_0^C(\hat{B}, \hat{f}, \{0\})\geqq\int_0^{r_0}\frac{ds}{\sqrt{a_2(s)}}\equiv\sigma_0(r_0). \tag{5.24}$$

Proof. By virtue of Lemma 5.4, to prove (5.24) it is sufficient to prove the t-analyticity of the solution $v(t, y)$ of Eq.(5.19) when $y = 0$ in the interval $]0, \sigma(r_0)[$.

The coefficients of Eq.(5.19) and the function $\hat{f}$ are analytic in the strip

$$P=\{t\in\mathbb{R}\}\times\{-r_0<y<r_0\}.$$

Consider the case when $a_2(y) \neq 0$ in the strip P. (The case when $a_2(y_0) = 0$ for some y_0, $-r_0 < y_0 < 0$ reduces to that considered using the substitution $y = y_0 + z^2/4$, and we shall consider this case later.)

When $a_2 \neq 0$ in P Eq.(5.19) is hyperbolic in P. The equations of its characteristics have the form

$$dy_{\pm}/dt = \pm\sqrt{a_2(y_{\pm})}. \tag{5.25}$$

Integrating this equation, we obtain:

$$\pm\sigma_0(y_{\pm}(t)) = t - s, \quad y_{\pm}(t) = \sigma_0^{-1}(\pm(t-s)), \tag{5.26}$$

where $\sigma_0(s)$ is determined in (5.24). It is obvious that the characteristics $y_{\pm}(t)$ pass throught the point $y=0$ when $t=s$. It follows from (5.25) that $y_+ < 0$ when $t < s$, $y_- > 0$ when $t < s$. If $t_1 < \sigma_0(r)$, then, assuming $s=t_1$, we obtain that the characteristics y_+ and y_- intersect the line $t=0$ at the points $\sigma_0^{-1}(-t_1)$ and $\sigma_0^{-1}(t_1)$ respectively, whilst under the nonzero angle. These two characteristics and the line $t=0$ form a curvilinear triangle Δ. The characteristics which pass through any point of this triangle intersect its base $t=0$ under the nonzero angles. The triangle Δ is contained in P, therefore the coefficients of Eq.(5.19) and its initial data are analytic in P. As is well-known (see, for example, Hadamard's book [1]), it follows hence that the solution $v(t, y)$ is analytic in Δ, and therefore $v(t, 0)$ is t-analytic when $0 < t < t_1$. The number t_1 is as close to $\sigma_0(r_0)$ as desired, and therefore $v(t)$ is analytic when $0 < t < t_0 = \sigma(r_0)$. The theorem is thereby proved when $a_2(y) \neq 0$ in P.

We shall now analyse the case when

$$a_2(y_0) = 0, \quad -r_0 < y_0 < r_0.$$

Since $a_2 \in A^+$ and, consequently, increases with respect to y, this is only possible when $-r_0 < y_0 < 0$.

Let us first consider the case $y_0 < 0$. If $a'_2(y_0) = 0$, it is then easy to see that the characteristic $y_+(t)$ does not intersect with the line $y = y_0$ and all the arguments proceed in exactly the same way as in the case $a_2 \neq 0$. We will therefore assume that $a'_2(y_0) \neq 0$. We will assume $y_0 = -\xi$, $0 < \xi < r_0$, and shall carry out the change $y = -\xi + z^2/4$ in Eq.(5.19). We will formulate $w(t, z) = v(t, -\xi + z^2/4)$. The function w satisfies equation

$$\partial_t w = a_{20}(y)\,\partial_z^2 w + a_{10}(y)\,\partial_z w + a_0(y)w, \tag{5.27}$$

where

$$\begin{gathered} y = y(z) = -\xi + z^2/4, \\ a_{10}(y, z) = (2a_1(y) - a_{20}(y))/z, \quad a_{20}(y) = a_2(y)/(y+\xi). \end{gathered} \tag{5.28}$$

It is obvious that the functions $a_0(y(z))$ and $\hat{f}(y(z))$ are analytic when $|z^2/4 - \xi| < r_0$, and since $a_2(-\xi) = 0$, then $a_{20}(y(\xi))$ is also analytic when $|z^2/4 - \xi| < r_0$. If

$$2a_1(-\xi) - a_{20}(-\xi) = 0, \tag{5.29}$$

then

$$a_{10}(y(z), z) = (2a_1(-\xi + z^2/4) - a_{20}(-\xi + z^2/4))/z$$

is analytic when $|z^2/4 - \xi| < r_0$.

We shall now show that we can achieve the satisfaction of condition (5.29) by replacing the coefficient a_1 in the appropriate way. We shall take the function $a_1^0(y)$ instead of $a_1(y)$

$$a_1^0(y) = a_{20}(y)/2 + \hat{a}_{20}(y)(1 + y/\xi) + a_1(y)(1 + y/\xi). \tag{5.30}$$

Here $a_{20}(y)$ is determined by (5.28), and $\hat{a}_{20}(y)$ is a majorant at zero of the function $a_{20}(y)$. The function $a_1{}^0(y)$, which is determined by (5.30), has the following properties:

$$a_1^0 \in \mathscr{A}^+, \tag{5.31}$$

$$a_1^0 \gg a_1, \tag{5.32}$$

$$r(a_1^0) \geqq r_0, \tag{5.33}$$

and condition (5.29) also holds, where $a_1 = a_1{}^0$.

We shall verify these properties of $a_1{}^0$. First of all, we shall verify condition (5.29). Substituting $y = -\xi$ into (5.30), we immediately show its validity. Since $\xi > 0$, then obviously

$$\hat{a}_{20}(y) \ll \bar{a}_{20}(y)(1+y/\xi) \qquad 0 \ll a_1(y) \ll (1+y/\xi)a_1(y).$$

Therefore all the derivatives at zero from $a_1{}^0(y)$, determined by (5.30), are not less than the derivatives of the function

$$a_{20}(y)/2+\hat{a}_{20}(y)+a_1(y).$$

The derivatives at zero of the function

$$a_{20}(y)/2+\hat{a}_{20}(y)$$

are nonnegative by virtue of the definition $\hat{a}_{20}$. Hence follow (5.31) and (5.32). In order to verify (5.33), note that $\hat{a}_{20}(y)$ decomposes at zero into a series with respect to y, whose coefficients equal the moduli of coefficients of the corresponding series $a_{20}(y)$. Therefore the radii of convergence of these series are equal,

$$r(\hat{a}_{20}) = r(a_{20}).$$

Since $a_2(-\xi) = 0$, then by virtue of (5.28)

$$r(a_{20}) = r(a_2) \geqq r_0.$$

Using this estimate, the definition of (5.30) and the fact that $r(a_1) > r_0$, we immediately derive (5.33).

Note that to prove (5.24) it is sufficient to prove the inequality

$$R_0^C(\hat{B}_0, \hat{f}, \{0\}) \geqq \sigma_0(r_0), \tag{5.34}$$

where $\hat{B}_0$ is the operator (5.13), where the function a_{10}, determined by (5.30), is substituted instead of the function a_1. Indeed, by virtue of (5.32) according to Lemma 5.3

$$M(\hat{B}_0, \hat{f}, \{0\}, R) \geqq M(\hat{B}, f, \{0\}, R). \tag{5.35}$$

Since the function $\sigma_0(y)$ is the same for $\hat{B}$ and $\hat{B}_0$, and the number r_0 is the same by virtue of (5.33), then, assuming $R = \sigma_0(r_0) - \varepsilon$ in (5.35), we obtain that

$$M(\hat{B}, \hat{f}, \{0\}, \sigma_0(r_0) - \varepsilon) < \infty, \qquad \forall \varepsilon > 0,$$

whence follows (5.24).

To prove (5.34), we shall consider Eq.(5.27). The point $z_0 = 2\sqrt{\xi}$ corresponds to the point $y=0$. The characteristics $z_\pm$ which pass through z_0 when $t=s$, are determined from the equation

$$\varphi(z_\pm) = \pm(t-s), \quad \varphi(z) = \int_{2\sqrt{\xi}}^{z} \frac{d\eta}{\sqrt{a_2(\eta^2/4 - \xi)/(\eta^2/4)}}. \tag{5.36}$$

Obviously,

$$z_{\pm}(t)=\varphi^{-1}(\pm(t-s)).$$

The function $\varphi(z)$ is monotonic and analytic when

$$-\xi < \eta^2/4 - \xi < r_0.$$

To determine the domain of definition φ^{-1}, we shall estimate the range of change $\varphi(z)$. If $z < -2\sqrt{\xi}$, we obtain, bearing in mind that the integrand expression in (5.36) is an even positive function:

$$\varphi(z)\leqq \int_{-2\sqrt{\xi}}^{z} \frac{d\eta}{\sqrt{a_2(\eta^2/4-\xi)/(\eta^2/4)}}=-\int_0^{z^2/4-\xi}\frac{dy}{\sqrt{a_2(y)}}=-\sigma_0(z^2/4-\xi).$$

Considering the positive values of z in a similar way, and bearing in mind that $|z^2/4 - \xi| < r_0$, we obtain that $\varphi^{-1}(z)$ is clearly defined in the interval

$$]-\sigma_0(r_0),\ \sigma_0(r_0)[$$

Hence follows the analyticity $w(t, z)$ in the triangle Δ_1, formed by the characteristics z_+, z_- and the line $t = 0$ when $t_1 = s < \sigma_0(r_0)$. Hence follows the analyticity with respect to t of $w(t, -2\sqrt{\xi})$ when $0 < t < \sigma_0(r_0)$ and, consequently, the analyticity of $v(t, 0)$ for these t. Hence, as shown at the beginning of the proof of the theorem, follows (5.34). Thus the theorem is also proved in the case when $a_2(y_0) = 0$, $y_0 < 0$.

It remained to consider the case $y_0 = H$. In this case we will assume

$$a_{2\varepsilon}(y)=a_2(y)+\varepsilon,\ \varepsilon>0.$$

Estimate (5.24) is proved for the operator $\hat{B}_\varepsilon$ with the coefficient $a_2 = a_{2\varepsilon}$. Since $a_2^0 > a_2$, from Lemma 4.3 we obtain that

$$R_0^C(\hat{B}, \hat{f}, \{0\}) \geq R_0^C(\hat{B}_\varepsilon, \hat{f}, \{0\}) \geq \int_0^{r_0} \frac{ds}{\sqrt{a_2(s)+\varepsilon}}.$$

Passing to the limit on the right-hand side of this inequality as $\varepsilon \to 0$, we also obtain (5.24) in the case $a_2(0) = 0$. The theorem is thus completely proved.

Theorem 5.4. Suppose the operator B is determined by (5.3),

$$\omega \subset \Omega,\ f \in (\mathscr{A}(\omega))^\kappa.$$

Suppose the functions $a_i(\omega, y)$ are determined by (5.12),

$$\hat{f}(y) = \hat{f}(\omega, y)$$

and the number $\sigma_0(r_0)$ is the same as in Formula (5.24). Then

$$R_0^C(B, f, \omega) > \sigma_0(r_0). \tag{5.37}$$

Proof. Inequality (5.37) directly follows from (5.15) and (5.24).

Remark 5.1. It is obvious from (5.37) that when the function f is an entire function, there is a lower bound for $R_0^C(B, f, \omega)$ which does not depend on f. Therefore (at least in the bounded domain Ω), according to (5.9) and (5.5) there is an lower bound for $R_0(B, f)$ which does not depend on the entire function f.

§ 6. Estimate of the parameter $R_0(B, f)$ for the model operator B.

Consider the operator B, which acts on the functions $v \in A(T^1)$:

$$Bv = -\partial_x(\cos^2 x\,\partial_x v) + a(x)v, \tag{6.1}$$

where $a \in A(T^1)$ is an entire function.

Obviously, the operator (6.1) is the simplest example of the operators considered in Example 3.1 in §3. The number of variables $m=1$ and the dimensions $\varkappa = 1$, i.e. B is a scalar operator.

Theorem 6.1. If $f \in A(T^1)$ is an entire function, then $R_0(B,f) > \pi/2$.

Remark 6.1. It follows from the results of §§4 and 5 of the following chapter (see Theorem 5.4.1 and Remark 5.5.1), that for many $a(x)$ and $f(x)$ $R_0(B,f) < \pi/2$ and, consequently, $R_0(B,f) = \pi/2$

The proof of Theorem 6.1 is based on Theorem 5.1. To apply this theorem, let us consider Cauchy's problem (5.1), where B is determined by (6.1):

$$\partial_t^2 v = -\partial_x \cos^2 x\,\partial_x v + a(x)v, \quad v|_{t=0} = f,\ \partial_t v|_{t=0} = 0, \tag{6.2}$$

where $a(x)$ is an entire function, $a \in A(T^1)$. The specific form $a(x)$ henceforth will not be essential.

From the Cauchy-Kovalevska theorem directly follows:

Proposition 6.1. Problem (6.2) has a unique solution $v(t, x)$ which is analytic with respect to t and x when

$$|t| < \varepsilon,\ \varepsilon > 0,\ x \in \mathbb{R},$$

and periodic with respect to x with the period 2π.

Proposition 6.2. If $\delta > 0$ is sufficiently small, the solution $v(t, x)$ extends to the solution analytic on the sets Ω_1 and $\Omega_2 \subset \mathbb{R}^2$:

$$\Omega_1=\{(x,t):x\in[-\pi,\pi]:|x-\pi/2|\leqq\delta,|t|<\pi\}$$
$$\Omega_2=\{(x,t):x\in[-\pi,\pi],|x+\pi/2|\leqq\delta,|t|<\pi\}. \tag{6.3}$$

Proof. We shall prove that $v(t, x)$ can be extended to Ω_1. For this, by virtue of Formula (5.4), it is sufficient to prove that

$$R_0^C(B,f,\omega_1^\delta)\geqq\pi, \tag{6.4}$$

where B is an operator on the right-hand side of (6.2),

$$\omega_1^\delta=\{x:|x-\pi/2|\leqq\delta\}. \tag{6.5}$$

We shall use Theorem 5.4 to estimate

$$R_0^C(B,f,\omega_1^\delta)$$

since the coefficients of the operator B and the function f are entire, in (5.24) $r_0 = +\infty$. Therefore

$$R_0^C(B,f,\omega_1^\delta)\geqq\int_0^\infty\frac{ds}{\sqrt{a_2(s)}}, \tag{6.6}$$

where

$$a_2(y)=a_2(y,\omega_1^\delta)$$

is a majorant on ω_1^δ of the function $\cos^2 x$. According to (5.10)

$$a_2(y,\omega_1^\delta)=\sum_{k=0}^\infty\frac{y^k}{k!}\sup_{|x-\pi/2|<\delta}|\partial_x^k\cos^2 x|. \tag{6.7}$$

Obviously, when $k=1,2,,,$

$$b_k=\sup_{|x-\pi/2|<\delta}|\partial_x^k\cos^2 x|=\tfrac{1}{2}\sup_{|x|<\delta}|\partial_x^k(\cos 2x)|.$$

Therefore $b_0 = \sin^2\delta$

$$b_k = \begin{cases} 2^{k-1},\ k \text{ is even}, & k \geqq 2 \\ 2^{k-1}\sin(2\delta),\ k \text{ is odd}, & k \geqq 1 \end{cases}. \tag{6.8}$$

From Formulae (6.7) and (6.8) we derive that

$$a_2(y, \omega_1^\delta) = \sin^2\delta + (\mathrm{ch}(2y) - 1)/2 + \sin(2\delta)\,\mathrm{sh}(2y)/2. \tag{6.9}$$

We shall make an upper estimate of the right-hand side of (6.6). It is obvious that due to (6.9) when $0 < y < 1$

$$a_2(y) \leqq \sin^2\delta + C_1 y^2 + C_2 y \sin\delta. \tag{6.10}$$

Therefore

$$\int_0^\infty \frac{ds}{\sqrt{a_2(s)}} \geqq \int_0^1 \frac{ds}{\sqrt{a_2(s)}} \geqq \int_0^1 \frac{ds}{\sqrt{\sin^2\delta + C_2 \sin\delta s + C_1 s^2}}. \tag{6.11}$$

Making the substitution $s = t\sin\delta$ we obtain hence and from (6.11)

$$R_0^C(B, f, \omega_1^\delta) \geqq \int_0^{1/\sin\delta} \frac{dt}{\sqrt{1 + C_2 t + C_1 t^2}} \to +\infty,\ (\delta \to 0) \tag{6.12}$$

Hence we obtain that for small δ

$$R_0^C(B, f, \omega_1^\delta) > \pi,$$

and Proposition 6.2 is proved for Ω_1. (Ω_2 is considered in a similar way.)

Proposition 6.3. The solution $v(t, x)$ of problem (6.2) extends to a solution which is analytic in the set

$$\Omega = \{x \in \mathbb{R}\} \times \{t \in \mathbb{R} : 0 \leqq t < \pi/2\}. \tag{6.13}$$

Proof. Bearing in mind Proposition 6.2 and the periodicity of all the functions in (6.2) with respect to x, it is sufficient to prove that $v(t, x)$ extends to the sets

$$\Omega_3 = \{x \in [-\pi, \pi] : |x| < \pi/2 - \delta, 0 < t < \pi/2\},$$
$$\Omega_4 = \{|x - \pi| < \pi/2 - \delta\} \times \{0 < t < \pi/2\}.$$

Consider Ω_3. We make the change of variable

$$x = 2 \operatorname{arctg} \operatorname{th}(z/2).$$

Then

$$\partial_x = \operatorname{ch} z\, \partial_z, \cos x = 1/\operatorname{ch} z$$

and problem (6.2) will take the form

$$\partial_t^2 u = -\partial_z^2 u + \operatorname{th} z\, \partial_z u + a_1(z) u, \tag{6.14}$$

$$u|_{t=0} = f_1, \quad \partial_t u|_{t=0} = 0, \tag{6.15}$$

where

$$a_1(z) = a_2(x(z)), \quad f_1(z) = f_2(x(z)),$$

and f_2 are entire functions. The function

$$x = 2 \operatorname{arctg} \operatorname{th}(z/2)$$

is holomorphic in the strip $|\operatorname{Im} z| < \pi/2$. In fact, the function

$$w = \operatorname{th}(z/2) = (e^z - 1)/(e^z + 1)$$

maps this strip into the circle $|w| < 1$. The function

$$x = 2 \operatorname{arctg} w$$

is the primitive of the function $2(1 + w^2)^{-1}$ which, in turn, is holomorphic in this circle. The coefficients and initial data thz, $a_1(z)$ and $f_1(z)$ of problem (6.14), (6.15) are thus holomorphic and bounded in the strip $|\mathrm{Im} z| < q$ when $q < \pi/2$.

We shall now prove that the solution (6.14), (6.15) exists and is analytic with respect to z and t for all real z and for complex t, $|t| < \pi/2$.

We will assume $z = \xi + i\eta$. It follows from (6.14), (6.15) that the function

$$u_2(t, \xi, \eta) = u_1(t, \xi + i\eta)$$

satisfies for each $\xi \in \mathbb{R}$ the equation

$$\partial_t^2 u_2 = \partial_\eta^2 u_2 - i \operatorname{th}(\xi + i\eta)\, \partial_\eta u_2 + a_1(\xi + i\eta) u_2 \tag{6.16}$$

with the initial condition

$$u_2|_{t=0} = f_1(\xi + i\eta), \quad \partial_t u_2|_{t=0} = 0. \tag{6.17}$$

Eq.(6.16) is hyperbolic. Its coefficients and initial data are analytic with respect to η when $-\pi/2 < \eta < \pi/2$ for each $\xi \in \mathbb{R}$. Consider (6.16) in the triangle

$$\Delta_l = \{t \geqq 0\} \cap \{t + \eta < l\} \cap \{t - \eta < l\}, \qquad (l < \pi/2), \tag{6.18}$$

formed by the characteristics of Eq.(6.16). The unique solution

$u_2(t, \xi, \eta)$ of Eq.(6.14) with the initial data (6.17) exists in Δ_l. This solution is analytic with respect to η and with respect to t. (See, for example, Hadamard's book [1].) The solution (6.16), (6.17) in Δ_l smoothly depends on ξ. Applying the operator $\partial/\partial z = \partial_\eta - i\partial_\xi$ to (6.16), (6.17), we obtain that

$$u_3 = (\partial_\eta - i\,\partial_\xi)u_2(t, \xi, \eta)$$

satisfies Eq.(6.16) with zero initial data, i.e. $u_3(t, \xi, \eta) = 0$. Therefore the function $u_2(t, \xi, \eta)$ satisfies the Cauchy-Riemann system and is an analytic function of the complex argument $z = \xi + i\eta$. Consequently the function $u(t, z)$ satisfies (6.14) for complex z.

If $|\eta| < \varepsilon$, the solution (6.14) in Δ_l is analytic with respect to t when $0 < t < l - \varepsilon$. Therefore, bearing in mind that we can take l as close to $\pi/2$ as desired, we obtain that a solution of (6.14), (6.15) which is analytic with respect to t and z when

$$0 \leqq t < \pi/2, \quad -\infty < z < +\infty,$$

exists. Consequently, the solution $v(t, x)$ of problem (5.10) extends analytically with respect to t to the interval $-\pi/2 < t < \pi/2$ when $-\pi/2 < x < \pi/2$, i.e. $v(t, x)$ analytically extends to Ω_3. Considering the function $v(t, x - \pi)$, which also satisfies a problem of the form (6.2), we obtain that $v(t, x)$ extends to Ω_4. Bearing in mind that by virtue of Proposition 6.2 $v(t, x)$ extends to Ω_1 and Ω_2, we obtain a statement of Proposition 6.3.

Proof of Theorem 6.1. Suppose $v(t, x)$ is a solution of problem (6.2) which is analytic in $[-\pi/2, \pi/2]$ and periodic with respect to x with the period 2π. This solution exists by virtue of Proposition 6.2. It is obvious that

$$v(t,\cdot)\in\mathscr{A}(T^1)=\mathscr{D}_0$$

when $0 < |t| < \pi/2$, and $B^k v(t, \cdot)$ are analytic with respect to t in $H = L_2(T^1)$. Lemma 5.1, according to which

$$v(t,\cdot)=\mathrm{ch}(t\sqrt{B})f$$

when $0 < t < \pi/2$ and $R_0(B, f) > \pi/2$, is therefore applicable.

Remark 6.2. A rather more detailed analysis of problem (6.14), (6.15) shows that $u(t, z)$ is analytic with respect to t when $|\mathrm{Re}\, t| < \pi/2$, $z \in \mathbb{R}$. Therefore

$$R_0^K(B,f)\geqq\pi/2.$$

Remark 6.3. It follows from Remark 6.1 that the function $v(t)$ - the solution of (6.1) - has a singularity when $t = \pi/2$. In paragraph 8 of Chapter 5 we obtain an upper estimate of the increase $\|v(t)\|$ when $t \to \pi/2$.

§ 7. Operators in unbounded domains.

Suppose $\Omega \subset \mathbb{R}^m$, and $\mathcal{O}_\delta^{\,c}(\Omega)$ is its complex δ-neighbourhood (see 1.12). We shall use $A_\delta^{\,c}(\Omega)$ to denote a set of functions bounded in $\mathcal{O}_\delta^{\,c}(\Omega)$, i.e. a set of u, such that

$$\sup\{r:r=|u(z)|,\, z\in\mathcal{O}_\delta^C(\Omega)\}\leqq C=C(u). \tag{7.1}$$

If

$$u(z)\in\mathscr{A}_\delta^C(\Omega),$$

we will assume when $x \in \Omega$

$$\check{u}_\delta(x)=\sup\{r:r=|u(x+z)|,\, z\in\mathbb{C}^m, |z|<\delta\}. \tag{7.2}$$

The majorant

$$\check{u}_\delta(x)$$

is determined in a similar way for the vector u.

Theorem 7.1. Suppose $\mathcal{D}_0$ is a set of functions v from $\mathcal{A}_\delta{}^C(\Omega)$, such that the function

$$\check{v}_\delta(x) \in L_2(\Omega).$$

Suppose the operator B is determined in $\mathcal{D}_0$ by Formula (3.1), whilst the coefficients B belong to $\mathcal{A}_\delta{}^C(\Omega)$, and suppose conditions (3.1), (3.2), (3.3) and (3.5) hold. (It is assumed that $\partial\Omega$ is piecewise-smooth and for almost all r the domain $\Omega \cap \{\langle x, x\rangle = r\}$ has a piecewise-smooth boundary.) The operator B then satisfies conditions (2.2)-(2.5).

Proof. Conditions (2.2)-(2.4) are verified in the same way as in Sect.3. The verification of condition (2.5) is based on the following theorem.

Theorem 7.3. Suppose the self-adjoint operator B in $H = (L_2(\Omega))^\varkappa$ is determined in the set $\mathcal{D}_0$ using Formula (3.1) and satisfies conditions (2.2)-(2.4). Suppose $f \in \mathcal{D}_0$ and the function $S(x) \in L_2(\Omega)$ exists, such that $\forall x \in \Omega$

$$M^C(B, f, \{x\}, R) \leqq S(x). \tag{7.3}$$

Then

$$\|\mathrm{ch}(R\sqrt{B})f\|^2 \leqq \kappa \int_\Omega S^2(x)\, dx. \tag{7.4}$$

Proof. We shall expand chz in a series and shall use $Q_n(z)$ to denote the partial sum of this series. Obviously, since $f \in \mathcal{D}_0 \subset \mathcal{D}_\infty(B)$

$$\|Q_n(R\sqrt{B})f\|^2 = \int_\Omega \langle Q_n(R\sqrt{B})f(x), Q_n(R\sqrt{B})f(x)\rangle\, dx$$
$$\leqq \kappa \int_\Omega |Q_n(R\sqrt{B})f|^2\, dx.$$

Since, according to (5.6),

$$|Q_n(R\sqrt{B})f(x)| \leqq M^C(B, f, \{x\}, R),$$

then by virtue of (7.3)

$$\|Q_n(R\sqrt{B})f\|^2 \leqq \kappa \|S\|^2.$$

We shall now pass to the limit as $n \to \infty$. Since $Q_n(z) \to$ chz monotonically increases with respect to n, Estimate (7.4) holds by virtue of Fatu's theorem and Formula (1.3.3).

Using the fact that the coefficients of the operator B are bounded in $\mathcal{O}_\delta{}^C(\Omega)$, and using the results of §5 for estimates at the point $x \in \Omega$, we obtain

$$M^C(B, f, \{x\}, R) \leqq C \check{f}_\rho(x)$$

for some $\rho > 0$. Hence and from Theorem 7.2 it follows that condition (2.5) holds.

§ 8. The use of polynomial representations to solve differential equations numerically.

In this paragraph the polynomial representation of the form (2.20) of the solution of the equation $Bu = f$ is constructed in the form of an iteration process. A comparison is then made of the number of arithmetical operations necessary to calculate the approximate solutions of the equation $Bu = f$ using finite-difference methods and using a polynomial representation of the solutions. Cases are considered when B are operators from Examples 3.1, 3.2 and 3.3, whose coefficients are trigonometric or algebraic polynomials, and the right-hand side is also a polynomial. The end of the paragraph considers polynomial representations of the Cauchy problems (2.16) and (2.17), considers the question of the number of arithmetical operations necessary to attain the specified accuracy, and makes a comparison with finite-difference methods. The basic results of this paragraph were published in the author's paper [11].

According to sect.1) of Theorem 2.3, the polynomial representation of the form (2.20) holds for the solution u of the equation $Bu = f$, and at the same time the approximate solution is represented in the form

$$u_n = P^{00}_{0n}(B)f, \qquad P^{00}_{0n}(B) = P^{00}_{n}(B - bI)$$

and estimate (2.8) holds. To use it for numerical calculations, however, it is important to point out a convenient method of calculating the polynomial $P^{00}_{0n}(B)$. When the operator is bounded, iteration processes of the following form are frequently used in computational mathematics:

$$u_0 = 0 \quad u_{k+1} = u_k - \tau_{k+1}(Bu_k - f), \qquad k = 0, \ldots, n-1. \tag{8.1}$$

(see Bakhvalov [1], Godunov and Ryaben'kii [1], Marchuk [1], Samarskii [1], Samarskii and Nikolaev [1].)

It is obvious that u_n is the result of applying some polynomial $P_n(B)$ in the operator B to the right-hand side f. The numbers τ_j are called parameters of the iteration process. When the operator B is bounded, and its spectrum is arranged on the segment $[\rho^2, M]$, the minimum error is attained with a Chebyshev set of parameters. At the same time the numbers τ_j are expressed in a simple way in terms of the root of the Chebyshev polynomial of the n-th degree and in terms of the numbers ρ^2 and M. In the above case the spectrum B is bounded only from below, and instead of the boundedness of B we require the boundedness of

$$\|\mathrm{ch}(R\sqrt{B})f\|, \; R>0.$$

Theorem 8.1. Suppose the self-adjoint operator B satisfies condition (2.4):

$$B \geqq bI, \; b=\rho^2>0,$$

and

$$f \in \mathscr{D}(\mathrm{ch}(R\sqrt{B})), \; R>0.$$

Suppose the parameters τ_j are determined by the equations

$$1/\tau_j=\rho^2+x_j^2\pi^2/(4R^2) \qquad (j=1,\dots,n), \tag{8.2}$$

where x_j are the solutions of Eq.(3.1.20) of Chapter 3. Suppose u_j, $j = 1, \dots, n$ is determined by the scheme (8.1), and u is the solution of the equation $Bu = f$. Then

$$\rho^4\|u-u_n\|^2+\|A(u-u_n)\|^2 \leqq Cn^{-2\sigma}\|\mathrm{ch}(R\sqrt{A})f\|^2, \tag{8.3}$$

where $A=B-\rho^2 I$, $\sigma=2R\rho/\pi$.

Proof. We will assume $z_k = u_k - u$. Using the fact that

$$u = u - \tau_{k+1}(Bu - f), \tag{8.4}$$

and subtracting (8.4) from (8.1), we will obtain:

$$z_0 = u = A^{-1}f, \qquad z_{k+1} = z_k - \tau_{k+1}Bz_k \tag{8.5}$$

when $k = 0, \ldots, n - 1$. Hence we obtain that

$$z_n = \prod_{j=1}^{n} (I - \tau_j B) z_0. \tag{8.6}$$

We shall formulate

$$K_n(B) = \prod_{j=1}^{n} (I - \tau_j B). \tag{8.7}$$

Since $z_n = u - u_n$, then

$$\|Bz_n\| = \|BK_n(B)B^{-1}f\| = \|K_n(B)f\|. \tag{8.8}$$

Denoting $A = B - \rho^2 I$, $A > 0$, we obtain, in a similar way to inequality (2.1.5):

$$\|K_n(B)f\| \leq \mu^+(K_n(\lambda + \rho^2), 0, \Phi_n(R\sqrt{\lambda}))\|\Phi_n(R\sqrt{A})f\|, \tag{8.9}$$

where $\Phi_n(x)$ is determined by Formula (3.1.5), and $\mu^+(P_n, g, \Phi)$ is determined by Formula (2.1.6). Now note that when $\lambda = -\rho^2$, $K_n(\lambda + \rho^2) = 1$. Replacing $\lambda = x^2$, we obtain that the polynomial

$$K_n(x^2 + \rho^2) \in \mathcal{N}_n(\rho)$$

(the class $\mathcal{N}_n(\rho^2)$ is determined in §1 of Chapter 3. See (3.1.12)). According to Theorem 3.1.1. of Chapter 3 $\mu(Q, 0, \Phi_n(Rx))$ attains its minimum in the set $\mathcal{N}_n(\rho)$ in the element

$$Q^0, \quad Q^0 = T_n(rx)/T_n(ir\rho).$$

We shall now prove that the polynomial $K_n(\rho^2 + x^2)$ agrees with $Q^0(x)$. Indeed, it follows from Lemma 3.1.4 and from (8.2) that $1/\tau_j = y_j^2 + \rho^2$, where y_j are roots $T_n(rx)$, $r = 2R/\pi$. Since $1/\tau_j$ are the roots of $K_n(\lambda)$ by virtue of (8.7), then $\lambda = y^2 = 1/\tau - \rho^2$ are the roots of $K_n(\lambda + \rho^2)$. Bearing in mind that the even polynomials $K_n(x^2 + \rho^2)$ and $Q^0(\lambda)$ equal 1 when $x = \pm i\rho$, and also have n common positive roots y_j, we obtain that these polynomials are identical. Therefore, by virtue of Formula (3.1.14) of Chapter 3

$$\mu^+(K_n(\lambda+\rho^2), 0, \Phi_n(R\sqrt{\lambda})) =$$
$$= \mu(K_n(x^2+\rho^2), 0, \Phi_n(Rx)) = \mu(Q^0, 0, \Phi_n(Rx)) = 1/T_n(ir\rho).$$

Using (3.1.6) and (3.2.38), where j=1, from (8.9) we obtain that

$$\|K_n(B)f\| \leqq Cn^{-r\rho}\|\text{ch}(R\sqrt{A})f\|,$$

where C does not depend on n. Hence and from (8.8), bearing in mind that

$$\|Bv\|^2 \geqq \rho^4\|v\|^2 + \|Av\|^2,$$

we obtain (8.3), and the theorem is proved.

Remark 1.1. Using estimate (3.2.26) and Remark 3.2.1 instead of estimate (3.2.38), we can refine the value of the constant C

in (8.3). Namely, (8.3) takes the form

$$\rho^4\|u-u_n\|^2+\|B(u-u_n)\|^2\leqq 4(1+2n+\sigma)^{-2\sigma}(1+\sigma)^{2\sigma}\|\mathrm{ch}(R\sqrt{B})f\|^2.$$

Remark 1.2. To calculate the parameters τ_j using Formula (8.2), it is required to calculate the roots of Eq.(3.1.20). This can require, for large n, a large number of calculations, in connection with there being in (3.1.20) a sum with respect to j from 1 to n.

We shall present a more economical method of constructing the parameters τ_j, which have all the properties necessary for the convergence of (8.1).

We will assume

$$\varphi(y)=\mathrm{arctg}\, y-(\ln(1+y^2))/(2y) \tag{8.10}$$

and shall consider the equation

$$\varphi(y_k)=\pi(2k-1)/(4n), \qquad k=1,\dots,n. \tag{8.11}$$

This equation has the unique root $y_k > 0$ for each k. We will assume

$$x_k=2n/y_k. \tag{8.12}$$

A detailed analysis shows that if τ_k is determined by Formula (8.2), where x_j are determined by (8.11), (8.12), estimate (8.3) will remain valid, for the function u_n, determined by (8.1). Calculations carried out for some examples on a computer show that the errors u-u_n differ little for x_j, determined by (3.1.20) and for x_j, determined by (8.11) and (8.12), i.e. the constant C in (8.3) differs little for these two modifications.

A function of the form (8.10) naturally arises in connection with the fact that the left-hand side of (3.1.20) has the form

$$2\sum_{j=1}^{n}\operatorname{arctg}\frac{x}{2j-1}=\pi n-2\sum_{j=1}^{n}\operatorname{arctg}\frac{2j-1}{x}.$$

Making the substitution $x = 2n/y$, we note that

$$\frac{1}{n}\sum_{j=1}^{n}\operatorname{arctg}\left(y\left(\frac{2j-1}{2}\right)\right)$$

is an integral sum of the integral

$$\varphi(y)=\int_{0}^{1}\operatorname{arctg}(yt)\,dt,$$

which, as can easily be seen, is identical with (8.10).

Let us now consider the case when B is a differential operator from Examples 3.1, 3.2, 3.3, whose coefficients are trigonometric (in Example 3.1) or algebraic polynomials, the right-hand side of which is also a polynomial. In this case the calculations using Formula (8.1) reduce to the differentiation and multiplication of polynomials, i.e. to purely algebraic operations on the coefficients of polynomials. These calculations can be easily implemented on a computer.

We shall estimate the number of calculations necessary to achieve the specified accuracy of ε as a function of ε. Suppose the coefficients A are polynomials of a degree no higher than p, and the function f is a polynomial of a degree no higher than q. Then $u_{k+1}(x)$ is a polynomial of a degree no higher than $q + kp$. This polynomial of m variables $x_1, \ldots, x_m$ has about $C(\varkappa, m)(q + kp)^m$ coefficients, where $\varkappa$ is the size of the vectors f and u. The calculation Bu_{k+1} reduces to the multiplication of polynomials of the power $q + kp$ and polynomials of the power p, and also to the differentiation of polynomials. The number of arithmetical operations on the coefficients of the polynomials is about

$$C_1(q+kp)^m p^m \quad C_1=C_1(\kappa, m).$$

Thus for the common number of N operations necessary to calculate n iterations using Formula (8.1), the following estimate holds:

$$N \leqq C_1 p^m \sum_{k=0}^{n-1} (q+kp)^m = C_2(q+np)^{m+1}, \tag{8.13}$$

where $C_2 = c_2(\varkappa, m)$. We shall now note that if it is required to obtain a solution correct to ε, i.e.

$$\|u_n - u\| \leqq \varepsilon, \tag{8.14}$$

then, according to inequality (8.3), it is sufficient to take n, such that

$$Cn^{-\sigma} < \varepsilon. \tag{8.15}$$

Therefore for (8.14) to be satisfied, it is sufficient to take

$$n = C_0 \varepsilon^{-1/\sigma}, \tag{8.16}$$

where $C_0 = C_0(R, \rho, f)$. Substituting this value into (8.13), we obtain that the number of $N(\varepsilon)$ arithmetical operations required to achieve estimate (8.14) will not exceed

$$N(\varepsilon) = C_3 \varepsilon^{-(m+1)/\sigma}, \qquad \sigma = 2R\rho/\pi, \tag{8.17}$$

where $C_3 = C_3(R, \rho, p, q, f)$ does not depend on ε. It follows from (8.17) that the increase in $N(\varepsilon)$ as $\varepsilon \to 0$ essentially depends on the parameter $\sigma = 2R\rho/\pi$, which, in turn, strongly depends on the analytic properties of the operator A.

We shall now compare the most common features of the above method with difference methods. To be specific, we shall consider the two-dimensional case $m=2$. When solving equations using net-point methods we take a net with the step h. At the same time the number of nodes is proportional to h^2. The corresponding system of net-point equations contains orders of $K_0 h^{-2}$ unknowns - the values of the function at the net points. No fewer than $K_1 K_0 h^{-2}$ arithmetical operations $K_1 > 1$ are required to solve this system of net-point equations. Moreover, usually $K_1 \to \infty$ as $h \to 0$. (See Godunov and Ryaben'kii [1], Marchuk [1], Samarskii [1]). Note, further, that the net step h is chosen as sufficiently small, in order to make a sufficiently small error of approximation. Usually, for schemes of the second order of accuracy, the error of approximation equals $\mathcal{O}(h^2)$ on sufficiently smooth solutions. Therefore for the specified accuracy of ε, $h = \mathcal{O}(\sqrt{\varepsilon}$, and the number of arithmetical operations when using difference approximations is not less than $C\varepsilon^{-1}$

$$N(\varepsilon) \geqq K_3 \varepsilon^{-1}. \tag{8.18}$$

(At the same time $N(\varepsilon)$ weakly depends on the analytic properties and does not usually increase faster than $C\varepsilon^{-1}\ln(\varepsilon^{-1})$). Thus, comparing (8.17) and (8.18), we note that for small $R\rho$ it can be more advantageous to use scheme (8.1) directly than to proceed to difference approximations.

Remark 8.3. For the error of approximation to decrease as $h \to 0$ when proceeding to net-point equations, some smoothness of the solution of the initial differential equation is required. In particular, the boundedness of the four derivatives of the solution is required to obtain second-order approximation schemes (see Godunov and Ryaben'kii [1], Marchuk [1], Samarskii [1]). When estimating the error of method (8.1), no assumptions are made about the smoothness of the solution $u(x)$, and the analyticity of only the right-hand side $f(x)$ is required. When the operator B is not elliptic, the solution of the equation $Bu = f$ can be non-smooth, see, for example, §4 of the following chapter. The solution $u(x)$ of Eq.(5.4.11) of this paragraph does not belong to the Sobolev space $W_2^{\sigma_0}(T^1)$ (with the corresponding value of the parameter β in (5.4.11), where $\sigma_0 = \beta - 1/2 = 2R\rho/\pi$, whilst by virtue of Theorem 8.1 estimate (8.3) holds and, consequently, (8.17) for any $\sigma < \sigma_0$. It is obvious from Formula (5.4.14) that the second derivative of the solution of Eq.(5.4.11) is continuous when $\beta > 3$, the third derivative is bounded when $\beta > 4$, and the fourth is bounded when $\beta > 5$. Even if we require a weaker condition instead of the boundedness of $\partial_x^4 u$, namely the boundedness of

$$\partial_x^2(\sin x\, \partial_x^2 u)$$

(it seems sufficient for a second order of approximation), this holds only when $\beta > 3$. Thus, applying difference schemes of the second order of accuracy to Eq.(5.4.11) is only justified when $\beta > 3$, and with these values $\sigma_0 = \beta - 1/2 > 2.5$. Comparing (8.17), where $m=1$, $\sigma = 2.5 - \delta$, δ is as small as desired, and (8.18), we show that in the case of degenerating equations the use of method (8.1) can be more effective for any - not necessarily large - values of $R\rho$.

Remark 8.4. The approximate solutions u_n, $n = 1, 2, \ldots$, determined by (8.1) form a sequence of polynomials which approximate the accurate solution of Eq.(5.4.11) with the maximum possible rate (in order). Indeed, if approximation were possible using polynomials of the n-th degree with an error of order $n^{-\gamma}$, $\gamma > \sigma_0$, the solution u would belong to

$$W_2^{\sigma_0+\delta}(T^1),\ \delta>0,$$

according to Bernshtein's inverse theorem, and this contradicts Lemma 5.4.4. It is thus impossible to obtain in any way - for example, using Galerkin's method - polynomials u_n of the n-th degree which approach the exact solution faster (in order) as $n\to\infty$ than those that are determined by (8.1).

We shall now indicate cases when the parameter $R\rho$ is small.

Example 8.1. Consider an equation with an operator of the form (3.1), containing the small parameter ν, $\nu > 0$:

$$\mathrm{Au}=-\nu\sum_{i,j=1}^{m}\partial_i(a_{ij}(x)\,\partial_j u)+a_{00}(x)-f(x), \tag{8.19}$$

and it is assumed that all the conditions imopsed in Example 3.1 hold, and the matrix $a_{00}(x)$ is positive definite:

$$a_{00}(x)\geqq q_0 I, \qquad q_0>0. \tag{8.20}$$

Obviously, in this case we can take the number

$$\rho_0=\sqrt{q_0}\,.$$

as the parameter ρ. It is also assumed that the coefficients a_{ij} and the function f are trigonometric polynomials. We shall use Theorem 5.3 for an lower bound of R. Since the functions a_{ij} and f are integral, then in Formula (5.24) $r_0 = +\infty$. It follows from the construction of the majorant $a_2(y)$ that

$$a_2(y)=a_2(y,\nu)=\nu a_{20}(y).$$

Therefore

$$R_0^C(\hat{A},\hat{f},\{0\})\geqq\int_0^\infty\frac{ds}{\sqrt{\nu a_{20}(s)}}=\frac{1}{\sqrt{\nu}}R_0'. \tag{8.21}$$

From (5.5), (5.9) and from (8.21) we obtain that

$$R_0(A,f)\geqq R_0'\nu^{-1/2}.$$

Assuming $R = R_1\nu^{-1/2}$, where $R_1 < R_0'$, we obtain that we can take the following as the number σ in (8.17)

$$\sigma=2\rho_0R_1/(\pi\sqrt{\nu}), \tag{8.22}$$

and inequality (8.17) takes the form

$$N(\varepsilon)\leqq C_3\varepsilon^{-S_0\sqrt{\nu}}. \tag{8.23}$$

It is obvious that for small ν the rate of increase $N(\varepsilon)$ in (8.23) will be less than in (8.18).

Example 8.2. Consider Eq.(8.19), whilst it is assumed that all the conditions imposed in Example 3.2 hold, and also condition (8.20), whilst the components of the matrices a_{ij} and of the vector f are algebraic polynomials. Repeating the arguments of Example 8.1 verbatim, we obtain that estimate (8.23) holds for a number of arithmetical operations $N(\varepsilon)$, i.e. $N(\varepsilon)$ slowly increases as $\varepsilon \to 0$, if ν is fairly small. We consider an equation of the form (3.7) with a small parameter ν for the second derivatives in a similar way.

Example 8.3. Consider Eq.(8.19) with zero boundary conditions (4.13): $u\big|_{\partial\Omega} = 0$. We will assume that all the conditions imposed in Example 4.2 hold. Suppose, in addition, condition (8.20) holds. Suppose the coefficients a_{ij} of the operator B, the right-hand side of f and the function G are algebraic polynomials. According to Theorem 4.1 inequality (2.4) holds, where $b > 0$. Since, by virtue of (8.19), $C_0 = C_{00}\nu$ in inequality (4.10), from (4.26) follows an estimate for the number $\rho = \sqrt{b}$

$$\rho \geqq C\nu^{1/4}, \qquad C > 0. \tag{8.24}$$

It follows from this inequality that the following estimate of the number of arithmetical operations needed to achieve the accuracy ε is valid (see (8.14))

$$N(\varepsilon) \leqq C\varepsilon^{-S_0\sqrt[4]{\nu}}, \qquad S_0 > 0. \tag{8.25}$$

Indeed, we shall use estimate (8.17). According to this estimate

$$N(\varepsilon) \leqq \varepsilon^{\pi(m+1)/(2R\rho)}.$$

According to (8.22), the number R can be taken in the form $R = R_1\nu^{-1/2}$, and according to estimate (8.24) $\rho = C\nu^{1/4}$. Therefore $R\rho = C_1\nu^{-1/4}$, whence follows (8.25). It follows from this estimate that for small ν $N(\varepsilon)$ slowly increases as $\varepsilon \to 0$.

Remark 8.5. The exponent in (8.25) for small ν is greater than in (8.23). This is connected with the fact that with the substitution $u = gv$ the matrix $a_{00}g$ ceases to be positive definite. Therefore we are obliged to use the positiveness of the principal part of the operator A to get lower bound of $\rho = \rho(\nu)$. Since the principal part is multiplied by the small parameter, $\rho = \rho(\nu)$ decreases as ν decreases, more slowly, however, than $R = R(\nu)$ increases.

Remark 8.6. The algorithm for calculation using Formula (8.1) does not depend on the form of the domaon Ω, on the coefficients B or on f. All the calculations reduce to the multiplication, differentiation and summation of polynomials, and are easily implemented on a computer in the form of a program that is suitable for the solution of any equation of the form (3.1), where m=2 or m=3 with polynomial coefficients and a right-hand side. Another situation arises when solving equations using net-point methods. The net equation strongly depends on the form of the domain Ω, which creates additional difficulties during programming. Therefore it can sometimes be advantageous to use method (8.1) in comparison with traditional methods, due to the programming savings even when there is no gain in consumption of computer time.

Remark 8.7. To estimate the error $\|u - u\|_k$ it is most convenient to use formula

$$\|u-u_k\| \leq \|B^{-1}\| \, \|z_k\|, \qquad z_k = Bu_k - f. \tag{8.26}$$

It is in many cases easy to obtain a lower bound of the norm B^{-1} for example

$$\|B^{-1}\| \leq q_0^{-1},$$

where q_0 is the number in Formula (4.15). The discrepancy z_k is calculated at each step of scheme (8.1), and it is easy to obtain the approximate estimate $\|z_k\|$.

The value of the error $\|u - u_k\|$ for finite k strongly depends on the constant C and on the multiplier

$$\|\text{ch}(R\sqrt{A})f\|$$

in (8.3). To give an impression of the value of this error, we shall give a specific example of a computer calculation.

Example 8.4. In the domain $\Omega \subset \mathbb{R}^2$, consisting of the square $|x_1| < 1$, $|y_1| < 1$ with the excluded circle $|x_1|^2 + |y_1|^2 < 1/4$ we consider the equation

$$\begin{aligned} Bu \equiv 0.1\, \partial_1((1-x_1^2)(0.25-x_1^2-x_2^2)\, \partial_1 u) \\ +0.1\, \partial_2((1-x_2^2)(0.25-x_1^2-x_2^2)\, \partial_2 u)+u=f(x), \end{aligned} \tag{8.27}$$

where

$$f(x)=0.05+0.75x_1^2-0.2x_2^2+x_1^4+0.06x_1^2x_2^2.$$

The unique bounded solution of this equation equals $u(x) = x_1^2$. The approximate solution was sought using scheme (8.1), for which $\rho = 1$, $R = 1.1$, $n = 50$ are taken.

We present a table which indicates the values of the function $u_{50}(x_1, x_2)$ at the points

$$(x_{1i}, x_{2j})=(0.2i, 0.2j),\ i,j \in \mathbb{Z},\ 0 \leqq i, j \leqq 5.$$

$x_2 \backslash x_1$	0	0.2	0.4	0.6	0.8	1.0
0	0.053	0.053	0.124	0.332	0.638	1.008
0.2	0.038	0.044	0.126	0.336	0.640	1.008
0.4	0.007	0.028	0.131	0.346	0.643	1.006
0.6	-0.020	0.015	0.138	0.354	0.644	1.003
0.8	-0.031	0.012	0.144	0.357	0.641	0.998
1.0	-0.031	0.015	0.147	0.356	0.638	0.995

The maximum deviation from the exact solution $u=x_1^2$ at the points belonging to Ω, was equal to 0.031. The calculated values of the discrepancy $z_{50} = Bu_{50}-f$ at these points do not exceed 0.05, the function $u_{50}(x_1, x_2)$ is an even polynomial in x_1 and x_2 of the degree 104, most of the coefficients of which exceed 1, and the coefficients that are maximum in value achieve 6000. Moreover this polynomial approximates the exact solution $u(x_1, x_2) = x_1^2$ in Ω well.

It is important to note that from the fact that $u_n \to u$ as $n \to \infty$ it generally does not follow that the coefficients of the polynomials u_n generally converge to coefficients of the polynomial u.

The results of the calculations show that the sequence of polynomials u_k, $k = 1, ..., 50$, constructed using the scheme (8.1) for Eq.(8.27), is quickly stabilised. For example, all the functions u_k when $k > 10$ differ from u_{50} at the points (x_{1i}, x_{2j}) by not more than 0.03. We shall present a table of values $u_5(x_{1i}, x_{2j})$.

$x_2 \backslash x_1$	0	0.2	0.4	0.6	0.8	1.0
0	0.050	0.054	0.123	0.327	0.638	1.010
0.2	0.038	0.046	0.124	0.332	0.640	1.009
0.4	0.008	0.028	0.127	0.343	0.645	1.007
0.6	-0.022	0.012	0.134	0.354	0.646	1.003
0.8	-0.037	0.007	0.142	0.360	0.642	0.998
1.0	-0.037	0.011	0.148	0.358	0.637	0.995

The corresponding values of the discrepancies for u_5 do not exceed 0.07, and do not exceed 0.05 for u_k, $k > 5$.

The polynomial u_5 has the degree 14, and moduli of its coefficients do not exceed 52. 4 mins. 30 secs. of computer time is required to calculate u_{50} when using an ES-1033 computer, and it takes several seconds to calculate u_5. Thus to calculate the approximate solution of Eq.(8.17) which degenerates on the boundary Ω with an error of 0.04 requires that 5 steps be satisfied using scheme (8.1), $k = 1, ..., 5$. All the operations hold on the coefficients of polynomials of a degree from 4 to 14. Bearing in mind that these polynomials are even with respect to x_1 and x_2, the maximum number of coefficients which it is required to store in the computer memory is 42 coefficients for the function $u_5(x_1, x_2)$.

Remark 8.8. In Example 8.4 we can estimate the value of the error accurately, since the exact value of u is known. Note that the use of estimate (8.26) gives close results. The number q_0 in (4.15) equals 1 for Eq.(8.27) from Example 8.4, and therefore $\|B^{-1}\| < 1$. The value of the discrepancy z_k at the points where it is calculated does not exceed 0.05 when $k > 5$, By virtue of (8.26) this indicates the value of the error $u - u_k$ of order 0.05, which agrees with the calculation data.

Remark 8.9. According to (8.26) the smallness of $u - u_k$ follows from that of $z_k = Bu_k - f$. Therefore it makes sense to run the calculations using Formula (8.1) not up to $k = n$, but up to the value $k=k_0$, after which the discrepancy z_k stabilises in value. As Example 8.4 shows, this k_0 can be much smaller than n. If the value of the stabilising discrepancy is insufficiently small, we can increase n. At the same time estimate (8.3) shows that we need to increase n by a factor of $p^{1/\sigma}$ to reduce the error by a factor of p (naturally, the last argument is not strict, but it accords well with the calculation data).

Remark 8.10. It is required to specify the parameters τ_k for calculations using Formula (8.1), and for this we required lower bounds of the parameters R and ρ. The lower bound for R is given in §4. Note that it can be disadvantageous to take the value R, close to the maximum $R = R_0$, as R, since the multiplier

$$\|\mathrm{ch}(R\sqrt{A})f\|$$

can have a strong effect on the value of the error $\|u - u_n\|$ in (8.3). This multiplier can become very large when R is close to $R_0 = R_0(A, f)$.

Remark 8.11. Since calculations using Formula (8.1) have a local character by virtue of the locality of the differentiation operation, then when it is required to obtain a solution not in the whole domain Ω, but in the neighbourhood of some point $x_0 \in \Omega$, we can expand the coefficients and the right-hand side of the equation into polynomials with respect to $(x - x_0)$ and we can calculate the values $u(x)$ only at points that are close to x_0.

Remark 8.12. The volume of calculations when solving the equation $Bu = f$ in the multi-dimensional case $m=2$ or $m=3$ is much larger than when $m=1$. Therefore, to estimate the number of iterations necessary to attain the specified value of error for the initial multi-dimensional equation, it is useful to make a model study of a one-dimensional equation. For example, we can take the following equation for Eq.(8.27) as the model:

$$0.2\,\partial_1[(1-x_1^2)(0.25-2x_1^2)\,\partial_1 u]+u=f,$$
$$f=0.05+0.5x_1^2+1.6x_1^4.$$

The results of the calculations show the agreement in order of the discrepancy $\|z_k\|$ for the initial equation when $5 \leqslant k \leqslant 100$.

Let us now consider Cauchy's problem (2.16) and (2.17), where the operator B is determined by (3.1). When solving Cauchy's problem using finite-difference methods we use discretisation both with respect to x, and with respect to t with the step Δx and Δt respectively. For schemes of a second order of approximation the error of approximation is proportional to $(\Delta t)^2 + (\Delta x)^2$. Therefore, if it is required to achieve accuracy of order ε, it is required to assume

$$\Delta t = O(\sqrt{\varepsilon}),\ \Delta x = O(\sqrt{\varepsilon}).$$

In this case the number of mesh points will increase like $\varepsilon^{-1/2}\varepsilon^{-m/2}$ even in the one-dimensional case $m=1$. The number of arithmetical operations necesary to solve the corresponding equations will increase no slower than $\varepsilon^{-1/2}\varepsilon^{-m/2}$.

Let us now consider finding approximate solutions of problems (2.16) and (2.17) in the form (2.18). We will assume that the coefficients of the operator B, determined by (3.1), and the function f, are trigonometric or algebraic polynomials. As indicated above, the number of arithmetical operations necessary to calculate

$$p_n^{ij}(B-bI)f_j,$$

is proportional to n^{m+1}, where m is the number of spatial variables. By virtue of estimate (2.10), we obtain that to achieve the accuracy ε the power n of the polynomial P_n^{1j} must be subject to the condition

$$n^{-\sigma_1}\ln^n \leqq \varepsilon/C,$$

i.e. conditions of the following form must hold:

$$n(\varepsilon) = \exp[(1/\sqrt{\sigma_1})\sqrt{\ln 1/\varepsilon + \ln C}]. \tag{8.28}$$

The number of arithmetical operations necessary to achieve this accuracy is proportional to

$$\exp[((m+1)/\sqrt{\sigma_1})\sqrt{\ln 1/\varepsilon + \ln C}]. \tag{8.29}$$

When ε approaches zero this number increases more slowly than any power of the number $1/\varepsilon$ with a positive indicator. When using (2.18) for solutions of the hyperboilc Cauchy problem (2.17), according to (2.12) the error will not exceed ε, if the following condition holds:

$$e^{-\sigma_2 n} \leqq \varepsilon/C,$$

i.e. $n(\varepsilon)$ can be taken in the form

$$n(\varepsilon) = (\ln 1/\varepsilon + \ln C)/\sigma_2. \tag{8.30}$$

The number of arithmetical operations will be proportional to

$$[(\ln 1/\varepsilon + \ln C)/\sigma_2]^{m+1}, \tag{8.31}$$

i.e. it will increase as $\varepsilon \to 0$ far more slowly than $\varepsilon^{-1/2-m/2}$.

The above facts show that the use of formulae of the form (2.18) can be useful for solving problems of the form considered.

Remark 8.13. It is important to note that the numbers σ_1 and σ_2 in Formulae (2.1) and (2.12), and consequently also in (8.29) and (8.30), strongly depend on t and on the parameter R. Estimates (2.10) and (2.12) are derived from estimates (3.5.1), (3.5.5) and (3.5.14) of Chapter 3. Consideration of these estimates shows that for large values of t, σ_1 and σ_2 become small. It therefore makes sense to use Formula (2.18) when t is not too large in comparison with R. We shall present examples of the numbers σ_1 and σ_2 for illustration (possibly rather understated):

$$\sigma_1 > \tfrac{1}{3} \quad \text{when} \quad 0 \leqq t < 2R^2/\pi^2$$

in Formula (8.28) and (8.29), and

$$\sigma_2 \geqq -2\ln(1-e^{-\pi/2}) > 0.225 \quad \text{when} \quad |t| \leqq R,$$

$$\sigma_2 \geqq -2\ln(1-e^{-\pi}) > 0.085 \quad \text{when} \quad |t| \leqq 2R.$$

in Formulae (8.30) and (8.31).

Remark 8.14. If it is required to calculate the solution of problem (2.16) $u_1(t)$ for large $t = t_N$, it is feasible to divide $[0, t_N]$ into intervals of length Δt, which is not large in compared with R (smallness in comparison with ε and R is not required). We can use Formulae (2.18) to calculate

$$u(t_k),\ t_k = k\,\Delta t, \quad k = 1, 2, \ldots, N,$$

The function $u(t_k)$ is a polynomial of high degree. Approximating $u(t_k)$ using a polynomial of not very large fixed degree

$$n_0 = n_0(\varepsilon)\ \hat{u}(t_k),$$

taking

$$f_k = \hat{u}(t_k)$$

and solving the resulting problem in the next segment, we will obtain $u(t_{k+1})$, etc.. When the operator B is elliptic in T^m, the error of approximation $u(t_k)$ using the polynomial $\hat{u}(t_k)$ is not great, since the exact solution of the parabolic equation is uniformly analytic when $0 < t < t_N$, and consequently is very rapidly approximated in L_2 by trigonometric polynomials.

The solution given by (2.18) is close to the accurate one in L_2, and is consequently also rapidly approximated using polynomials. The accumulation of errors proceeds slowly due to the finiteness of the step Δt and the correctness in L_2 of Cauchy's problems for the parabolic equation.

CHAPTER 5. ESTIMATES OF THE SMOOTHNESS OF FUNCTIONS OF SELF-ADJOINT DIFFERENTIAL OPERATORS

In the previous chapter we gave examples of differential operators B with analytic coefficients, the functions $g(B)$ of which can be constructed in the form of the limit of the polynomials in the operator B using Formula (4.1.1), where $A = B$. The operator $g(B)$, applied to the analytic function $f(x)$, is also some function $u(x) = g(B)f(x)$ of variable x. In this chapter we shall use the results obtained earlier to examine the connection between the properties of the function $g(\lambda)$ and the smoothness with respect to x of the function $u(x)$. The results obtained are used to estimate the smoothness of the solutions of degenerating differential equations with partial derivatives. These estimates are precise.

§ 1. Formulation of the theorems on smoothness.

The differential operators which will be considerred in this chapter will satisfy the following condition.

Condition 1.1. The operator B is defined on functions from $C^{\infty}(\overline{\Omega})$ using Formula (4.5.3), and its coefficients belong to $\mathcal{A}(\overline{\Omega})$. In addition, the set $\mathcal{D}_0$ of functions exists in the space $H = L_2(\Omega, G)$ (see §2, Ch.4), such that $\mathcal{D}_0 \subset \mathcal{A}(\overline{\Omega})$ and conditions (4.2.2)-(4.2.5) hold.

Examples of differential operators which satisfy condition 1.1 are given in §3, §4 and §7 of Chapter 4.

As usual, we shall determine the norm $\| \ \|_s$ in the Sobolev spaces $W_2{}^s(\mathbb{R}^m)$ using the formula

$$\|f\|_2^s = \int_{\mathbb{R}^m} (1+|\xi|^2)^{2s} |\tilde{f}(\xi)|^2 \, d\xi, \tag{1.1}$$

where

$$\tilde{f}(\xi) = \mathcal{F} f(\xi)$$

is a Fourier transform of the function f, determined by Formula (1.2.3).

We shall use $\varphi_0(x)$ to denote a function of the class

$$C^\infty(\mathbb{R}^m),\ \varphi_0(x)=1$$

when

$$|x|\leqq 1/2,\ \varphi_0(x)=0$$

when $|x| > 1$. We shall introduce local Sobolev classes $W_2^s{}_{loc}(\Omega)$, where $\Omega \subset \mathbb{R}^m$ is an open domain. The function $f\in W^s_{2\,loc}(\Omega)$, if

$$\forall x_0\in\Omega\ \exists\delta>0$$

such that the function $\varphi_0((x - x_0)/\delta)f(x)$ belongs to $W_2^s(\mathbb{R}^m)$. As is known, this definition does not depend on the choice φ_0.

If $K \subset \mathbb{R}^m$ is a compactum, then $u \in W_2^s(K)$ if $u(x)$ is determined in some neighbourhood of the compactum K and belongs to $W_2^s{}_{loc}$ in this neighbourhood.

If $g(z)$, $z \in \mathbb{R}$ is an even function, we will assume

$$g^+(\lambda)=g(\sqrt{\lambda}),\qquad \lambda\in\mathbb{R}_+. \tag{1.2}$$

If the function $g(z)$ of the class $A(\mathcal{J}_\beta)$ or the class $A(\mathcal{J}_\infty) = A_0(\mathcal{J}_\infty)$ (see §3 and §5 of Ch.3), the function $g^+(\lambda)$ is of the class $A(\mathcal{J}^+_\beta)$ $(\beta < +\infty)$ (see §6, Ch.3).

We shall introduce the following majorant for the functions g and g^+, connected by Eq.(1.2):

$$g^*(s)=\sup\{p:p=|g(z)|,\ |\operatorname{Im} z|\leqq s\}.$$

It follows from the definition of the class $A(\mathcal{J}_\beta)$, that when $\beta < +\infty$, $g^*(s)$ is finite when $0 < s < \beta$. It is obvious that $g^*(s)$ monotonically increases with respect to s, $0 < s < \beta$. We will assume $g^*(s) = +\infty$ for $s > \beta$. When $\beta = +\infty$, if $g(z)$ differs from the constant, then $g^*(s) \to +\infty$ when $s \to +\infty$.

Besides the majorant $g^*(s)$ we shall introduce the regular majorant $g^{**}(s)$, which is connected to $g^*(s)$ by the relations:

$$g^*(s) < Cg^{**}(s) \ \forall s > 0, \ g^{**}(s) = \exp(s\nu^*(s)) \tag{1.3'}$$

At the same time it is assumed that $\nu^*(s)$ is a monotonically non-decreasing function, $\nu^*(0) > 0$, $\nu^*(s) = +\infty$ when $s > \beta$. Everywhere henceforth, it is also assumed when $\beta = +\infty$, that $\nu^*(s)$ satisfies one of the conditions - A or B - formulated below.

Condition A. $s_0 > 0$, $C > 0$ exist for any $\varepsilon > 0$, such that for all $s > s_0$ the inequality $-\varepsilon\nu^*(s) + \ln\ln s < C$ holds.

Condition B. $s_0 > 0$, $C > 0$ exist for any $\varepsilon > 0$, $t_0 > 0$, such that for all $t > t_0$, $s > s_0$ the following inequality holds:

$$\ln\ln\nu^*(st) < \varepsilon(\nu^*(s) + \ln t) + C$$

It is obvious that the functions $g^{**}(s)$ and $\nu^*(s)$ are defined in more than one way with respect to g, but they are convenient for applications, since it is much easier to construct some majorant g^{**} than to calculate g^* accurately.

Note that the requirement on ν^*, which consists of the satisfaction of one of the two conditions, A or B, is slightly limiting. Indeed, if $\nu^*(s)$ increases more rapidly than $\ln\ln s$, condition A holds. If $\nu^*(s)$ increases more slowly than $\exp(s^\varepsilon)$ $\forall \varepsilon > 0$ and has fairly regular behaviour, condition B holds. In particular, condition B holds if the function

$$\varphi(s) = \ln\ln\nu^*(\exp s)$$

has a bounded derivative φ', whilst $\varphi'(s) \to 0$ as $s \to +\infty$. Thus condition B holds for the functions $\nu^*(s)$ of the form

$$\exp\exp\ln^q s \ (q < 1),\ s^p,\ (\ln s)^p,\ (\ln\ln s)^p (p > 0)$$

and so forth , and also for $\nu^*(s)$ = const.

We shall formulate the basic results of this chapter.

Theorem 1.1. Suppose the operator B satisfies condition 1.1,

$$f \in \mathscr{D}_0,\ R_0 = R_0(B, f) > 0$$

(see (4.1.2)). Suppose

$$g^+ \in \mathscr{A}(\mathscr{J}_\beta^+), \quad \beta > 0.$$

Suppose b is a number, satisfying (4.2.4), $A = B - bI$, $u = g^+(A)f$. Then

$$u \in W^s_{2\,\mathrm{loc}}(\Omega),\ \forall s: s < r\beta, \quad r = 2R_0/\pi. \tag{1.4}$$

If K is a compactum, $K \subset \Omega$, we will assume

$$\|v\|_K^2 = \int_K |v(x)|^2\, dx. \tag{1.5}$$

Theorem 1.2. Suppose the conditions of Theorem 1.1 hold, where $\beta = +\infty$. Then for any compactum $K \subset \Omega$ and for any $\varepsilon >$ 0 the numbers C and R_1 exist, such that

$$\|\partial^\alpha U\|_K \leqq C R_1^{|\alpha|} |\alpha|!\, g^{**}(|\alpha|/(r-\varepsilon)), \tag{1.6}$$

where $r = 2R_0/\pi$.

Theorem 1.1 is a corollary of Theorem 6.1, which will be proved in §6, and Theorem 1.2 is the corollary of Theorem 7.1 from §7. §§2 and 3 prove Theorems 1.1 and 1.2 in the special, simpler case, when $\Omega = T^m$ is a torus. The general case $\Omega \subset \mathbb{R}^m$ is then reduced, in §§6 and 7, to the case $\Omega = T^m$.

Let us now consider the domain Ω, whose boundary $\partial\Omega$ satisfies the following condition in the domain Ω_1.

Condition 1.2. Suppose $\Omega_1 \subset \mathbb{R}^m$. For any point

$$y_0 \in \bar{\Omega} \cap \Omega_1$$

the neighbourhood $\mathcal{O}(y_0)$ exists, such that the analytic mapping Y of the cube

$$K_1 = \{|x_i| \leqq 1,\ i = 1, \ldots, m\}$$

into

$$\Omega_1, \qquad Y(K_1) \subset \bar{\Omega},$$

exists, whilst

$$\mathcal{O}(y_0) \cap \bar{\Omega} \subset Y(K_1).$$

At the same time Y is one-to-one, and the differential Y' is non-degenerate: det $Y' \neq 0$ to K. It is also assumed that the matrix G in the definition $L_2(\Omega, G)$ is positive definite in $\bar{\Omega} \cap \Omega_1$.

It is obvious that condition 1.2, imposed on Ω, is in fact a condition on $\partial\Omega$. It indicates that $\partial\Omega$ is a piecewise-analytic hypersurface in Ω_1, whilst the singularities of $\partial\Omega$ must be locally the same as on the boundary of the cube K_1. These singularities are possible on the generic piecewise-analytic hypersurface.

If condition 1.2 holds, we have estimates of the smoothness of $u(x)$ in Ω_1 up to the boundary.

Theorem 1.1'. Suppose the conditions of Theorem 1.1 hold and condition 1.2 holds. Then for any compactum $K \subset \Omega_1 \bigcap \bar{\Omega}$

$$u \in C^l(K), \forall l < (r\beta - 3m/2)/2, \qquad l \subset \mathbb{Z}_+.$$

Theorem 1.2'. Suppose the conditions of Theorem 1.2 hold, and condition 1.2 holds. Then for any compactum $K \subset \bar{\Omega} \bigcap \Omega_1$ C and R_1 exist, such that

$$\|\partial^\alpha u\|_{C(K)} \leqq CR_1^{|\alpha|}|\alpha|!g^{**}((2|\alpha| + 3m/2)/(r-\varepsilon)). \tag{1.6'}$$

We shall now derive from Theorems 1.1, 1.1', 1.2 and 1.2' theorems on the smoothness of the solutions of partial differential equations. We shall introduce the necessary notation.

If B is a self-adjoint upper-bounded operator, we shall use $b_0(B)$ to denote the lower bound of the operator B

$$b_0(B) = \inf\{(Bu, u) : u \in \mathscr{D}(B), \|u\| = 1\}. \tag{1.7}$$

We shall use $b_1(B)$ to denote the lower bound of the continuous spectrum B, i.e.

$$b_1(B) = \sup\{(s \in \mathbb{R} : (\text{spectrum of } B) \cap]-\infty, s] \quad \text{is finite}\}. \tag{1.7'}$$

Obviously, $b_1(B) \geqslant b_0(B)$. For an operator B of the form (4.5.3), we shall use $R_0^K(B, f)$ to denote the radius of the convergence in $L_2(\Omega, G)$ of the Cauchy-Kowalewska series (4.5.4), and shall use $R_0(B, f)$ to denote the number, determined by (4.1.2), where A is the Friedrichs extension of the operator B. By virtue of (4.5.5)

$$R_0(B, f) \geqq R_0^k(B, f).$$

We shall determine the smoothness indicator for the function $u \in L_2(\Omega, G)$

$$s_2(u) = \sup\{s \in \mathbb{R}_+ : u \in W^s_{2\,\mathrm{loc}}(\Omega)\}. \quad (1.8)$$

Note that the operators B, which satisfy condition 1.1, can be degenerating operators. Indeed, condition (4.3.3) permits arbitrary degeneracy of the principal part of the operator B, preserving its nonnegativeness. We shall estimate the smoothness of the solutions of the equations $Bu = f$ with a generally degenerating operator B.

Theorem 1.3. Suppose the operator B satisfies condition 1.1 and is invertible, $b_1(B) > 0$. Suppose $f \in \mathfrak{D}_0$. Then

$$s_2(B^{-1}f) \geqq 2/\pi\sqrt{b_1(B)}R_0(B, f) = s_{20}. \quad (1.9)$$

If condition 1.2 holds, then $B^{-1}f$ of the class

$$C^l(\bar{\Omega} \cap \Omega_1)$$

when

$$2l + 3m/2 < s_{20}.$$

Proof. Suppose $b = b_1(B) - \varepsilon > 0$. Only a finite number of points $\lambda_1, \ldots, \lambda_K$ of the spectrum of the operator B, is contained in the segment $[b_0(B), b]$. Suppose $Q(\lambda)$ is an interpolation polynomial of the $K-1$-th degree of the function λ^{-1} with interpolation nodes $\lambda_1, \ldots, \lambda_K$. We put

$$u_1 = Q(B)f,\ u_2 = B^{-1}f - u_1.$$

Then

$$Bu_2 = f_1,\ f_1 = f - BQ(B)f.$$

We shall formulate $Q_1(B) = I - BQ(B)$. We will assume $\mathcal{D}_{01} = Q_1(B)\mathcal{D}_0$. It is obvious that $\mathcal{D}_{01} \subset A(\bar{\Omega})$, and that $\mathcal{D}_{01}$ satisfies condtions (4.2.2)-(4.2.4). According to Proposition 4.1.3

$$R_0(B, f) = R_0(B, Q_1(B)f),$$

therefore $\mathcal{D}_{01}$ satisfies condition (4.2.4). Note that

$$(BQ_1(B)f, Q_1(B)f) = \int_{b_0}^{\infty} \lambda |Q_1(\lambda)|^2\, d(E_\lambda f, f)$$

$$= \int_{b_0}^{b} \lambda |Q_1(\lambda)|^2\, d(E_\lambda f, f) + \int_{b}^{\infty} \lambda |Q_1(\lambda)|^2\, d(E_\lambda f, f).$$

Since the points of increase $(E_\lambda f, f)$ coincide with $\lambda_1, \ldots, \lambda_K$ in the segment $[b_0, b]$ and at these points $Q_1(\lambda) = 0$, the integral from b_0 to b equals zero. Therefore

$$(BQ_1(B)f, Q_1(B)f) \geqq b\int_{b}^{\infty} |Q_1(\lambda)|^2\, d(E_\lambda f, f) = b\|Q_1(B)f\|^2.$$

We can thus take $b = b_1(B) - \varepsilon$ for $\mathcal{D}_{01}$ as the constant b in inequality (4.2.4).

Consequently, the operator B satisfies all the conditions of Theorem 1.1, where $\mathcal{D}_0 = \mathcal{D}_{01}$, $b = b_1(B) - \varepsilon$. Obviously

$$B^{-1}f_1=(A+b)^{-1}f_1,\ A=B-bI.$$

The function

$$g^+(\lambda)=(\lambda+b)^{-1}\in\mathscr{A}(\mathscr{I}^+_\beta)$$

when $\beta < \sqrt{b}$. Therefore, by virtue of Theorem 1.1

$$u_2=g^+(A)f_1\in W^s_{2\,\mathrm{loc}}(\Omega)$$

$$\forall s<2R_0\sqrt{b_1(B)-\varepsilon}/\pi.$$

Since $u_1 \in A(\Omega)$, and ε is as small as desired, the following inclusion holds for $u = u_2 + u_1$:

$$u\in W^s_{2\,\mathrm{loc}}(\Omega),\ \forall s<2R_0(B,Q_1(B)f)\sqrt{b_1(B)}/\pi.$$

Since

$$R_0(B,Q_1(B)f)=R_0(B,f),$$

hence follows (1.9).

The smoothness $u(x)$ up to the boundary, when condition 1.2 is satisfied, is derived instead of Theorem 1.1 from Theorem 1.1′.

Remark 1.0. If the boundary $\partial\Omega$ does not have singularities in the neighbourhood of the point $x_0 \in \partial\Omega$, there exist derivatives which are continuous in this neighbourhood along the boundary up to the order l when $l + m/2 + 1 < s_{20}$, where s_{20} is the same as in (1.9), and the derivatives with respect to the normal to $\partial\Omega$ up to order l_1 when $2l_1 + m/2 + 1 < s_{20}$ (see Remark 6.4).

Remark 1.1. An example which shows the accuracy of estimate (1.9) will be constructed in §4 for the case $\Omega = T^m$.

Remark 1.2. Other methods of estimating the smoothness of the solutions of equations of the form $Bu = f$ with arbitrary degeneracy, permitted by (4.33), have been obtained by Oleinik [1] (see also the monograph by Oleinik and Radkevich [1]) and Cohn and Nirenberg [1]. The conditions on the boundary, which are imposed in these works, differ from those here, and the estimates of the smoothness of the solutions are formulated in terms completely different from our own.

Remark 1.3. §8 investigates the case when the function $u = B^{-1}f$ belongs to a space with a limit indicator of smoothness $s = 2R_0\sqrt{b}$. It appears that the limit class of smoothness to which u belongs is the Nikol'skii space $H_2{}^s$, and not the Sobolev space $W_2{}^s$. This example, constructed in §4, shows that it is impossible to take the narrower Besov space $B^s_{2,\theta}$ when $\theta < \infty$ instead of the wider (for fixed s) Nikol'skii space $H_2{}^s = B^s_{2,\infty}$

We shall now give estimates of the smoothness of solutions of parabolic and hyperbolic equations with a generally degenerating operator B.

Theorem 1.4. Suppose the operator B satisfies condition 1.1,

$$f_0, f_1 \in \mathscr{D}_0,\ R_0 = \min(R_0(B, f_0), R_0(B, f_1)).$$

Suppose $u(t, x)$ is the solution of problem (4.2.16). Then

$$u(t,\cdot) \in C^\infty(\Omega),\ \forall t \geq 0,\ u(t,\cdot)$$

extends up to a function that is analytic with respect to t when $\operatorname{Re} t > 0$. If

$$|t|^2/\operatorname{Re} t \leqq \theta, \quad \operatorname{Re} t > 0 \qquad (\theta > 0),$$

then $\forall \varepsilon > 0$, $\forall$ of the compactum $K \subset \Omega$ exists $R_2 = R_2(\Theta, \varepsilon)$, such that

$$\|\partial_x^\alpha u(t,x)\|_{C(K)} \leqq C_\varepsilon R_2^{|\alpha|} |\alpha|!\, e^{\sigma_1 |\alpha|^2}, \tag{1.10}$$

where

$$\sigma_1 = \theta/(r_0 - \varepsilon)^2, \quad r_0 = 2R_0/\pi,$$

If condition 1.2 holds, estimate (1.10), where

$$\sigma_1 = 4\theta/(r_0 - \varepsilon)^2$$

holds for any $K \subset \Omega_1 \cap \overline{\Omega}$.

The functions $\partial_x^{\alpha} u(t, \cdot)$ are infinitely differentiable with respect to t when $t > 0$.

Proof. The solution of problem (4.2.16) is given by Formula (4.2.19), where the functions $g_{1j}(\lambda)$ $j = 0, 1$ are determined by (1.3.8):

$$u(t) = g_{10}(t, b + A) f_0 + g_{11}(t, b + A) f_1.$$

As shown in the proof of sect.2) of Theorem 4.2.3, for t, satisfying (4.2.9), the functions

$$g_{1j}(z^2 + b) \in \mathscr{A}_{2,\theta}(\mathscr{I}_\infty),$$

i.e.

$$g_{1j}^*(s) \leqq Ce^{\theta s^2+s} = Cg^{**}(s). \tag{1.11}$$

According to Sobolev's imbedding theorem, if the domain $\omega \subset \mathbb{R}^m$ the norm in the space of continuous functions $C(\omega)$ is estimated in terms of the norm in the Sobolev space $W_2^l(\omega)$ when

$$l > m/2, \qquad \|u\|_{C(\omega)} \leqq C\|u\|_l.$$

Therefore by virtue of the determination of the norm in $W_2^l(\omega)$ (see (7.4.1)),

$$\max_{|\alpha|=k} \|\partial^\alpha u\|_{C(K)} \leqq C \max_{|\alpha| \leqq k+l} \|\partial^\alpha u(x)\|,$$

where $\| \ \|$ is the norm in $L_2(\omega)$. Using the above remark, we obtain from (1.6) and (1.11) estimate (1.10) when $K \subset \Omega$. When

$$K \subset \bar{\Omega} \cap \Omega_1$$

we obtain (1.10) from (1.11) and (1.6'). The function $u(t, \cdot)$ is analytic with respect to t in $L_2(\Omega)$ when $\operatorname{Re} t > 0$ (see Proposition 1.3.1), and since it is bounded in $C^l(\omega)$, $\omega \subset \Omega$, it is also analytic in $C^l(\omega)$ when $\operatorname{Re} t > 0$. Thus $u(x, t)$ is analytic with respect to $t > 0$ and infinitely differentiable with respect to $x \in \Omega$.

We shall now prove that $u(t, x)$ is infinitely differentiable with respect to t when $t > 0$. Indeed, by virtue of Eq.(4.2.16) $\partial_t u = -Bu + f_0$, and therefore $\partial_t^j u$ satisfies the equation

$$\partial_t \partial_t^j u = -B\partial_t^j u, \quad j \geqq 1, \partial_t u|_{t=0} = -Bf_1 + f_0, \ \partial_t^j u|_{t=0} = (-B)^{j-1}(-Bf_1 + f_0) = f_j.$$

Using the already-proved statement of Theorem 1.4 on the smoothness with respect to x for problem (4.2.16) with the initial condition $u|_{t=0} = f_j$ and with the right-hand side $f_0 = 0$, we obtain that $\partial_t^j u$ are bounded in $W_2^s(K)$ when $0 < t < \theta$ $\forall j \forall s$. Using Sobolev's imbedding theorem, we obtain that $u(t, x)$ is bounded in $C^l(K \times [0, \theta]), \forall l$, and the theorem is proved.

Remark 1.4. Using the form of the dependence of the constants R_1 and M_0 in (1.6) on f, we can obtain the following estimate:

$$\|\partial_t^j \partial^\alpha u(t, x)\|_K \leqq CR_2^{j+|\alpha|}(2j)!\, e^{\theta|\alpha|^2/(r_0-\varepsilon)^2}(1 + j \ln j)^{|\alpha|}$$

when $0 < t < \theta$, $K \subset \Omega$.

Remark 1.5. The functions $v(x)$, whose derivatives $\partial^\alpha v(x)$ increase like

$$|\alpha|!\, R^{|\alpha|} e^{\kappa|\alpha|^2},$$

form the class

$$C(|\alpha|!\, e^{\kappa|\alpha|^2})$$

of infinitely differentiable functions, which is much wider than Gevrait's known classes (see Chapter 2 Remark 2.2.3). In §5 we construct an example of a degenerating parabolic equation, the solutions of which $u(t, \cdot)$ do not belong to a class of functions that are more regular than

$$C(|\alpha|!\, e^{\kappa|\alpha|^2}).$$

The value of the parameter

$$\kappa = \kappa_0 t, \ \kappa_0 = 1/r_0^2 + \varepsilon,$$

is also accurate.

Oleinik proved the infinite differentiability with respect to x of $u(t, x)$ (without estimating the increase in the derivatives) in the scalar case ($\varkappa = 1$ in (4.5.3)) using another method.

We shall now consider Cauchy's problem for a hyperbolic equation.

Theorem 1.5. Suppose B satisfies condition 1.1,

$$f_0, f_1, f_2 \in \mathscr{A}(\bar{\Omega}), \ \Omega$$

Ω is a bounded domain, the boundary of which $\partial\Omega$ satisfies condition 1.2 in Ω_1. Suppose $u(t, x)$ is the solution (4.2.17). Then $u(t, \cdot) \in A(K)$ for any compactum $K \subset \Omega$, or

$$K \subset \bar{\Omega} \cap \Omega_1.$$

In addition, $u(t, x)$ is analytic with respect to t, $u(t, x) \in A(K \times \mathbb{R})$.

Proof. The solution of problem (4.2.17) is given by Formula (4.2.19). It is obvious that by virtue of (1.3.13) the functions

$$g_{20}^*(s), g_{21}^*(s),$$

and $g_{22}{}^*(s)$, corresponding to the functions

$$g_{20}(z^2 + b), \ g_{21}(z^2 + b),$$

in (4.2.19), permit the estimate

$$g_{2j}^*(s) \leqq C e^{|t|s+s} = Cg^{**}(s). \tag{1.12}$$

From this estimate and from (1.6′) follows the estimate

$$\|\partial_x^\alpha u(t, x)\|_{C(K)} \leqq C_1 C_0^{|\alpha|} e^{|t||\alpha|/(r-\varepsilon)} |\alpha|! \tag{1.13}$$

for $K \subset \Omega_1$. Bearing in mind that condition 1.2 obviously holds for any domain $\Omega_1 \subset \Omega$, we obtain that the analyticity of $u(t, x)$ in Ω and on $\Omega \cup (\Omega_1 \cap \Omega)$ follows from (1.13).

We shall prove the analyticity of $u(t, x)$ with respect to t. Since $Bu(t, x)$ is analytic with respect to x, by virtue of Eq.(4.2.17) $\partial_t^2 u(t, x)$ is also analytic with respect to $x \in K$ when $t \in [0, t_0]$. The function $\partial_t u$ is analytic with respect to x by virtue of the formula

$$\partial_t u(t, x) = \partial_t u(0, x) + \int_0^t \partial_t^2 u(t, x)\, dt,$$

since estimate (1.6′) is uniform with respect to t when $t \in [0, t_0]$ by virtue of (1.12). According to the Cauchy-Kowalewska theorem, the solution (4.2.17)

$$u^1(t, x),\ |t - t_1| < \varepsilon,$$

which is analytic with respect to t and x, exists for specified $f_{01} = u(t_1)$, $f_{02} = \partial_t u(t_1)$, $|t| \leqq t_0$, which are analytic in K. Since $u(t, \cdot)$ is analytic with respect to t in $L_2(\Omega)$, then

$$u^1(t, x) = u(t, x)$$

when $x \in K$ by virtue of the uniqueness of the analytic function with specified derivatives at the point $t = t_1$. Since the numbers

t_0 and t_1 are arbitrary, hence follows the analyticity of $u(t, x)$ with resepct to x and t for all t.

Remark 1.6. The question of the analyticity of the solutions of nonstrict hyperbolic equations was examined by Bony and Schapira using another method [1-3]. Note that the method used here enabled us to prove analyticity up to the boundary, which is difficult to do using their methods.

Remark 1.7. All the investigations in this chapter which are carried out for the functions $g(z)$ from $A(\mathcal{J}_\beta)$ and $A(\mathcal{J}_\beta^{+})$ could be carried out for the functions $g \in A_q(\mathcal{J}_\beta)$ and $A_q(\mathcal{J}_\beta^{+})$, $q > 0$, which have a power increase with respect to $|\mathrm{Re}\, z|$. We confined ourselves to the case $q=0$ for brevity of discussion, and in connection with the fact that the functions $g(\lambda)$, corresponding to problems for differential equations, belong to $A(\mathcal{J}_\beta)$ and $A_q(\mathcal{J}_\beta^{+})$.

§2. Functions of the class $A(\mathcal{J}_\beta^{+})(\beta < \infty)$ of the self-adjoint operator A on a torus.

In this paragraph we shall consider the operator B from Example 4.3.1. We shall prove the theorem from which follows Theorem 1.1 for the case when $\Omega = T^m$ - m is another m-dimensional torus. This case is simpler than the general case $\Omega \subset \mathbb{R}^m$, which we consider later in §6.

Theorem 2.1. Suppose

$$g^{+} \in \mathcal{A}(\mathcal{J}_\beta^{+}),\ \beta > 0,\ f \in \mathcal{A}(T^m),$$

B is an operator from Example 4.3.1,

$$R_0 = R_0(B, f),\ A = B - bI,$$

where b is a number from (4.2.4). Suppose $u = g^{+}(A)f$. Then

$$u \in W_2^s(T^m) \text{ when } s < 2\beta R_0/\pi. \tag{2.1}$$

To prove Theorem 2.1, we need certain lemmas.

Lemma 2.1. Suppose A is a self-adjoint operator in the Hilbert space H, $A > 0$. Suppose

$$f \subset \mathscr{D}(\mathrm{ch}(R\sqrt{A})),$$

Suppose

$$g^+ \in \mathscr{A}(\mathscr{I}_\beta^+),\ P_n(\lambda) = \Pi_r^{+n} g^+(\lambda)$$

is an interpolation polynomial of the function $g^+(\lambda)$ (see 3.6.4). Supopse $u = g^+(A)f$. Then

$$u = u_0 + w, \tag{2.2}$$

where

$$u_0 = P_0(A)f = g^+(\lambda_1)f,\ g^+(\lambda_1) \in \mathbb{C}$$

is a numerical multiplier. The vector w is the sum of series

$$w = \sum_{N=0}^{\infty} y_N, \tag{2.3}$$

$$y_N = u_{N+1} - u_N,\quad u_N = P_n(A)f,\quad n = 2^N. \tag{2.4}$$

At the same time the following estimate holds:

$$\|y_N\| \leqq (\mu_R^+(P_n, g^+) + \mu_R^+(P_{2n}, g^+))\|\mathrm{ch}(R\sqrt{A})f\|. \tag{2.5}$$

Proof. Using Lemma 4.1.1., where $j=0$, from inequality (4.1.14), where $j=0$, we obtain the estimate

$$\|g^+(A)f - P_n(A)f\| \leqq \mu_R^+(P_n, g^+)\|\mathrm{ch}(R\sqrt{A})f\|. \tag{2.6}$$

Since

$$\mu_R^+(P_n, g) \to 0$$

as $n \to \infty$ (Theorem 3.6.1), series (2.3) converges in H. Obviously

$$\|u_{N+1} - u_N\| = \|u_{N+1} - g^+(A)f + g^+(A)f$$
$$- u_N\| \leqq \|g^+(A)f - u_{N+1}\| + \|g^+(A)f - u_N\|.$$

Bearing in mind (2.4) and (2.6) we obtain hence (2.5).

Lemma 2.2. Suppose

$$g^+ \in A(\mathcal{J}_\beta^{\ +}),\ 0 < \beta < +\infty,\ P_n = \Pi_r^{\ +n} g^+$$

Then $P_n(\lambda) = q_0 + Q_1\lambda + \ldots + Q_n\lambda^n$,

$$|Q_k| < C_0 e^n C_1^{\ 2k} \min(1,\ (\nu^*(2k))^{2k}(2k)^{-2k}) \tag{2.7}$$

where C_0 and C_1 depend only on $\nu^*(s)$.

Proof. When replacing $\nu^*(s)$ by $c\nu^*(s)$ the form (2.7) does not change, therefore we will assume $\nu^*(0) > 1$. We will assume $n > 3$ (when $n < 3$ (2.7) is obvious). Obviously, $P_n = P_n - g + g$. To estimate the coefficients Q_k it is sufficient to estimate the derivatives at zero of order j, $j < 2n$, of the even functions $g(x^2)$ and $\varphi(x) = P_n(x^2) - g(x^2)$. By virtue of Formula (3.3.4)

$$\varphi(x) = T_n(rx)\Psi(x),\ \Psi(x) = \frac{i}{2\pi} \int_\Gamma \frac{g^+(\zeta^2)}{T_n(r\zeta)} \frac{d\zeta}{\zeta - x},$$

$\Gamma = \{\zeta = \xi \pm i\rho,\ \xi \in \mathbb{R}\}$. Since $|g^+(z^2)| \leqslant g^*(\rho)$ when $|z| < \rho$, then by virtue of the Cauchy formula

$$|d^k g^+(x^2)/dx^k|_{x=0}| \leqslant q_k\ k!,\ g_k = \inf_{\rho > 0} q^*(\rho)\rho^{-k} \tag{2.8}$$

In a similar way, using (2.39'), we obtain:

$$\left|\frac{d\,T_n(rx)}{dx^k}\right|_{x=0}\Big| \leqslant C_1 e^n \inf_{\rho > 0} e^{3\rho r}\rho^{-k}k! \leqslant C_2 e^n C_3{}^k.$$

Without loss of generality we can assume $C_3 > 2$. Calculating the derivative at zero of the function $\Psi(x)$, determined by the integral with respect to $\Gamma=\Gamma_\rho$, and estimating its modulus using the inequality $|T_n(r\zeta)| \geqslant \delta_1(1 + x^2)$, $(\delta_1 > 0)$ when $|\mathrm{Im}\,\zeta| \geqslant \delta > 0$, which follows from (3.2.38), we obtain

$$|d^k\Psi(x)/dx^k|_{x=0}| \leqslant C_n(\delta)g^*(\rho)\rho^{-k}\,k!$$

Bearing in mind that the infinum in (2.8) is attained when $\rho = \rho_0 > 0$, we obtain hence that

$$|d^k\Psi(0)/dx^k| \leqslant C_5 q_k\ k!$$

Estimating the derivative of $\varphi = T_n\Psi$ using the Leibnitz formula, we obtain:

$$|d^k\varphi(0)/dx^k| \leqslant C_2C_5 2^{k+1}C_3{}^k \max(q_i i!,\ i = 0, \ldots, k)e^n. \tag{2.9}$$

We shall now make an upper estimate of q_i in terms of ν^*. By virtue of (2.8) and (1.3′)

$$\ln q_i \leqslant \rho\nu^*(\rho) - i\ln\rho + C_6$$

Assuming here $\rho = i/\nu^*(i)$, we obtain that when $\nu^*(i) < +\infty$

$$\ln q_i \leqslant i\nu^*(i/\nu^*(i))/\nu^*(i) - i\ln i + i\ln\nu^*(i) + C_6$$

Since $\nu^*(s)$ increases and $\nu^*(i) \geqslant \nu^*(0) \geqslant 1$, then $\nu^*(i/\nu^*(i)) \leqslant \nu^*(i)$. Therefore

$$\ln q_i \leqslant i - i\ln i + i\ln\nu^*(i) + C_6$$

Obviously, this formula also holds when $\nu^*(i) = +\infty$. When $i \leqslant k$ we have

$$q_i i! \leqslant e^i\nu^*(i)^i \leqslant e^k\nu^*(k)^k$$

Using this estimate, from (2.9) we obtain inequality (2.7) when $\nu^*(2k)^{2k}(2k)^{-2k} \leqslant 1$. Otherwise we will take $\rho = \delta_0$, $0 < \delta_0 < \beta$. Obviously,

$$\ln q_i \leqslant \delta_0\nu^*(\delta_0) - i\ln\delta_0 + C_6 \leqslant C_7 + C_8 i.$$

We obtain (2.7) from this estimate and from (2.9).

Lemma 2.3. Suppose the polynomial P_n is the same as in Lemma 2.2, and the operator B is determined by Formula (4.3.1). Suppose the compactum $K \subset \Omega$, $f \in A(K)$, and the number $\delta > 0$ is so small that f extends analytically to $\mathcal{O}_\delta(K)$. Suppose

$$C_0(f) = \sup\left\{t : t = |f(z)|,\ z \in \mathcal{O}_\delta(K)\right\}.$$

Then R_1 exists, depending only on δ and on B, such that

$$\sup_{x\in K}|\partial^\alpha(P_n(B)f)(x)| \leqslant CC_0(f)R_1^{|\alpha|+2n}|\alpha|!\min((2n)^{2n},\ \nu^*(2n)^{2n}). \tag{2.10}$$

Proof. According to Lemma 1.1.2 the function $f(x)$ satisfies (1.1.1), where $C_0 = C_0(f)$, $C_1 = 2/\delta$. According to Theorem 1.1.1, where $p = 2$, the following estimate holds

$$|\partial^\alpha B^k f(x)| \leqslant C_0 R^{2k+|\alpha|}(|\alpha| + 2k)! \leqslant C_0(2R)^{2k+|\alpha|}|\alpha|!(2k)!, \tag{2.11}$$

$C_0 = C_0(f)$, $x \in K$. Therefore, by virtue of (2.7)

$$|\partial^\alpha P_n(B)f(x)| \leqslant C_2 C_0 \sum_{k=0}^{n} |Q_k| C_1^{2k+|\alpha|}(2k)!|\alpha|! \tag{2.12}$$

Using the estimate $|Q_k|$, given in (2.7), we obtain hence (2.10).

We shall now give some statements on the properties of functions that are periodic with respect to each variable, i.e. functions determined on the torus T^m.

If the function $f(x)$ is defined on the m-dimensional torus T^m, its Fourier coefficients $\tilde{f}(\xi)$, $\xi \in \mathbb{Z}^m$ are determined by the formula

$$\tilde{f}(\xi) = (2\pi)^{-m} \int_{[0,2\pi]^m} f(x)e^{-i\xi\cdot x}\,dx, \tag{2.13}$$

where

$$\xi \cdot x = \xi_1 x_1 + \cdots + \xi_m x_m.$$

The Fourier expansion $f(x)$ is given by the formula

$$f(x) = \Sigma \tilde{f}(\xi)\, e^{i\xi \cdot x}, \tag{2.14}$$

where the summation is carried out over $\xi \in \mathbb{Z}^m$.

We shall specify the norm $\| \; \|_s$ in the Sobolev space $W_2^{\,s}(T^m)$ using the equality

$$\|f\|_s^2 = \Sigma (1 + |\xi|)^{2s} |\tilde{f}(\xi)|^2. \tag{2.15}$$

The following Parseval equality holds:

$$\|f\|_0 = (2\pi)^{-m} \int_{[0,\, 2\pi]^m} |f(x)|^2 \, dx. \tag{2.16}$$

It follows directly from Formula (2.14) that

$$(\partial^\alpha f)(\xi) = i^{|\alpha|} \xi^\alpha \tilde{f}(\xi). \tag{2.17}$$

Here ∂^α and ξ^α are notation introduced at the beginning of the first chapter. The modulus $| \; |$ of the vector in $\mathbb{R}^m$ is determined by Formula (4.5.7) (where $\varkappa = m$).

We introduce the notation

$$\|E_\gamma f\| = (\Sigma\, e^{2\gamma|\xi|} |\tilde{f}(\xi)|^2)^{1/2} \tag{2.18}$$

(it is obvious that $\|E_\gamma f\| = \infty$, if $|\tilde{f}(\xi)|$ decrease significantly).

We denote the following set by Π_ε, $\varepsilon > 0$:

$$\Pi_\varepsilon=\{z\in\mathbb{C}^m:|\operatorname{Im} z_1|+\ldots+|\operatorname{Im} z_m|<\varepsilon\}. \tag{2.19}$$

Lemma 2.4. Suppose the function $f(z)$ is holomorphic and bounded in Π_ε:

$$|f(z)| < M \text{ when } Z \in \Pi_\varepsilon, \tag{2.20}$$

and periodic in Π_ε with respect to the real part z with the period 2π:

$$\begin{aligned}&f(x_1+iy_1,\ldots,x_j+2\pi+iy_j,\ldots,x_m+iy_m)\\&=f(x_1+iy_1,\ldots,x_j+iy_j,\ldots,x_m+iy_m),\qquad j=1,\ldots,m.\end{aligned} \tag{2.21}$$

Then when $\gamma < \varepsilon$

$$\|E_\gamma f\|\leqq CM, \tag{2.22}$$

where C does not depend on f.

Proof. Since the integrand expression on the right-hand side of (2.13) is holomorphic with respect to $z = x + iy$ in Π_ε and periodic with respect to x with the period 2π, the integral in (2.13) does not change if we replace each segment $[0, 2\pi]$ by the segment $[0, 2\pi] + y_j i$. Therefore, denoting $y = (y_1, \ldots y_j$, from (2.13) we obtain:

$$\tilde{f}(\xi)=(2\pi)^{-m}\int_{[0,2\pi]^m} f(x+iy)\,e^{-i\xi\cdot x}\,dx\,e^{\xi\cdot y}. \tag{2.23}$$

Estimating the integral on the right-hand side of this formla using (2.20), we obtain that

$$|\tilde{f}(\xi)|\,e^{-\xi\cdot y}\leq M \tag{2.24}$$

If

$$|\xi_{j_0}| = \max |\xi_j|,$$

we will put

$$y_{j_0} = -\lambda \operatorname{sign} \xi_{j_0},\ y_j = 0$$

when $j \neq j_0$, where $\lambda > 0$, $\lambda < \varepsilon$. From (2.24) we directly obtain estimate

$$|\tilde{f}(\xi)| e^{\lambda|\xi|} \leqq M. \tag{2.25}$$

Assuming

$$\lambda = \gamma + (\varepsilon - \gamma)/2,\ 0 < \gamma < \varepsilon$$

(obviously, $0 < \lambda < \varepsilon$), from (2.18) and (2.25) we obtain:

$$\|E_\gamma f\|^2 \leqq M^2 \Sigma\, e^{2(\gamma - \lambda)|\xi|} = M^2 \Sigma\, e^{(\gamma - \varepsilon)|\xi|} < \infty,$$

whence directly follows (2.22).

Lemma 2.5. Suppose $\|f\|_0 < q$ and for some $\gamma > 0$

$$\|E_\gamma f\| \leqq M < \infty. \tag{2.26}$$

Then when $s > \gamma$

$$\|f\|_s \leqq 2q\mathscr{D}(s)^s, \quad \mathscr{D}(s) = \max(\mathscr{D}^1(s), \mathscr{D}^2), \tag{2.27}$$

where

$$\mathscr{D}^1(s)=s/\gamma, \quad \mathscr{D}^2=1+1/\gamma\ln(M/q). \tag{2.28}$$

Proof. Obviously, for any $r > 0$

$$\begin{aligned}\|f\|_s^2 &= \sum_{|\xi|<r}(1+|\xi|)^{2s}|\tilde{f}(\xi)|^2 \\ &+ \sum_{|\xi|\geqq r}(1+|\xi|)^{2s}e^{-2\gamma|\xi|}e^{2\gamma|\xi|}|\tilde{f}(\xi)|^2 \\ &\leqq (1+|r|)^{2s}q^2+\sup_{|\xi|\geqq r}[(1+|\xi|)^{2s}e^{-2\gamma|\xi|}]M^2.\end{aligned} \tag{2.29}$$

Note that when $r+1 > s/\gamma$

$$\sup_{|\xi|\geqq r}[(1+|\xi|)^{2s}e^{-2\gamma|\xi|}]=(1+r)^{2s}e^{-2\gamma r} \tag{2.30}$$

Therefore, assuming $1+r=t$, from (2.29) we obtain that for any $t > s/\gamma$

$$\|f\|_s^2 \leqq t^{2s}[q^2+e^{-2\gamma t}e^{2\gamma}M^2]. \tag{2.31}$$

We will put

$$t_0=1/\gamma\ln(M/q)+1. \tag{2.32}$$

Substituting this value $t = t_0$ into the right-hand side of (2.31), we obtain that when $t_0 > s/\gamma$ the following inequality holds:

$$\|f\|_s^2 \leqq 2q^2t_0^{2s}=2q^2[1/\gamma\ln(M/q)+1]^{2s},$$

and (2.27) holds. If $t_0 < s/\gamma$, then $M < qe^{s-\gamma}$. Using this estimate, from (2.29) we obtain the estimate

$$\|f\|_s^2 \leqq q^2[(1+r)^{2s}+\sup_{|\xi|\geqq r}[(1+|\xi|)^{2s}e^{-2\gamma|\xi|}]e^{2s-2\gamma}]. \tag{2.33}$$

Assuming here $r=s/\gamma-1$ and using (2.30), we obtain that the right-hand side of (2.33) when $r=s/\gamma-1$ equals $2q^2(s/\gamma)^{2s}$. Hence it follows that inequality (2.27) also holds when $t_0 < s/\gamma$.

Lemma 2.6. Suppose the polynomial P_n is the same as in Lemma 2.2, and B is an operator from Example 4.3.1. Suppose $f \in A(T^m)$, f satisfies inequality (1.1.6) for some $\delta > 0$, where $K = [0, 2\pi]^m$. Then the numbers γ, C and R exist, depending only on B and δ, such that $\forall n \in \mathbb{N}$

$$\|E_\gamma P_n(B) f\| \leqq C C_0(f) M_n,\ M_n = R_1^n \min((2n)^{2n}, v^*(2n)^{2n}). \tag{2.34}$$

Proof. According to Lemma 2.3, where $K = [0, 2\pi]^m$, the function $f_n = P_n(B)f$ satisfies estimate (2.10), i.e. estimate (1.1.1), where

$$f_n,\ C_0 = C C_0(f) M_n,\ C_1 = P$$

is substituted instead of f. According to Lemma 1.1.1 and bearing in mind that $f_n(x)$ is a periodic function, we obtain that $f_n(x)$ extends to a function which is analytic on Π_ε for some $\varepsilon > 0$ (see (2.19)), and which satisfies inequality (1.1.3) on Π_ε. Inequality (2.20), where

$$M = C' C_0(f) M_n$$

thus holds. Using Lemma 2.4, we obtain inequality (2.22) for f_n, from which directly follows (2.34).

Proof of Theorem 2.1. When $0 < R < R_0(B, f)$, according to Proposition 4.1.2,

$$\|\mathrm{ch} R\sqrt{A} f\| < M_0 < \infty.$$

According to Lemma 2.1 $u = g(A)f$ is represented in the form $u = u_0 + w$, $u_0 \in A(T^m)$ and w is determined by (2.3), where y_N is determined by (2.4). Estimate (2.5), in which according to Theorem 3.4.1, where $\omega=0$,

$$\mu_R^+(P_n, g) = \mu_{0,R}^+(P_n, g) < C_1 n^{-r\rho} \text{ when } \rho < \beta \tag{2.35}$$

holds for $\|y_N\|$. Therefore we have the following inequality from (2.5):

$$\|y_N\| < q_N, \; q_N = C_2 2^{-Nr\rho} \tag{2.36}$$

Estimating $\|E_\gamma y_N\| = \|E_\gamma(P_{2n}(A)f - P_n(A)f)\|$, $n = 2^N$ using Lemma 2.6, from (2.34) we obtain

$$\|E_\gamma y_N\| < C_3 M_N, \; M_N = (2n)^{4n} R_1^{2n} C_0(f), \; n = 2^N \tag{2.37}$$

We shall use Lemma 2.5, where $q = q_N$, $M = M_N$ (we shall consider the case when $\mathcal{D} = \mathcal{D}^2$ in (2.27), and the case $\mathcal{D} = \mathcal{D}^1$ is simpler). From (2.27), (2.28) we obtain, bearing in mind (2.36) and (2.37):

$$\|y_N\|_s < 2C_2 2^{-Nr\rho}(1 + 1/\gamma \ln[C_4 (2n)^{4n} R_1^{2n} 2^{Nr\rho}])^s <$$

$$C_5 2^{-Nr\rho}(C_6 + 2^{N+2}(N+1)\ln 2 + 2^{N+1}\ln R_1 + Nr\rho\ln 2)^s$$

Obviously, when $r\rho > s$ hence we obtain the estimate

$$\|y_N\|_s < C_7 2^{-N\varepsilon} \tag{2.38}$$

where $\varepsilon > 0$ $(\varepsilon < r\rho - s)$, C_7 does not depend on N. Summing this estimate over N, we obtain that $w \in W_2^s(T^m)$ when $s < r\rho$.

Since ρ is as close to β as desired, R is as close to $R_0(B, f)$, and $u = u_0 + w$, where $u_0 \in A(T^m)$, hence follows the statement of Theorem 2.1.

§3. Functions of an operator on a torus of the class $A(\mathcal{J}^+_\infty)$

Theorem 3.1. Suppose B is an operator from Example 4.3.1,

$$f \in \mathscr{A}(T^m),\ R_0 = R_0(B, f),\ r_0 = 2R_0/\pi.$$

Suppose $A = B - bI$,

$$b < b_0(B),\ u(x) = g(A)f(x),\ g^+ \in A(\mathcal{J}^+_\infty).$$

The constants C and C_1 then exist for any $\varepsilon > 0$ such that when $s > 0$

$$\|u\|_s < CC_1{}^s s^s g^{**}(s/r_0 - \varepsilon)), \tag{3.1}$$

and the constants C and C_1 depend only on ε, g^{**}, r_0, on the number $C_0(f)$ from Formula (1.1.6), where $K = [0, 2\pi]^m$ and on the number M_0, which satisfies the condition

$$M_0 > \|\mathrm{ch}((R_0 - \varepsilon/2)\sqrt{A})f\|.$$

The following lemma is needed to prove Theorem 3.1.

Lemma 3.1. Suppose $g^+ \in \mathscr{A}(\mathcal{J}^+_\infty)$, $\omega \geqq 0$, let $r = 2R/\pi$. Then the constants Θ and C exist, such that if

$$P_k = \Pi_r^{+k} g(\lambda),\ k \in \mathbb{N},$$

then for any $n \in \mathbb{N}$

$$\mu^{+}_{\omega,R}(P_n, g^{+}) + \mu^{+}_{\omega,R}(P_{2n}, g^{+}) \leqq C \inf_{t \geqq r} [g^*(t/r+\omega) n^{-t} t^t t^{\theta}], \tag{3.2}$$

where g^* is determined in (1.3).

Proof. According to Lemma 3.6.1

$$\mu^{+}_{\omega,R}(\Pi_r^{+k} g^{+}, g^{+}) = \mu_{\omega,R}(\Pi_r^{2k+1} g, g), \qquad g(z) = g^{+}(z^2). \tag{3.3}$$

Since $g(z) \in A(\mathcal{J}_{\infty})$ then, according to (3.5.1), the following estimate holds for $P_{2n+1} = \Pi_r^{2n+1} g$:

$$\mu_{\omega,R}(P_{2n+1}, g) \leqq C n^{r\omega} \sigma_{n+1}(r, \omega, g). \tag{3.4}$$

We shall use the estimate σ_{n+1}, given in (3.5.2), where we will assume $q=0$. The exponent in the exponential in (3.5.2) has the form

$$-(2n+y)\ln(1+2n+y) + 2n\ln(1+2n) + (8+y)\ln(1+y)$$

$$\leqq -y\ln(1+2n+y) + 8\ln(1+y) + y\ln(1+y)$$

$$\leqq -y\ln n + 8\ln y + y\ln y + C.$$

(the inequalities hold when $y > r > 0$, and the constant C depends on r). From this inequality and from (3.5.2) we obtain:

$$\sigma_{n+1}(r, \omega, g) \leqq C_1 \inf_{y \geqq r + r\omega} [g^*(y/r) n^{-y} y^y y^8].$$

Writing $y - r\omega = t$, we obtain hence:

$$\sigma_{n+1}(r, \omega, g) \leqq C_1 \inf_{t \geqq r} [g^*(t/r+\omega) n^{-r\omega \cdot t} (t+r\omega)^{t+r\omega} (t+r\omega)^8. \tag{3.5}$$

Writing $8 + r\omega = \theta$, when $t > r$ we obtain

$$(t+r\omega)^{t+r\omega}(t+r\omega)^8 = (t+r\omega)^{\theta} t^t (1+r\omega/t)^t \leqq C_2 t^{\theta} t^t e^{r\omega}.$$

Setting $C=2C_1C_2e^{r\omega}$, we obtain estimate (3.2) hence, from (3.5), (3.4) and (3.3). (If we replace n by $2n$ in (3.5), the right-hand side of (3.5) will only reduce)

Lemma 3.2. Suppose the conditions of Lemma 3.1 hold, and suppose

$$q_n = \inf_{t>0}[g^*(t/r + \omega)n^{-t}t^te^t] \tag{3.6}$$

Then a constant C exists, such that

$$\mu^+_{\omega,R}(P_n, g^+) + \mu^+_{\omega,R}(P_{2n}, g^+) \leq Cq_n \tag{3.7}$$

In addition, for any $\varepsilon > 0$, $\varepsilon \leq 1$, the constants C_0 and C_1 exist such that for any s and δ, satisfying the conditions $(1 + 2\varepsilon)s/r + \omega < \beta$, $0 \leq \delta \leq \varepsilon$, the following inequality holds:

$$q_n \leq C_0C_1{}^s(s/n)^{s+\delta s}\exp(-\varepsilon s\nu^*(s/r + \omega))g^{**}((1 + 2\varepsilon)(s/r + \omega)). \tag{3.8}$$

Proof. Note that $t^\Theta \leq Ce^t$ when $t > 0$. Since when $0 \leq t \leq r$

$$g^*(t/r + \omega))n^{-t}t^te^t > g^*(0)/e,$$

then for large n the infimum on the right-hand side of (3.6) is attained when $t > r$. (3.7) therefore follows from (3.2).

We shall proceed to prove (3.8). We shall take under the sign of the infimum in (3.6)

$$t = \eta s,\ \eta = 1 + \delta,\ 0 \leq \delta \leq \varepsilon \leq 1.$$

We will obtain:

$$q_n < Cg^*(\eta s/r + \omega)n^{-\eta s}s^{\eta s}\eta^{\eta s} < C_1C_2{}^s g^{**}(\eta(s/r + \omega))\ (s/n)^{\eta s} \tag{3.8'}$$

Note that when $\eta > 1$, $\varepsilon > 0$ according to (1.3')

$$g^{**}(\eta y) < g^{**}((\eta + \varepsilon)y)\exp[-\varepsilon y \nu^*(y)]$$

by virtue of the monotony of $\nu^*(y)$. Estimating the right-hand side of (3.8') using this inequality, we obtain (3.8).

Lemma 3.3. Suppose the numbers q_n are determined by Formula (3.6), and the numbers M_n are determined by (2.34). Then numbers $R > 0$, $C_1 > 0$, $\delta_1 > 0$, $s_0 > 0$ exist, such that when $s > s_0$

$$q_n(\ln(M_{2n}/q_n))^s < \xi_n g^{**}((1 + 2\varepsilon)(s/r + \omega))R^s s^s \tag{3.9}$$

where $\xi_n < C_1$ when $2n < s^2$, $\xi_n < C_1 n^{-\delta_1}$ when $4n > s^2$.

Proof. Since $g^*(s) > g$, then according to (3.6)

$$\ln q_n > \ln g^*(0) + n\inf[tn^{-1}\ln(tn^{-1})] > \ln g^*(0) - C_2 n.$$

Therefore $M_{2n}/q_n < C_3C_4{}^n M_{2n}$, and according to (2.34)

$$\ln(M_{2n}/q_n) < C_5 + nC_6 + 4n \min(\ln(4n), \ln \nu^*(4n)). \tag{3.10}$$

By virtue of (3.8) and (3.10) we have:

$$q_n(\ln(M_{2n}/q_n))^s < C_7 g^{**}((1 + 2\varepsilon)(s/r + \omega))s^s G^s C_8{}^s, \tag{3.11}$$

where

$$G = (s/(4n))^{\delta}\exp(-\varepsilon\nu^*(s/r + \omega))(C_g + \min(\ln(4n), \ln\nu^*(4n))).$$

We shall prove the boundedness of G. When $4n \leqslant s/r + \omega$ we shall take $\delta=0$ and, bearing in mind that

$$\ln\nu^*(4n) \leqslant \ln\nu^*(s/r + \omega)$$

and that

$$\exp[-\varepsilon\nu^*(y)]\ln\nu^*(y) \leqslant C_{10}$$

we obtain that $G \leqslant C_{10}'$ for n. When $4n \geqslant s/r + \omega$ we will assume $4n = ts$, $t \geqslant 1/r$, and shall assume $\delta = \varepsilon$. If Condition A holds (see §2 of Chapter 3), then when $t \geqslant 1/r$

$$t^{-\delta}\exp[-\varepsilon\nu^*(s/r + \omega)](\ln t + \ln s) \leqslant C_{11} \tag{3.12}$$

Since $s^{\delta}/(4n)^{\delta} \leqslant C_{11}'t^{-\delta}$, then $G \leqslant C_{12}$ and estimate (3.9) holds. If Condition B holds (see §2 of Chapter 3), then

$$t^{-\delta}\exp[-\varepsilon\nu^*(s/r + \omega)]\ln\nu^*(rts/r) \leqslant C_{12}' \tag{3.13}$$

We obtain that $G \leqslant C_{12}''$ from this inequality and from the estimate $s^{\delta}/(4n)^{\delta} \leqslant C_{11}'t^{-\delta}$. We shall now show that if $4n \geqslant s^2$, then $G \leqslant C_{13}n^{-\delta_0}$, $\delta_0 > 0$. We shall take

$$\delta_0 = \varepsilon/4, \ s^{\varepsilon}/(4n)^{\varepsilon} \leqslant (4n)^{-\varepsilon/2} \leqslant (4n)^{-\varepsilon/4}t^{-\varepsilon/4}$$

when $s = 4n/t \geqslant 1$. From estimates (3.12) and (3.13), where $\delta=\delta_0$, we obtain the estimate $G < C_{14}n^{-\varepsilon/4} = \xi_n$. (3.9) follows from the estimates of G obtained and from (3.11).

Lemma 3.4. Suppose w is determined by (2.3), where $y_N \in A(T^m)$ whilst the estimates

$$\|E_\gamma y_N\| < C_0 M_{2n}, \ \|y_N\| < C_1 q_n$$

hold for y_N, where q_n are determined in (3.6) and M_n in (2.34). Then $\forall\, \varepsilon > 0$, $C(\varepsilon)$ and $C_1(\varepsilon)$ exist such that $\forall\, s > 0$ and the following inequality holds:

$$\|w\|_s < C(\varepsilon)C_1(\varepsilon)^s s^s g^{**}((1 + \varepsilon)(s/r + \omega)). \tag{3.14}$$

Proof. If s is such that $g^{**}((1 + \varepsilon)(\omega + s/r)) < \infty$, the left-hand side of (3.14) is bounded (see the proof of Theorem 2.1). It is therefore sufficient to prove (3.14) for large s. We shall use lemma 2.5. According to (2.27)

$$\|y_N\|_s < 2C_1 q_n s^s \gamma^{-s} + 2C_1 q_n[1 + 1/\gamma \ln(C_0 M_{2n}/(C_1 q_n))]^s. \tag{3.15}$$

Obviously,

$$(1 + 1/\gamma \ln(C_0 M_n/(C_1 q_n)))^s < C_2{}^s \ln^s(M_{2n}/q_n).$$

Therefore, by virtue of (3.9) and (2.3)

$$\|w\|_s < 2C_1\gamma^{-s}\sum_N q_n + C_1 C_2{}^s s^s \sum_N \xi_n g^{**}((1 + 2\varepsilon)(s/r + \omega))R^s$$

Since $n = 2^n$, $q_n < C_3 n^{-\delta}$, $\xi_n < C_4 n^{-\delta_1}$ when $n > s^2$, the series converge

$$\sum_N q_n < C_5, \ \sum_N \xi_n < C_6 \ln s + C_7$$

and therefore

$$\|w\|_s < C_8 C_g{}^s(1 + g^{**}((1 + 2\varepsilon)(s/r + \omega)).$$

Hence, bearing in mind that $g^{**}(y) > g^{**}(0) > 0$, that ε is arbitrarily small and that r is arbitrarily close to r_0, we obtain estimate (3.14).

Proof of Theorem 3.1. Since f is analytic, then

$$\|u_0\|_s < C\|f\|_s < Cs^s C_1{}^s$$

Therefore by virtue of (2.2), bearing in mind that $g^{**}(s) > C_0 > 0$, it is sufficient to estimate $\|w\|_s$. Suppose $R < R_0(A, f)$, $R > 0$. Then the inequalities (2.34) and (2.5) hold. Bearing in mind estimate (3.7) we obtain that the conditions of Lemma 3.4 hold, and inequality (3.14) holds, where ω=0. Since R is arbitrarily close to R_0, and ε is as large as desired, hence we obtain (3.1).

§4. The accuracy of the estimate of the smoothness of solutions of the stationary equation.

We shall use B_+ to denote a set of differential operators of the form (4.3.1) which are invertible in $L_2(T^m)$, with analytic coefficients for which the lower bound of the continuous spectrum $b_1(B)$ (see (1.7')) is positive: $b_1(B) > 0$. Theorem 1.3 gives an estimate of the smoothness (1.9) of functions $u(x) = B^{-1}f(x)$ where

$$f \in \mathscr{A}(T^m),\ B \in B_+.$$

We can rewrite this estimate in the form

$$\inf_{B\in B_+,\, f\in \mathscr{A}(T^m)} \left\{ \frac{s_2(B^{-1}f)}{\sqrt{b_1(B)}R_0(B,f)} \right\} \equiv \gamma_m \geqq \frac{2}{\pi}. \tag{4.1}$$

The basic result of this paragraph is Theorem 4.1, which shows the accuracy of estimate (4.1).

Theorem 4.1. Suppose the number γ_m is determined by Formula (4.1). Then

$$\gamma_m = 2/\pi \qquad \forall\, m \in \mathbb{N}. \tag{4.2}$$

The proof of Theorem 4.1 is based on the following theorem.

Theorem 4.2. Suppose $\beta > 1/2$, and the operator B_β is determined in $A(T^m)$ by the formula

$$B_\beta v = -\partial_1((\sin^2 x_1)\,\partial_1 v) + (\sin^2 x_1)v + \beta \cos x_1 v + \beta^2 v. \tag{4.3}$$

The following statements then hold: 1) if $\beta > 1/2$, then

$$b_0(B_\beta) \geqq (\beta - 1/2)^2 \tag{4.4}$$

(see (1.7) for the definition of $b_0(B)$); 2) if $f \in A(T^m)$ is an entire function and $f(x)$ does not depend on $x_2, \ldots, x_m$, then

$$R_0(B_\beta, f) \geqq \pi/2; \tag{4.5}$$

3) if

$$f(x) = -\sin x_1 \cos x_1 + \beta \sin x_1, \qquad \beta > 1/2,$$

then

$$B_\beta^{-1} f \notin W_2^{\beta-1/2}(T^m). \tag{4.5'}$$

Theorem 4.2 will be proved later. We shall now derive Theorem 4.1 from Theorem 4.2.

Proof of Theorem 4.1. It follows from (4.5') that

$$s_2(B_\beta^{-1} f) \leqq \beta - 1/2.$$

Using (4.4) and (4.5) we obtain hence that when $\beta > 1/2$

$$s_2(B_\beta^{-1} f)/(\sqrt{b_0(B_\beta)} R_0(B_\beta, f)) \leqq 2/\pi.$$

Since $b_0(B) < b_1(B)$ (see (1.7) and (1.7')), it follows hence that $\gamma_m < \pi/2$. Bearing in mind inequality (4.1), we obtain that (4.2) holds, and Theorem 4.1 is proved.

We shall prove a number of preliminary statements in order to prove Theorem 4.2.

Lemma 4.1. Suppose B_β is determined by Formula (4.3), $\beta > 1/2$. Then

$$(B_\beta v, v) \geqq (\beta - 1/2)^2 \|v\|^2 \qquad \forall v \in \mathscr{A}(T^m), \tag{4.6}$$

where (,) and $\| \ \|$ are the scalar product and norm in $L_2([-\pi,\pi]^m)$ and $\mathscr{A}(T^m)$ is a set of periodic functions with the period 2π that are real-analytic in R^m.

Proof. Consider a first-order differential operator

$$A_1 v \equiv \partial_1((\sin x_1)v) + \beta v. \tag{4.7}$$

The operator A_1^*, which is adjoint to A_1, is determined by formula

$$A_1^* v \equiv -(\sin x_1)\,\partial_1 v + \beta v. \tag{4.8}$$

Direct calculations show that the operator B_β, which is determined by (4.3), is the superposition of A_1 and A_1^*:

$$B_\beta = A_1^* A_1. \tag{4.9}$$

By virtue of (4.9), for (4.6) to hold it is sufficient that the following estimate holds:

$$(A_1 v, A_1 v) \geqq (\beta - 1/2)^2 \|v\|^2, \qquad \forall\, v \in \mathscr{A}(T^m). \tag{4.10}$$

We shall now prove that (4.10) holds. Obviously,

$$\begin{aligned}\operatorname{Re}(A_1 v, v) &= \operatorname{Re}\int \partial_1(\sin x_1 v)\bar v\, dx + \beta \int |v|^2\, dx \\ &= -\operatorname{Re}\int \sin x_1 v\, \partial_1 \bar v\, dx + \beta\|v\|^2 = -\tfrac{1}{2}\int \sin x_1\, \partial_1 |v|^2\, dx + \beta \|v\|^2 \\ &= \tfrac{1}{2}\int |v|^2 \cos x\, dx + \beta\|v\|^2 \geqq (\beta - \tfrac{1}{2})\|v\|^2\end{aligned}$$

(the integral will be taken over $[-\pi, \pi]^m$). Therefore

$$\|A_1 v\|\,\|v\| \geqq |\operatorname{Re}(A_1 v, v)| \geqq (\beta - \tfrac{1}{2})\|v\|^2.$$

Therefore

$$\|A_1 v\| \geqq (\beta - \tfrac{1}{2})\|v\|,$$

whence we obtain that (4.10) holds and, consequently, (4.6).

Lemma 4.2. Suppose the operator B_β is determined by Formula (4.3), $f \in A(T^m)$ is an entire function, and $f(x)$ does not depend on $x_2, \ldots, x_m$. Then (4.5) holds.

Proof. By virtue of Propositions 4.1.1 and 4.1.2

$$R_0(B_\beta, f) = R^H(A, f),$$

where

$$H = L_2(T^m),\ A = B_\beta - (\beta - \tfrac{1}{2})I$$

Since $f(x)$ depends only on x_1, and A acts only with respect to variable x_1, it is obvious from (4.13) that

$$M^H(A, f, R) \leqq CM^{H_1}(A, f, R)$$

where $H_1 = L_2(T^1)$. Therefore

$$R^H(A, f) \geqq R^{H_1}(A, f).$$

Using Theorem 4.6.1, we obtain that $R_0(A, f) > \pi/2$ (in $H_1 = L_2(T^1)$). Therefore, by virtue of (4.16), $R^{H_1}(A, f) > \pi/2$. Consequently $R^H(A, f) > \pi/2$ and $R_0(A, f) > \pi/2$ (in $H = L_2(T^m)$). By virtue of Proposition (4.1.2) (4.5) holds and the lemma is proved.

Lemma 4.3. For any $\beta > 0$ equation

$$\partial_x(u(x) \sin x) + \beta u(x) = \sin x, \tag{4.11}$$

$x \in \mathbb{R}$, has a unique solution, $u(x) = u_\beta(x)$, belonging to $L_1(-\pi, \pi)$. This solution is periodic with respect to x with the period 2π and is analytic at all points, with the exception of the point $\pm\pi$. The function

$$\varphi_\beta(z) = u_\beta(2 \operatorname{arcctg} z), \quad \psi_\beta(z) = \varphi_\beta(z)/(1+z^2) \tag{4.12}$$

has singularities at the point $z = 0$. Namely, suppose

$$\beta=l+\delta, \quad l\in\mathbb{Z}_+, \quad 0\leqq\delta<1. \tag{4.13}$$

Then

$$\psi_\beta(z)=\begin{cases} C_\beta|z|^{\beta-1}\operatorname{sign} z+\varphi_{1\beta}(z) & (\beta\neq 2k, \quad k\in\mathbb{Z}), \\ C_\beta|z|^{\beta-1}\ln z+\varphi_{1\beta}(z) & (\beta=2k, \quad k\in\mathbb{Z}), \end{cases} \tag{4.14}$$

where $\psi_{1\beta}(z)$ are functions that are analytic on $\mathbb{R}$, $C_\beta \neq 0$.

Proof. The solution v_0 of the linear homogeneous equation corresponding to (4.11) has the form

$$v_0=C|\operatorname{ctg}(x/2)|^\beta \sin x$$

and does not belong to $L_1(-\pi, \pi)$ when $C \neq 0$. A unique soultion of the inhomogeneous equation (4.11), which belongs to $L_1(-\pi, \pi)$, is given by the formula

$$u(x)=\frac{|\operatorname{ctg}(x/2)|^\beta}{\sin x}\int_0^x \sin t|\operatorname{ctg}(t/2)|^{-\beta}\,dt, \quad -\pi<x<\pi, \tag{4.15}$$

and periodically extends to all x. Obviously, $u(x) = -u(x)$. It is obvious that $u(x)$ is analytic when $x \neq \pi k$, $k \in \mathbb{Z}$. The singularity at zero is removable.

We shall consider the function $u(x)$ in the neighbourhood of the point $\pm\pi$ in more detail. We make the substitution $\tau = \operatorname{ctg}^2(t/2)$ in the integral (4.15) when $t > 0$:

$$u(x)=-2\frac{|\operatorname{ctg}(x/2)|^\beta}{\sin x}\int_\infty^{\operatorname{ctg}^2 x/2}\frac{\tau^{-\beta/2}}{(\tau+1)^2}\,d\tau, \qquad 0<x<\pi.$$

Carrying out the substitution $z = \operatorname{ctg}(x/2)$ when $0 < x < \pi$, we shall rewrite the formula obtained in the form

$$\varphi(z)=z^{\beta-1}(1+z^2)\int_{z^2}^\infty \frac{\tau^{-\beta/2}}{(\tau+1)^2}\,d\tau, \tag{4.16}$$

where $\varphi(z) = \varphi_\beta(z)$ is determined by (4.12). It follows from the oddness of the function $u: u(-x) = u(x)$ and from the periodicity of u that $\varphi(-z) = -\varphi(z)$. Thus, oddly extending $\varphi(z)$, which is determined when $z > 0$ using Formula (4.16), we shall determine $\varphi(z)$ for all $z \neq 0$. Note that the point $\pm\pi$ becomes the point $z = 0$ with the substitution $z = \mathrm{ctg}(x/2)$. It is obvious that this substitution is an analytic diffeomorphism of the neighbourhoods of the point $x = \pm\pi$ and the point $z = 0$. Therefore the singularities of the function $\varphi(z)$ when $z = 0$ are the same as for $u(x)$ at the point $x = \pm\pi$.

We shall proceed to examine

$$\psi(z) = \varphi(z)/(1+z^2)$$

in the neighbourhood of zero. We shall carry out integration by parts on the right-hand side of Formula (4.16) k times. Let us first consider the case when $\beta/2$ is not integral, and let us take $k > \beta/2$. We obtain integrated terms of the form

$$z^{\beta-1}C_j \frac{\tau^{-\beta/2+j}}{(\tau+1)^{1+j}}\bigg|_{z^2}^{\infty} = \frac{-C_j z^{2j-1}}{(z^2+1)^{j+1}}, \qquad z>0, \tag{4.17}$$

where $j = 1, \ldots, k$. These functions extend, oddly on $z < 0$ to functions which are analytic at zero, and do not thereby contribute to the singularity at zero. After integration by parts an integral of the following form will remain:

$$\begin{aligned} z^{\beta-1}C'_k \int_{z^2}^{\infty} \frac{\tau^{-\beta/2+k}}{(\tau+1)^{2+k}}\,d\tau &= C'_k z^{\beta-1} \int_0^{\infty} \frac{\tau^{-\beta/2+k}}{(\tau+1)^{2+k}}\,d\tau \\ &- C'_k z^{\beta-1} \int_0^{z^2} \frac{\tau^{-\beta/2+k}}{(\tau+1)^{2+k}}\,d\tau, \qquad C'_k \neq 0. \end{aligned} \tag{4.18}$$

Expanding the function $(\tau+1)^{-2+k}$ in a series in powers of τ in the integrand expression of the second integral on the right-hand side of (4.18), and integrating from 0 to z^2, we obtain that this integral is expanded in a series in terms of the form

$$b_i(z^2)^{i+k-\beta/2}.$$

Multiplying this series by $C'_k z^{\beta-1}$, we obtain that the second term on the right-hand side of (4.18) is expanded in a series in integer odd powers of z. This term, which is extended with respect to oddness, is also a function which is analytic at zero. Only the first term in Formula (4.18), which has the form $Cz^{\beta-1}$ when $z > 0$, remains. Extending this term with respect to oddness to $z < 0$, we obtain

$$\psi(z)=\psi_\beta(z)=C_\beta|z|^{\beta-1}\operatorname{sign} z+\psi_{1\beta}(z), \tag{4.19}$$

where

$$C_\beta=C'_k\int_0^\infty \frac{\tau^{-\beta/2+k}}{(\tau+1)^{2+k}}d\tau\neq 0,$$

The function $\psi_{1\beta}(z)$ is analytic at zero. (4.14) follows from (4.19) for noninteger $\beta/2$. Thus the function $\psi_\beta(z)$ is not infinitely smooth at zero.

We shall now consider the case when $\beta/2$ is integral. In this case we shall also integrate by parts k times the integral on the right-hand side of (4.16), taking $k=\beta/2$, i.e. $\beta=2k$. The first k-1 integrated terms have the form (4.17), and are consequently odd and analytic at zero. The following integral remains before the last integration ($j = k - 1$):

$$-z^{2k-1}C_{k-1}\int_{z^2}^\infty \tau^{-1}d(\tau+1)^{-k}=kC_{k-1}z^{2k-1}\int_{z^2}^\infty \frac{\tau^{-1}d\tau}{(\tau+1)^{k+1}}.$$

Integrating by parts for the last time, we transform this integral to the form

$$C_k z^{2k-1}\left[\frac{-\ln z^2}{(z^2+1)^{k+1}}+(k+1)\int_0^{\infty}\frac{\ln t\,dt}{(t+1)^{k+2}}-(k+1)\int_0^{z^2}\frac{\ln t\,dt}{(t+1)^{k+2}}\right]. \tag{4.20}$$

the second term in square brackets is a constant. Let us consider the third term in square brackets in more detail.

Obviously,

$$-(k+1)\int_0^{z^2}\frac{\ln t\,dt}{(t+1)^{k+2}}=\int_0^{z^2}\ln t\,d[(t+1)^{-(k+1)}-1]$$

$$=\ln t[(t+1)^{-(k+1)}-1]\Big|_0^{z^2}-\int_0^{z^2}\left[\frac{1}{(t+1)^{k+1}}-1\right]\frac{dt}{t}.$$

It is obvious that the integrand expression in the last integral is an analytic function at zero, and consequently this integral is an analytic even function at zero. Substituting the expression obtained into (4.20), we obtain that it equals

$$C_k z^{2k-1}\left[\frac{-\ln z^2}{(z^2+1)^{k+1}}+\ln z^2\left(\frac{1}{(z^2+1)^{k+1}}-1\right)+\psi_{00}(z)\right],$$

where $\psi_{00}(z)$ is an even analytic function. Extending this expression to negative z with respect to oddness, and bearing in mind the analyticity of (4.17), when $\beta = 2k$ we obtain

$$\varphi_{2k}(z)=Cz^{2k-1}\ln|z|+\psi_{1,2k}(z), \qquad C\neq 0, \tag{4.21}$$

where $\psi_{1,2k}(z)$ is a function that is analytic in the neighbouhood of zero. Since $\psi_{2k}(z)$ is analytic on $\mathbb{R}\backslash 0$, in the same way as the first term in (4.21), $\psi_{1,2k}(z)$ is also analytic on $\mathbb{R}\backslash 0$, and consequently on $\mathbb{R}$. (4.14) follows from (4.21) when $\beta = 2k$.

Lemma 4.4. Suppose the function $w(z)$ is identical with the function $\psi_\beta(z)$, which is determined by (4.14), when $|z| < 3$, and $w(z)$ is infinitely smooth when $|z| > 3$, $w(z) = 0$ when $|z| > 4$. Then $w(z)$ does not belong to the Besov space $B^s_{q,\Theta}(\mathbb{R}^1)$ if

$$s \geqq 0, \quad s = \beta - 1 + 1/q, \quad 1 \leqq \theta < \infty. \tag{4.22}$$

Proof. Suppose

$$s = l + \delta, \; l \in \mathbb{Z}_+, \; 1 \geqq \delta > 0.$$

As is well known (see Besov, Il'in, Nikol'skii [1] or Nikol'skii [1]), if

$$w \in B^2_{q,\theta}(\mathbb{R}^+),$$

the following seminorm is finite:

$$p_1(w) = \left(\int_{-1}^{1} |t|^{-1-\theta\delta} \|\Delta_t^2 \, \partial^l w\|^\theta_{L_q(0,1)} \, dt \right)^{1/\theta}, \tag{4.23}$$

where Δ_t^2 is a second difference operator,

$$\Delta_t^2 w(z) = w(z + 2t) - 2w(z + t) + w(z)$$

(Note that the analogous seminorm where $L_q(0, 1)$ is replaced in (4.23) by $L_q(\mathbb{R})$ is also finite.)

We shall prove that the seminorm (4.23) is infinite for the function w, described under the condition of the lemma, and for s, determined by (4.22):

$$p_1(w) = +\infty. \tag{4.24}$$

The term $\varphi_{1\beta}(z)$ in (4,.14) is analytic, therefore $p_1(\varphi_{1\beta}) < +\infty$. Consequently, to prove (4.24) it is sufficient to prove the infiniteness of p_1 in the first terms in (4.14), which we shall denote by w_1.

Let us first consider the case when $\beta = 2k$ in (4.14). Obviously

$$\partial_z^l w_1 = \partial_z^l(|z|^{\beta-1}\operatorname{sign} z) = C|z|^{\beta-1-l}(\operatorname{sign} z)^{1+l}. \tag{4.25}$$

When $z > 0$ and $t > 0$, signz and sign$(z + t)$ are positive, therefore for these z

$$\Delta^2\,\partial^l(|z|^{\beta-1}\operatorname{sign} z) = C((z+2t)^\varepsilon - 2(z+t)^\varepsilon + z^\varepsilon), \tag{4.26}$$

where $\varepsilon = \beta - 1 - l$. Since $l = s - \delta$, $1 > \delta > 0$, then by virtue of (4.22) $\varepsilon = \beta - 1/q$, $-1 < \varepsilon < 1$. When $0 < t < 1$ by virtue of (4.26)

$$\begin{aligned}\|\Delta_t^2\,\partial^l w_1\|^q_{L_q(0,1)} &= C_1\int_0^1 t^{\varepsilon q}\left|\left(\frac{z}{t}+2\right)^\varepsilon - 2\left(\frac{z}{t}+1\right)^\varepsilon + \left(\frac{z}{t}\right)^\varepsilon\right|^q dz\\ &\geqq C_1 t^{\varepsilon q+1}\int_0^1 |(x+2)^\varepsilon - 2(x+1)^\varepsilon + x^\varepsilon|^q\,dx \geqq C_2 t^{\varepsilon q+1} = C_2 t^{\delta q} \qquad (C_2>0).\end{aligned} \tag{4.27}$$

Substituting (4.26) into (4.23), where $w = w_1$, we obtain:

$$p_1(w_1) \geqq C_3\left(\int_0^1 t^{-1-\theta\delta}t^{\theta\delta}\,dt\right)^{1/\theta} = +\infty. \tag{4.28}$$

Since the function $\varphi_{1\beta}$ in (4.14) is analytic, then $p_1(\varphi_{1B}) < +\infty$, and (4.24) follows from (4.28).

Let us consider the case when β is an integer. Then $l = \beta - 1$, $\delta = 1/q$. Supopse β is even. Then, according to (4.14)

$$\partial^l\varphi_\beta(z) = C\ln|z| + \psi_{2\beta}(z), \qquad C \neq 0 \tag{4.29}$$

where $\psi_{2\beta}$ is an analytic function. Obviously,

$$\Delta_t^2 \ln(z) = \ln[(z+2t)z/(z+t)^2]$$

$$= \ln\left[\left(\frac{z}{t}+2\right)\frac{z}{t}\Big/\left(\frac{z}{t}+1\right)^2\right].$$

Therefore

$$\|\Delta_t^2 \ln|z|\|_{L_q(0,1)}^q \geqq t \int_0^{1/t} \ln^q[(x+2)x/(x+1)^2]\, dx \geqq C_4 t = C_4 t^{q\delta}.$$

Hence, as in (4.28), we obtain that $p_1(w_1) = +\infty$, and consequently (4.24) holds.

Let us now consider the last case, when β is an integer and odd. Then (4.25) holds, where $l = \beta - 1$. Obviously, when

$$-1 < t < 0 |\Delta_t^2 \operatorname{sign} z| = 2$$

when $0 < z < -2t$. Therefore

$$(p_1(w_1))^\theta \geqq \int_{-1}^{0} |t|^{-1-\theta/q} 2|2t|^{\theta/q}\, dt = +\infty$$

and (4.24) holds. It follows from (4.24) that

$$w \notin B_{q,\theta}^s(\mathbb{R}^1)$$

and the lemma is proved.

Proof of Theorem 4.2. If $\beta > 1/2$, (4.4) follows from Lemma 4.1. Formula (4.5) follows from Lemma 4.2. We shall proceed to prove sect.3) of Theorem 4.2. Suppose $u(x)$ is the solution of (4.11), where $x = x_1$. According to Lemma 4.3, $u(x)$ is analytic when $x \neq \pm\pi$ and when $\beta > 1/2$, according to (4.14), $u \in L_2(-\pi, \pi)$. Since Eq.(4.11) has the form $A_1 u = \sin x_1$, where the operator A_1 is the same as in (4.7), the function $u(x)$ is the solution of the equation

$$A_1^* A_1 u = A_1^* \sin x_1,$$

where A_1^* is determined by (4.8). Since

$$A_1^* \sin x_1 = -\sin x_1 \cos x_1 + \beta \sin x_1 = f(x_1),$$

then by virtue of (4.9) $u(x)$ is the solution of the equation $B_\beta u = f$. Since the solution is unique by virtue of (4.6), then $u = B_\beta^{-1} f$. Note, now, that

$$u(x) \notin W_2^{\beta - 1/2}(T^1).$$

Indeed, we will assume the opposite. Then

$$u \in W_{2\,\mathrm{loc}}^{\beta - 1/2}(\mathbb{R}^1).$$

The Sobolev classes are invariant with respect to changes of variables (see Besov, Il'in, Nikol'skii [1]). Therefore the function $\varphi_\beta(z)$, which is determined by Formula (4.12), will then belong to

$$W_{2\,\mathrm{loc}}^{\beta - 1}(\mathbb{R}^1).$$

If $\chi(z)$ is an infinitely smooth function, which equals zero when $z > 4$, and equals 1 when $|z| < 3$, then the function

$$w(z) = \chi(z)\varphi_\beta(z) \in W_q^{\beta - 1/2}(\mathbb{R}^1).$$

It follows from embedding theorems in Besov spaces (see Besov, Il'in, Nikol'skii [1]), that

$$w(z) \in B_{2,2}^{\beta - 1/2}(\mathbb{R}^1).$$

Since $s = \beta - 1/2$ satisfies (4.22), where $q = 2$, we arrive at a contradiction with the statement of Lemma 4.4. Therefore

$$u(x_1) \notin W_2^{\beta-1/2}(T^1).$$

By virtue of the definition (1.1) of the norm in $W_2{}^s(T^m)$ the membership of the function of one variable in $W_2{}^s(T^1)$ is equivalent to its membership in $W_2{}^s(T^m)$. Therefore

$$u = B_\beta^{-1} f \notin W_2^{\beta-1/2}(T^m),$$

and (4.5) is proved.

§5. Upper bound of the smoothness of solutions of Cauchy's problem for a degenerating parabolic equation.

Consider on the circle $T^1 = [0, 2\pi]$ Cauchy's problem

$$\partial_t u = \sin x\, \partial_x^2((\sin x)u)), \quad u|_{t=0} = f, \tag{5.1}$$

where $f \in A(T^1)$, i.e. f is an analytic periodic function.

Theorem 5.1. Suppose the solution $u(t, x)$ of problem (5.1), which is infinitely differentiable with respect to t and x, where $f \in A(T^1)$, is analytic on T^1 with respect to x for some $t_1 > 0$, i.e.

$$u(t_1, \cdot) \in \mathscr{A}(T^1).$$

Then $f \equiv 0$.

This theorem shows that the statement of Thoerem 1.4 on the infinite differentiability of $u(t, x)$ with respect to x cannot be replaced by the analyticity of $u(t, x)$ with respect to x. At the end of this paragraph we will prove Theorem 5.2, from which it follows that the estimate of the increase of the norms of the derivatives $\partial^\alpha u(t, x)$, given by (1.10) at the internal points, is sharp (see also Remark 5.1 at the end of the paragraph).

The proof of Theorem 5.1 is based on a number of lemmas. Consider the equation

$$\partial_t v = -\sin^2 x\, \partial_x^2 v. \tag{5.2}$$

The proof of Theorem 5.1 is based on the possibility of constructing a solution of this equation when $t > 0$ in the neighbourhood of the points $0, \pm\pi$.

Lemma 5.1. If $v(t, x)$ is a solution of Eq.(5.2) which is infinitely differentiable with respect to x and with respect to t when $|x| < \delta$, $t_0 < t < t_1$, the functions

$$v_n(t) = \frac{1}{n!}\partial_x^n v(t, x)\Big|_{x=0} \qquad (n = 0, 1, \dots) \tag{5.3}$$

are solutions of the equations

$$\partial_t v_n = -n(n-1)v_n + \sum_{k=3}^{n} b_k v_{n-k+2}(n-k+2)(n-k+1), \tag{5.4}$$

where

$$b_k = \frac{1}{k!}\partial_x^k(-\sin^2 x)\Big|_{x=0}.$$

Proof. We obtain Eqs.(5.4) by differentiating (5.2) n times with respect to x when $x = 0$, using the Leibnitz formula, and noting that when $k = 0$ and $k = 1$ the corresponding terms equal zero.

Lemma 5.2. Suppose

$$|v_n(0)| \leqq C_0 R_0^n/(n(n-1)), \quad R_0 > 0, \qquad n = 2, 3, .., \tag{5.5}$$

and $v_n(t)$ are solutions of Eq,.(5.4) with the initial conditions

$$v_n(t)|_{t=0} = v_n(0).$$

Then

$$v_0(t) = v_0(0),\ v_1(t) = v_1(0)$$

and $R > 0$ exists such that when $|\mathrm{Im}t| < \mathrm{Re}t$ the following estimate holds:

$$|v_k(t)| \leqq C_0 e^{-2\mathrm{Re}t} R^k/(k(k-1)), \qquad k = 2, 3, \ldots \ . \tag{5.6}$$

Proof. We shall prove (5.) by induction. When $k = 2$ (5.6) is obvious when $R > R_0$. Suppose (5.6) holds when $2 < k < n - 1$. We shall prove (5.6) when $k = n_t$. We shall write the solution of (5.4) in the form

$$v_n(t) = e^{-n(n-1)t}\left[v_n(0) + \int_0^t \sum_{k=3}^n b_k v_{n-k+2}(\tau)(n-k+2)(n-k+1)\, e^{n(n-1)\tau}\, d\tau\right], \tag{5.7}$$

where the integral is taken over the segment in $\mathbb{C}$, which joins 0 and t, $|\mathrm{Im}t| < \mathrm{Re}t$. Obviously $|b_k| < B_0$ $\forall k$ for some number B_0 and by virtue of the assumption of induction

$$|b_k v_{n-k+2}(\tau)(n-k+2)(n-k+1)| \leqq 2C_0 B_0 R^{n-k+2} e^{-2\mathrm{Re}\tau}.$$

Bearing this estimate in mind, from (5.7) and (5.5), assuming $R > 1$, we derive:

$$|v_n(t)| \leqq e^{-n(n-1)\operatorname{Re} t}\left[\frac{C_0 R_0^n}{n(n-1)} + \frac{C_0 B_0 R^{n-1}}{1-R^{-1}} \int_0^{\operatorname{Re} t} e^{-2s} e^{n(n-1)s} \sqrt{2}\, ds\right]$$

$$= \frac{C_0 e^{-n(n-1)\operatorname{Re} t} R^n}{n(n-1)}$$

$$\times \left[\left(\frac{R_0}{R}\right)^n + \frac{n(n-1)}{R-1} B_0 \frac{e^{(n(n-1)-2)\operatorname{Re} t} - 1}{n(n-1)-2} \sqrt{2}\right].$$

Bearing in mind that $n(n-1) - 2 > n(n-1)/2$ when $n > 3$, we hence obtain for fairly large R the estimate

$$|v_n(t)| \leqq \frac{C_0 R^n e^{-2\operatorname{Re} t}}{n(n-1)} \left[\left(\frac{R_0}{R}\right)^n + \frac{2\sqrt{2} B_0}{R-1}\right] \leqq \frac{C_0 R^n e^{-2\operatorname{Re} t}}{n(n-1)},$$

Q.E.D.

Lemma 5.3. Suppose $v_n(t)$ are solutions of Eqs.(5.4) with the initial conditions $v_n(0)$, satisfying (5.5). Then the function

$$v^0(t,x) = \sum_{k=0}^{\infty} v_k(t) x^k \tag{5.8}$$

is analytic with respect to x and with respect to t in the domain

$$\Omega_\delta = \{(t,x): \operatorname{Re} t > 0, |\operatorname{Im} t| < \operatorname{Re} t,\ |x| < \delta\},$$

where $\delta > 0$. At the same time $v^0(t, x)$ is a solution of Eq.(5.2). If $v_0 = v_1 = 0$, then when $(t, x) \in \Omega_\delta$ the following estimate holds:

$$|v_0(t,x)| + |\partial_x v_0(t,x)| \leqq C e^{-2\operatorname{Re} t}. \tag{5.9}$$

Proof. It is obvious that $v_k(t)$ - the solutions of (5.4) - are analytic with respect to $t \in \mathbb{C}$. We shall use Lemma 5.2. By virtue of estimate (5.6) series (5.8) uniformly converges to Ω_δ if

$\delta < R^{-1}$. Therefore $v^0(t, x)$ is analytic in Ω_δ. Since Eqs.(5.4) are obtained by substituting (5.8) into (5.2) and by equating the coefficients of equal powers of x, $v^0(t, x)$ is the solution of (5.2). Estimate (5.9) follows from (5.6).

Lemma 5.4. On the assumptions of Lemma 5.3 the function $v^0(t, x)$ constructed in Lemma 5.3, extends to a solution of Eq.(5.2) which is analytic with respect to x and t when $-\pi < x < \pi$, $t > 0$. If $v_0(0) = v_1(0) = 0$, then when $-\pi + 1 < x < \pi - 1$, $t > \varepsilon$, when ε is as small as desired, estimate (5.9) holds.

Proof. Suppose $\varepsilon > 0$ is an arbitrarily small number, and the number $\alpha > 0$ is so small that $2a < \delta$, δ is the same as in Lemma 5.3. The function $v^0(t, x)$ is analytic in the neighbourhood of the line

$$L_a = \{(t, x) : t \geqq \varepsilon, x = a\}.$$

To extend $v^0(t, x)$ with respect to x, we shall consider Cauchy's problem with data on L_a for Eq. (5.2):

$$\partial_x^2 v = -\frac{1}{\sin^2 x}\partial_t v, \; v\Big|_{x=a} = v^0(t, a), \; \partial_x v\Big|_{x=a} = \partial_x v^0(t, a). \tag{5.10}$$

Doubly integrating (5.10) with respect to x, we obtain:

$$v(t, x) = K_0\, \partial_t v(t, x) + g(t, x), \; K_0 \varphi(x) = \int_a^x dx_1 \int_a^{x_1} dx_2 \frac{-\varphi(x_2)}{\sin^2 x_2}, \tag{5.11}$$

where

$$g(t, x) = v^0(t, a) + (x - a)\, \partial_x v^0(t, a).$$

We shall seek the solution of (5.11) in the form of the series

$$v(t,x)=\sum_{j=0}^{\infty}(K_0\,\partial_t)^j g(t,x)=\sum_{j=0}^{\infty}K_0^j\,\partial_t^j g(t,x) \tag{5.12}$$

(we have used the fact that K_0 and ∂_t commute). When

$$a/2 \leqq \operatorname{Re} x \leqq \pi - a/2,\ |\operatorname{Im} x| \leqq 1,$$

the following estimate holds:

$$|K_0^j\varphi(x)| \leqq \frac{|x-a|^{2j}}{(2j)!}\,\frac{1}{\sin^{2j}(a/2)} \sup_{\substack{a/2<\operatorname{Re} x \leqq \pi - a/2 \\ |\operatorname{Im} x| \leqq 1}} |\varphi(x)|. \tag{5.13}$$

Since $v^0(t,a)$ and $\partial_x v^0(t,a)$ are analytic with respect to t when

$$\operatorname{Re} t > 0,\ |\operatorname{Im} t| < \operatorname{Re} t,$$

then when $\operatorname{Re} t > \varepsilon$, $|\operatorname{Im} t| < \varepsilon/2$ the functions $\partial_x v^0(t,a)$ and $v^0(t,a)$ are bounded, and therefore if $v_0 = v_1 = 0$, then

$$\sup_{|\operatorname{Im} x| \leqq 1,\ 0 \leqq x \leqq \pi} |\partial_t^j g(t,x)| \leqq C e^{-2\operatorname{Re} t} r^j j!, \tag{5.14}$$

when $t > 2\varepsilon$, where $r = r(\varepsilon)$ is sufficiently large, $j = 1, 2, \ldots$. (It is obvious that $\partial_t^j g(t, x)$ when $j > 1$ do not depend on v_0 and v_1, since $v_0(t)$ and $v_1(t)$ do not depend on t. Therefore (5.14) when v_0, $v_1 \neq 0$ holds for $j > 1$ as previously. The convergence of series (5.12) in the complex neighbourhood of the set

$$P = \{a \leqq x \leqq \pi - a\},\ t \geqq 2\varepsilon,$$

follows from (5.13) and (5.14), whilst estimate (5.9) holds on P. Bearing in mind that ε is arbitrarily small, we obtain a statement of the lemma for positive x. In a similar way, for negative x we shall take $a < 0$ in (5.10).

Lemma 5.5. Suppose $v(t, x)$ is a solution of Eq.(5.2) that is infinitely smooth on $[0, t_0] \times T^1$, and for which the numbers $v_n(0)$ and $v_n^1(0)$, determined by (5.3) (for $v_n^1(0)$ in (5.3) $x = \pi$), satisfy estimate (5.5). Suppose $v(t, x)$ belongs to $A(T^1)$ when $t = t_0 > 0$. Then $v(t, x)$ extends to a soultion of Eq.(5.2) which is analytic on $[t_0, \infty] \times T^1$.

Proof. Estimate (5.5) with some $R_0 > 0$ holds for the functions $v_n(t)$, defined by (5.3). By virtue of Lemmas 5.3 and 5.4, the solution $v^0(t, x)$ of Eq.(5.2) exists, which is analytic with respect to x and t. At the same time by virtue of the construction

$$\partial_x^k v^0(t,x)\big|_{x=0} = \partial_x^k v(t,x)\big|_{x=0} \qquad (t \geqq 0, k=1,2,\ldots). \tag{5.15}$$

Since the function $v(t, x)$ is analytic with respect to x on T^1 by virtue of the assumption of the lemma, it follows from (5.15) and from the analyticity of $v^0(t, x)$ that

$$v^0(t_0, x) = v(t_0, x).$$

Note that with the substitution $x \to x - \pi$ Eq.(5.2) becomes itself, and the point π becomes 0. It is therefore obvious that the solution $v^1(t, x)$ of Eq.(5.2), which is analytic with respect to x and t when $0 < x < 2\pi$, $t > 0$, exists. When

$$t = t_0 v^1(t_0, x) = v(t_0, x),\ 0 < x < 2\pi.$$

Therefore

$$v^0(t_0, x+2\pi) = v^1(t_0, x),\ v^0(t_0, x) = v^1(t_0, x)$$

where v^1 and v^0 are determined. Note that since both $v^1(t, x)$ and

$v^0(t, x)$ are solutions of Eq.(5.2), and are identical with $v(t_0, x)$ when $t = t_0$, by virtue of (5.2)

$$\partial_j^k v^i(t,x)\big|_{t=t_0} = (-\sin^2 x\, \partial_x^2)^k v(t_0, x), \qquad i=0,1.$$

Since the functions $v^1(t, x)$ and $v^0(t, x)$ are analytic with respect to t when $t > 0$ for each

$$x \in]0, \pi[\cup]\pi, 2\pi[,$$

the agreement of these functions are identical when $t > 0$ in view of the fact that all the derivatives with respect to t are identical when $t = t_0$. Thus the function $\hat{v}(t, x)$, which is periodic with respect to x and analytic with respect to x and t when $x \in T^1$, $t > 0$ and which is identical when $t = t_0$ with $v(t, x)$, exists.

Proof of Theorem 5.1. Suppose $u(t, x)$ is the solution of (5.1), which is analytic when $t = 0$ and $t = t_1$. We shall multiply (5.1) by $\sin x$ and shall replace t by $t_1 - t$. We will obtain that

$$v(t,x) = \sin x u(t_1 - t, x)$$

is a solution of Eq.(5.2), which is analytic when $t = 0$ and $t = t_1$, whilst

$$v(t,0) = 0, \ v(t,\pi) = 0.$$

According to Lemma 5.5 the extension $v(t, x)$ to $t > t_1$, which is analytic with respect to x and t, exists. When $t > 0$ this extension agrees in the segments $[-\pi + 1, \pi - 1]$ and $[1, 2\pi - 1]$ with $v^0(t, x)$ and $v^1(t, x)$ respectively. We will assume

$$v^{01}(t,x)=v^0(t,x)-q_0x, \quad q_0=\partial_x v^0(t,x)|_{x=0},$$

and the number q_0 is a constant by virtue of Lemma 5.1. Since q_0x is the solution of (5.2), $v^{01}(t, x)$ is also a solution of (5.2). Obviously, $v^{01}(t, x)$ for small x is represented by series (5.8), where $v_0 = v_1 = 0$. Therefore $v^{01}(t, x)$ when $-\pi + 1 < x < \pi + 1$ by virtue of Lemma 5.4 satisfies estimate (5.9). In a similar way, the function

$$v^{11}(t,x)=v^1(t,x)-q_1(x-\pi), \quad q_1=\partial_x v^1(t,x)|_{x=\pi},$$

satisfies estimate (5.9) when $1 < x < 2\pi - 1$. Since $v^0(t, x)$ and $v^1(t, x)$ agree with $\hat{v}(t, x)$ when $t > t_1$, from the estimates for v^{01} and v^{11} we obtain

$$|\hat{v}(t,x)-q_0x|\leqq Ce^{-2t} \qquad (-\pi+1<x<\pi-1),$$

$$|\hat{v}(t,x)-q_1(x-\pi)|\leqq Ce^{-2t} \qquad (1<x<2\pi-1)$$

when $t > t_1$. Hence we obtain that when $1 < x < \pi - 1$

$$|q_0x-q_1(x-\pi)|\leqq|\hat{v}(t,x)-q_0x| + |\hat{v}(t,x)-q_1(x-\pi)|\leqq 2Ce^{-2t}.$$

Passing to the limit as $t \to +\infty$, we obtain that $q_0x = q_1(x - \pi)$ when $1 < x < \pi - 1$ i.e. $q_0 = q_1 = 0$. Therefore from (5.9) follows estimate

$$|\partial_x v(t,x)|+|v(t,x)|\leqq Ce^{-2t} \qquad (t\geqq t_1). \tag{5.16}$$

From estimate (5.16) we obtain that the function

$$u(t,x)=v(t_1-t,x)/\sin x$$

– the solution of Eq.(5.1) – satisfies the estimate $|u(t, x)| < C_1 e^{2t}$ when $t < 0$. Multiplying Eq.(5.1) by $\bar{u}(t, x)$ and integrating by parts with respect to x, we obtain:

$$\tfrac{1}{2}\partial_t \int_{T^1} |u|^2\, dx + \int_{T^1} |\partial_x(\sin xu)|^2\, dx = 0,$$

whence, integrating from t to 0, we obtain:

$$\tfrac{1}{2}[\|u(0)\|^2 - \|u(t)\|^2] + \int_t^0 \|\partial_x((\sin x)u)\|^2\, dt = 0.$$

Directing t to $-\infty$ and bearing in mind that

$$\|u(t)\|^2 \leqq C_3 e^{4t},$$

we obtain that $\|u\|(0)\| = 0$; *Q.E.D.*

We shall now prove more than $u(t, x)$ cannot be analytic when $t > 0$, namely it cannot have derivatives which increase too slowly.

Theorem 5.2. Suppose $u(t, x)$ is an infinitely differentiable solution (5.1), where $f \in A(T^1)$. Suppose for some $t_0 > 0$ the following estimate holds:

$$|\partial_x^n u(t, x)|/n! \leqq C_0 R^n e^{\kappa n^2}, \quad \forall x \in T^1, \quad n \in \mathbb{Z}_+, \tag{5.17}$$

where $\varkappa < t_0$. Then $f = 0$.

The proof of the theorem is based on the following lemma.

Lemma 5.6. Suppose $v_n(t)$ are determined like solutions of Eqs.(5.4), where

$$|v_n(0)| \leqq C_0 R_0^n e^{\kappa n(n-1)}/(n(n-1), \qquad n = 2, 3, \ldots \tag{5.18}$$

$(\varkappa > 0)$. Suppose β is a number, $0 < \beta < 1$. Then $R > 0$ exist such that when $0 \leqslant t \leqslant \varkappa/\beta$ the following estimate holds:

$$|v_k(t)| \leqq C_0 R^k e^{(\kappa-\beta t)k(k-1)}/(k(k-1)), \qquad k=2,3,\dots. \tag{5.19}$$

Proof. When $k = 2$ estimate (5.19), where $R > R_0$, follows directly from (5.7). Suppose (5.19) holds when $2 \leqslant k \leqslant n - 1$. We shall prove it when $k = n$. We shall use Formula (5.7) and will obtain:

$$|v_n(t)| \leqq e^{-n(n-1)t}\Big[|v_n(0)|$$

$$+B_0 \int_0^t \sum_{k=3}^n |v_{n-k+2}(\tau)|(n-k+2)(n-k+1)\, e^{n(n-1)\tau}\, d\tau\Big]$$

$$\leqq e^{-n(n-1)t}\Big[|v_n(0)| + B_0 C_0 \int_0^t \sum_{k=3}^n$$

$$\times R^{n-k+2} e^{(\kappa-\beta\tau)(n-k+2)(n-k+1)+n(n-1)\tau}\, d\tau\Big]$$

$$=|v_n(0)|e^{-n(n-1)t} + B_0 C_0 R^n \sum_{k=3}^n \Big[R^{-k+2} e^{\kappa(n-k+2)(n-k+1)}$$

$$\times \frac{e^{[n(n-1)-\beta(n-k+2)(n-k+1)]t}-1}{n(n-1)-\beta(n-k+2)(n-k+1)} e^{-n(n-1)t}\Big]$$

$$\leqq |v_n(0)|\, e^{-n(n-1)t} + B_0 C_0 R^n \sum_{k=3}^n \frac{R^{-k+2} e^{(n-k+2)(n-k+1)(\kappa-\beta t)}}{(1-\beta)n(n-1)}.$$

Since $\varkappa - \beta t \geqslant 0$, $k \geqslant 3$ then, assuming $R > 1$, we obtain hence:

$$|v_n(t)| \leq |v_n(0)|\, e^{-n(n-1)t} + \frac{B_0 C_0 R^n}{n(n-1)(1-\beta)} e^{(\kappa-\beta t)n(n-1)} \sum_{k=3}^\infty R^{-k+2}$$

$$\leqq \frac{C_0 R_0^n}{n(n-1)} e^{(\kappa-t)n(n-1)} + \frac{B_0 C_0 R^{n-1}}{n(n-1)(1-\beta)(1-R^{-1})} e^{(\kappa-\beta t)n(n-1)}$$

$$= \frac{C_0 R^n}{n(n-1)} e^{(\kappa-\beta t)n(n-1)}\Big[\Big(\frac{R_0}{R}\Big)^n e^{(\beta-1)n(n-1)t} + \frac{B_0}{(1-\beta)(R-1)}\Big].$$

It is obvious that for sufficiently large R the expression in square brackets is not greater than 1, and we obtain (5.19) when $k = n$..

Proof of Theorem 5.2. The function

$$v(t,x) = \sin x u(t_0 - t, x)$$

is a solution of Eq.(5.2). We shall prove that $v(0, x)$ satisfies the following estimate by virtue of (5.17):

$$\frac{1}{n!}|\partial_x^n v(0,x)| \leq C_1 R_1^n e^{\varkappa n(n-1)}/(n(n-1)), \qquad n=2,3,\dots. \tag{5.19'}$$

Indeed, by virtue of the Leibnitz formula

$$\frac{1}{n!}|\partial_x^n(u\sin x)| \leq \sum_{k=0}^{n} \frac{1}{k!(n-k)!}|\partial_x^k u||\partial_x^{n-k}\sin x|$$

$$\leq \sum_{k=0}^{n} \frac{1}{k!}|\partial_x^k u| \leq C_0 e^{\varkappa n^2} \sum_{k=0}^{n} R^k \leq C_0 e^{\varkappa(n(n-1))} e^{\varkappa n}(1+R)^n.$$

Since $n(n-1) < C_2 2^n$, hence follows (5.19') with $R_1 = 2(1+R)e^{\varkappa}$.

The fulfilment of condition (5.18) follows from (5.19'), and Estimate (5.19) holds by virtue of Lemma 5.6. Suppose $\beta > 0$ is a number, such that $\varkappa/t_0 < \beta < 1$. Such β exists by virtue of the condition $\varkappa < t_0$. By virtue of estimate (5.19), where $t = t_1 = \varkappa/\beta$, we obtain:

$$\frac{1}{k!}|\partial_x^k v(t_1,x)|\Big|_{x=0} \leq CR^k/(k(k-1)). \tag{5.20}$$

We also have a similar estimate at the point $x = \pi$. Using Lemma 5.5, we obtain the existence of the solution of Eq.(5.2) which is periodic and analytic when $t > t_1$. Hence we derive, in exactly the same way as in the proof of Theorem 5.1, that $u(0, x) \equiv 0$.

Remark 5.1. Theorem 5.2 shows that estimate (1.10), given in Theorem 1.4, cannot be significantly improved. Indeed, for the operator

$$Bv = -\sin x\partial_x^2(\sin xv) \tag{5.21}$$

and for the entire function $f \in A(T^1)$, $R_0(B, f) > \pi/2$ by virtue of Theorem 4.6.1. Since all thepoints T^1 are internal, estimate (1.10) takes the form

$$\|\partial^\alpha u(t, 0)\|_{C(T^1)} \leqq C_\varepsilon R^{|\alpha|} |\alpha|!\, e^{\varkappa_0 t|\alpha|^2}, \tag{5.22}$$

where $\varkappa_0 = 1 + \varepsilon$, and $\varepsilon > 0$ is as small as desired. Note that by virtue of Theorem 5.2, if $f \neq 0$, we cananot assume $\varkappa_0 = 1 - \varepsilon$, where $\varepsilon > 0$ is as small as desired, in (5.22).

Thus Theorem 1.4 not only precisely indicates the class

$$C(k!\, e^{\varkappa_0 t k^2})$$

of functions that are infinitely differentiable with respect to x, to which $u(t, x)$ belongs for each $t > 0$, but also precisely indicates the value of the parameter $\varkappa_0$ for this class.

§6. Proof of Theorems 1.1 and 1.1′ in the general case.

In §§2 and 3 we considered the operators B of Example 4.3.1, i.e. those that act on the torus T^m. The functions $u(x)$, whose smoothness we estimated, were defined when $x \in T^m$. In this paragraph we will prove a theorem on smoothness which generalises the theorems from §§2 and 3 to the case $u(x)$, defined when x, lie in the domain $\Omega \subset \mathbb{R}^m$.

Later we shall need the well-known properties of the Sobolev spaces (see, e.g., Besov, Il'in, Nikol'skii [1], Nikol'skii [1]).

The Sobolev classes are invariant with respect to a change of variable. If $\Psi: \Omega_1 \to \Omega$ is a one-to-one mapping of the class C^l, $l \in \mathbb{N}$ with a Jacobian, which everywhere differs from zero, and $0 < s < l$ then if

$$f(x) \in W^s_{2\,\text{loc}}(\Omega), \qquad \text{then} \qquad f(u(y)) \in W^s_{2\,\text{loc}}(\Omega_1). \tag{6.1}$$

If $a(x)$ is a function of the class C^l, $l > s$, then if

$$f(x) \in W^s_{2\,\text{loc}}(\Omega), \qquad \text{then} \qquad a(x)f(x) \in W^s_{2\,\text{loc}}(\Omega). \tag{6.2}$$

If $f(x)$ is defined on the torus T^m, then

$$f \in W^s_{2\,\text{loc}}(T^m) \Leftrightarrow \|f\|_s < \infty. \tag{6.3}$$

where $\| \ \|_s$ is determined by Formula (2.15).

Theorem 1.1 is an obvious corollary of the following theorem.

Theorem 6.1. Suppose

$$g \in \mathscr{A}(\mathscr{I}^+_\beta),\ \beta > 0,$$

the operator satisfies condition 1.1, $f \in \mathcal{D}_0$, $A = B - bI$, where b is the same as in (4.2.4). Suppose $R_0 = R_0(A, f) > 0$. Suppose $u(x) = g(A)f(x)$. Then

$$u \in W^s_{2\,\text{loc}}(\Omega),\ \forall s: 0 \leq s < 2\beta R_0/\pi. \tag{6.4}$$

To prove Theorem 6.1, we need to introduce some notation and to prove two lemmas. We shall prove Theorem 1.1′ at the end of the paragraph.

Suppose the point $x_0 \in \Omega$. We shall use $K_h(x_0)$ to denote a cube with its centre at the point $x_0 = (x_{01}, \ldots, x_{0m})$

$$K_h(x_0)=\{x\in\mathbb{R}^m:|x_{0i}-x_i|\leqq h, \quad i=1,\ldots,m\}. \tag{6.5}$$

If h is small, then $K_h(x_0) \subset \Omega$.

Henceforth we will use a special change of variables in $K_h(x_0)$, which we shall now describe.

Suppose $X_h:\mathbb{R}^m \to K_h(x_0)$ is a mapping which is determined for $\varphi = (\varphi_1, \ldots, \varphi_m) \in \mathbb{R}^m$ by the formula

$$X_h(\varphi_1,\ldots,\varphi_m)=(x_{01}+h\sin\varphi_1,\ldots,x_{0m}+h\sin\varphi_m). \tag{6.6}$$

It is obvious that the mapping X_h induces the mapping X_h^* of functions defined on Ω into a set of functions that are defined and periodic on R^m. This mapping operates to the function $f(x)$, $x \in \Omega$, using formula

$$X_h^* f(\varphi)=f(X_h(\varphi)). \tag{6.7}$$

We will assume

$$\chi(\varphi)=\prod_{j=1}^{m}\cos\varphi_j, \qquad \varphi\in\mathbb{R}^m. \tag{6.8}$$

Since the functions $\sin\varphi_j$ and $\cos\varphi_j$ are periodic, the mapping χX_h^*, which is a composition of the mapping X_h^* and a multiplication by $\chi(\varphi)$, maps the functions defined on $\Omega \subset \mathbb{R}^m$ into functions defined on the torus T^m.

Lemma 6.1. If $K_h(x_0) \subset \Omega$, then for any function $f \in L_2(\Omega, G)$ (see (4.2.1) for the definition of $L_2(\Omega, G)$) the following estimate holds:

$$\|\chi X_h^* f\|_0 \leq C(\pi h)^{-m/2} \max_{x \in K_h(x_0)} |G(x)^{-1/2}| \|f\|_{L_2(\Omega, G)}. \tag{6.9}$$

Proof. It is sufficient to prove inequality (6.9) for functions that are continuous on Ω, since these functions are everywhere dense in $L_2(\Omega, G)$. We will assume

$$f^0 = X^* f,\ f^1 = \chi X^* f,\ X = X_h.$$

According to (2.15)

$$\|f^1\|_0^2 = (2\pi)^{-m} \int_0^{2\pi} \dots \int_0^{2\pi} \chi^2(\varphi)|f^0(\varphi)|_2^2\, d\varphi_1 \dots d\varphi_m,\quad |f|_2^2 = \langle f, f \rangle.$$

Since the functions $\chi^2(\varphi)$ and $f^0(\varphi)$ depend only on $\sin\varphi$, and

$$\sin(\pi/2 - \xi) = \sin(\pi/2 + \xi),$$

then

$$\|f^1\|_0^2 = \pi^{-m} \int_{-\pi/2}^{\pi/2} \dots \int_{-\pi/2}^{\pi/2} \chi^2(\varphi)|f^0(\varphi)|_2^2\, d\varphi_1 \dots d\varphi_m$$

Since $0 < \chi(\varphi) < 1$ when $-\pi/2 < \varphi < \pi/2$, then $\chi^2(\varphi) < \chi(\varphi)$ and therefore

$$\|f^1\|_0^2 \leq \pi^{-m} \int_{-\pi/2}^{\pi/2} \dots \int_{-\pi/2}^{\pi/2} \chi(\varphi)|f^0(\varphi)|_2^2\, d\varphi_1 \dots d\varphi_m \tag{6.10}$$

Since $\cos\varphi_j d\varphi_j = d\sin\varphi_j$, then, substituting

$$f^0(\varphi) = X^* f = f(\chi(\varphi))$$

into (6.10) and bearing in mind (6.6), we obtain:

$$\|f^1\|_0^2 \leqq \pi^{-m} \int_{-\pi/2}^{\pi/2} \cdots \int_{-\pi/2}^{\pi/2} |f(x_{01}+h\sin\varphi_1,\ldots,$$

$$x_{0m}+h\sin\varphi_m)|_2^2\, d\sin\varphi_1 \ldots d\sin\varphi$$

$$=(\pi h)^{-m} \int_{-h}^{h} \cdots \int_{-h}^{h} |f(x_{01}+x_1,\ldots,x_{0m}+x_m)|_2^2\, dx_1 \ldots dx_m$$

$$=(\pi h)^{-m} \int_{K_h(x_0)} \langle G^{-1/2}G^{1/2}f, G^{-1/2}G^{1/2}f\rangle\, dx.$$

Estimate (6.9) follows directly hence.

Lemma 6.2. Suppose

$$v\in\mathscr{A}(\bar\Omega),\ \omega\subset\bar\Omega,\ \omega=Y(K_h(x_0)),$$

where Y is a mapping that is analytic on $K_h(x_0)$. Suppose the majorant $\hat{v}(y) \in A^+$ of the function $v(x)$ is determined by Formula (4.5.10), and suppose δ is such that

$$\hat{v}(\delta)<\infty. \tag{6.11}$$

Suppose

$$v_1(x)=v(Y(x)),\ x\in K_h(x_0).$$

Then the number $\gamma > 0$, $\gamma = \gamma(\delta, h)$ and constant $C = C(\delta)$ will be obtained, such that

$$\|E_\gamma \chi X_h^* v_1\| \leqq C\hat{v}(\delta), \tag{6.12}$$

where $\|E_\gamma f\|$ is determined by (2.18).

Proof. It follows from (6.11) that estimate (1.1.1), where

$$C_1=\delta^{-1},\ C_0=\hat{v}(\delta).$$

holds for $f = v$. According to Lemma 1.1.1, $v(x)$ analytically extends to the complex neighbourhood $\mathcal{O}_{\delta/m}(\omega)$ of the compactum ω, defined by (1.1.2), where $K = \omega$. According to estimate (1.1.3)

$$|v(z)| \leqslant 2\hat{v}(\delta) \text{ when } z \in \mathcal{O}_{\delta/(2m)}(\omega). \tag{6.13}$$

The mapping Y analytically extends to the complex neighbourhood $\mathcal{O}_{\delta 1}(K_h)$, whilst

$$Y(\mathcal{O}_{\delta_1}(K_h)) \subset \mathcal{O}_{\delta/(2m)}(\omega).$$

The mapping X_h determined by Formula (6.6), extends to the complex domain Π_ε, defined by Formula (2.19). If $\varepsilon \leqslant \varepsilon_0$, where $\varepsilon_0 = \varepsilon_0(h, \delta)$ is sufficiently small, then $X(\Pi_\varepsilon) \subset \mathcal{O}_{\delta 1}(K_h(x_0))$. Therefore the modulus of the function

$$v_1^0(\varphi)=(X^*v_1)(\varphi)=v_1(X(\varphi))$$

is bounded in Π_ε by the number $2\hat{v}(\delta)$. Since $|\chi(\varphi)| \leqslant C'$ when $\varphi \in \Pi_\varepsilon$, the modulus of the function $v_1{}^1(\varphi) = \chi(\varphi)v^0(\varphi)$ satisfies estimate

$$|v_1^1(\varphi)| \leq 2C'\hat{v}(\delta) \qquad \text{when} \qquad \varphi\in\Pi_\varepsilon,\ \varepsilon<\varepsilon_0, \tag{6.14}$$

Inequality (6.14) has the form (2.20). Using Lemma 2.4, we obtain from (2.22), where $M = 2C'\hat{v}(\delta)$, inequality (6.12), where $\mathcal{O} < \gamma < \varepsilon_0$.

Proof of Theorem 6.1. Suppose $x_0 \in \Omega$, $K_h(x_0) \in \Omega$. According to Lemma 2.1 $u = g(A)f = u_0 + w$, where w is represented in the form of series (2.3), the norms of the terms of which in $L_2(\Omega, G)$, according to (2.35), satisfy the estimate

$$\|y_N\| \leqq CM_0 2^{-Nr\rho}.$$

It follows from this estimate and from Lemma 6.1 that

$$\|\chi X_h^* y_N\|_0 \leqq q_N M_0, \quad q_N = C_1 2^{-Nr\rho} \tag{6.15}$$

when $r < 2R_0/\pi$, $\rho < \beta$. Since $f \in A(\Omega)$, then according to Lemma 2.3, where $K = K_h$, estimate (2.10) holds. According to (4.5.10) and (2.10) when

$$\delta = R_1^{-1}/2,\ 2\delta < 1,$$

we obtain $\hat{u}_N(\delta) < C'G_n$ where

$$G_n = C_0(F)(2\delta)^{-2n}\min((2n)^{2n}, \nu^*(2n)^{2n}),\ u_N = P_n(B)f,\ n = 2^N.$$

Therefore, by virtue of (2.4) and bearing in mind that $2^{N+1} = 2n$, we obtain:

$$\hat{y}_N(\delta) \leqq \hat{u}_N(\delta) + \hat{u}_{N+1}(\delta) \leqq G_{2n}, \quad n = 2^N. \tag{6.16}$$

Using Lemma 6.2, where Y is an identical mapping, $\omega = K_h$, from (6.16) and (6.12) we obtain:

$$\|E_\gamma \chi X_h^* y_N\| < C_3 G_{2n},\ n = 2^N. \tag{6.17}$$

It follows from (2.3) that

$$\chi X_h^* w = \Sigma \chi X_h^* y_N. \tag{6.18}$$

It follows from the satisfaction of (6.15) and (6.17) for $\chi X_h{}^* y_N = y_N{}^1$, that $y_N{}^1$ satisfies conditions (2.36) and (2.37), where $y_N = y_N{}^1$. As in the proof of Theorem 2.1, from these inequalities we obtain that

$$\chi X_h^* w \in W_2^s(T^m) \text{ when } s < 2R_0\beta/\pi = s_0. \tag{6.19}$$

Since $\chi(\varphi) \neq 0$ when $|\varphi| < \pi/6$, by virtue of (6.2)

$$X_h^* w = \chi^{-1} \chi X_h^* w \in W_{2\,\mathrm{loc}}^s(\Omega_1),\ \Omega_1 = \{|\varphi| < \pi/6\}.$$

Since X_h is an analytic diffeomorphism between Ω_1 and $K_{h/2}(x_0)$, by virtue of (6.2)

$$w \in W_{2\,\mathrm{loc}}^s(K_{h/2}(x_0)).$$

Since the point x_0 is arbitrary, hence follows the statement of Theorem 6.1.

We shall now prove some lemmas which we need to prove Theorem 1.1′.

We shall assume the following for the set $\Omega \subset \mathbb{R}^m$:

$$\|u\|_{C^l(\Omega)} = \max_{|\alpha| \leq l} \sup_{x \in \Omega} [|\partial^\alpha u(x)|/\alpha!], \tag{6.20}$$

where, as usual,

$$\alpha! = \alpha_1! \ldots \alpha_m!,\ \partial^\alpha = \partial_1^{\alpha_1} \ldots \partial_m^{\alpha_m}.$$

Remark 6.1. note that if the mapping $Y : K \to \overline{\Omega}$ is the same as in Condition 1.2, the constant C exists, such that for any $f \in L_2(\overline{\Omega})$

$$\|f(Y(x))\|_{L_2(K)} \leqq C\|f(y)\|_{L_2(\Omega)}. \tag{6.21}$$

The derivation of this inequality uses the fact that $|\det Y'| > 0$ on K, from which it follows that $|\det Y'| > C > 0$ on K.

Remark 6.2. Inequailty (6.17) is derived from Lemma 6.2 not only when $y_N = y_N(x)$ are defined on $K_h(x_0) \subset \overline{\Omega}$, but also when $Y : K \to \overline{\Omega}$ is not an identical mapping, but a mapping from Condition 1.2 and $y_N(x) = y_N(Y(x))$.

Lemma 6.3. Suppose

$$u \in C^{l+m}(T^m),\ \chi$$

is determined by (6.8). Then

$$\|u\|_{C^l(T^m)} \leqq C(l+m)^m \|\chi u\|_{C^{l+m}(T^m)} \qquad (T^m = [-\pi, \pi]^m), \tag{6.22}$$

where C does not depend on l.

Proof. We shall make the change of coordinates $x_j = \varphi_j - \pi/2$. Then $\cos\varphi_j = \cos(x_j + \pi/2) = -\sin x_j$ and $\chi(x)$ vanishes when $x_j = 0$, $x_j = \pi$. Obviously, it is sufficient to estimate $|\partial^\alpha u|$ in terms of $|\partial^\beta(\chi u)|$ when $|x_l| \leqslant \pi/2$, since we can reduce the estimate to this case using a translation by π. We will assume

$$K = \{x \in [-\pi, \pi]^m : |x_j| \leqq \pi/2\}.$$

Suppose $f(x) = 0$ when $x_j = 0$. Then

$$\|f/\sin x_j\|_{C^l(K)} \leqq C(l+1)\|f\|_{C^{l+1}(K)}, \tag{6.23}$$

where C does not depend on f and l. We shall prove (6.23) when j=1. Obviously,

$$f(x)=x_1\int_0^1 \partial_1 u(\theta x_1, x_2, \ldots, x_m)\,d\theta = x_1 f_1(x). \tag{6.24}$$

Obviously,

$$f(x)/\sin x_1 = x_1/\sin x_1 f_1(x).$$

According to the Leibnitz formula,

$$\partial^\alpha\left(\frac{f(x)}{\sin x_1}\right)=\sum_{k=0}^{\alpha_1}\partial_1^k\left(\frac{x_1}{\sin x_1}\right)\frac{\alpha_1!}{(\alpha_1-k)!\,k!}\partial_1^{\alpha_1-k}\partial^{\alpha'}f_1(x), \tag{6.25}$$

where $\partial^{\alpha'}$ contains only ∂_2, ..., ∂_m. Since the function $x_1/\sin x_1$ is analytic in the domain $|\mathrm{Re}\,x_1| < \pi/2$ and $\pi/2 > 3/2$, then

$$\left|\partial_1^k(x_1/\sin x_1)\right| \leqq C_1(2/3)^k k!. \tag{6.26}$$

By virtue of the definition of f_1 in (6.24) we obtain:

$$\left|\partial_1^{\alpha_1-k}\partial^{\alpha'}f_1(x)\right| \leqq \sup_\theta\left|\partial_1^{\alpha_1-k+1}\partial^{\alpha'}f(\theta x_1, x_2, \ldots, x_k)\right|. \tag{6.27}$$

Using (6.26) and (6.27), from (6.25) we obtain:

$$\left\|\frac{f}{\sin x_1}\right\|_{C^{|\alpha|}(K)} \leqq (|\alpha|+1)C_1\sum_{k=0}^{|\alpha|}\left(\tfrac{2}{3}\right)^k\|f\|_{C^{|\alpha|+1}(K)}. \tag{6.28}$$

Obviously, (6.23) follows hence when j=1. The case $j \neq 1$ reduces to that considered.

Since χu vanishes when $x_j = 0$, $j = 1, \ldots, m$, we obtain estimate (6.22) by successively applying estimate (6.23) to $f = u_1 = \sin x_1 u$, $u_2 = \sin x_2 u_1$ etc.

Lemma 6.4. Suppose

$$f \in C^l(\Omega_1),\ \Omega_1 \subset \mathbb{R}^m,\ \varphi: \Omega_2 \to \Omega_1,\quad \Omega_2 \subset \mathbb{R}^m_x,\ \varphi \in C^l(\Omega_2).$$

Suppose $\alpha \in \mathbb{Z}^m_+$, $1 \leqslant |\alpha| \leqslant l$. Then

$$\frac{1}{\alpha!}\partial^\alpha_x f(\varphi(x)) = \sum_{\beta \leqq \alpha} \frac{1}{\beta!}\partial^\beta_\varphi f(\varphi) \quad \times \sum_{\sigma(\beta,\gamma)=\alpha} \prod_{i=1}^{m} \prod_{j=1}^{\beta_i} \frac{1}{\gamma_{ij}!}\partial^{\gamma_{ij}}_x \varphi_i(x), \tag{6.29}$$

$$\sigma(\beta,\gamma) = \gamma_{11} + \ldots + \gamma_{1\beta_1} + \ldots + \gamma_{m1} + \ldots + \gamma_{m\beta_m}$$

(if $\beta_i = 0$, the corresponding factor equals unity, if $\beta_i \neq 0$ then $|\gamma_{ij}| \geqslant 1$).

Proof. In order to derive (6.29), it is sufficient to consider the case when f, φ are analytic, $\varphi(0) = 0$, and to derive (6.29) at the point $x=0$. We shall expand $f(\varphi)$ in a series in powers of φ using the Taylor formula:

$$f(\varphi) = \sum \frac{1}{\beta!}\partial^\beta_\varphi f(0)\varphi^\beta. \tag{6.30}$$

We shall substitute the expansion $\varphi(x)$ into this formula instead of $\varphi = \varphi_1, \ldots, \varphi_m$ in powers of x. Equating the coefficients when x^α in the resulting expansion and in the expansion $f(\varphi(x))$ using Taylor's formula, we obtain (6.29).

Remark 6.3. The number α_l can be represented as the sum of $|\beta|$ nonnegative integer numbers in no more than

$$(\alpha_l + |\beta|)!/(\alpha_l!\,|\beta|!) \leqq 2^{\alpha_l + |\beta|}$$

different ways. Therefore the sum over $\sigma(\beta, \gamma) = \alpha$ in Formula (6.29) contains nor more than

$$2^{|\alpha| + m|\beta|} \leqq 2^{|\alpha|(m+1)}$$

terms. In a similar way, the sum over $|\beta| < |\alpha|$ contains not more than $2^{|\alpha|(m+1)}$ terms.

Lemma 6.5. Suppose the function φ in Lemma 6.4 is analytic on $\overline{\Omega}_2$. Then

$$\|f(\varphi(x))\|_{C^l(\Omega_2)} \leqq C_0 C_1^l \|f(\varphi)\|_{C^l(\Omega_1)}, \tag{6.31}$$

where C_0, C_1 do not depend on f and l.

Proof. Since $\varphi(x)$ is analytic on $\overline{\Omega}_2$, then

$$|\partial^\gamma \varphi_i / \gamma!| \leqq C R^{|\gamma|},$$

where we can assume $R > 1$. Estimating the derivatives of φ_i on the right-hand side of (6.29) in this way, and using Remark 6.3, from (6 .29) we obtain:

$$|\partial^\alpha f(\varphi(x))| / \alpha! \leqq C 2^{2|\alpha|(m+1)} |f(\varphi)|_{C^{|\alpha|}(\Omega_1)} R^{|\alpha|}.$$

Hence follows (6.31).

Lemma 6.6. Suppose $X_h(\varphi) = X(\varphi)$ is determined by (6.6), where $x_{0i} = 0$, $h = 1$. Suppose

$$X_h^* f(\varphi) \in C^{2l}(T^m).$$

Then $f(x) \in C^l(K)$, where $K = K_h(x_0)$, $h = 1$, $x_0 = 0$. At the same time

$$\|f(x)\|_{C^l(K)} \leqq CC_1^l \|(X^*f)(\varphi)\|_{C^{2l}(T^m)}, \tag{6.32}$$

where C and C_1 do not depend on f and l.

Proof. In order to prove (6.32), it is sufficient to prove the analogous inequality, in which

$$K = K_1 = \{x \in \mathbb{R}^m : 0 < x_i < 1, \forall i\} \qquad T^m = T_0^m = \{\varphi : 0 < \varphi_j < \pi/2, \forall j\}.$$

Indeed, the cube $K = [-1, 1]^m$ divides into 2^m cubes of the form $K_1 = [0, 1]^m$, and having proved (6.32) for one such cube K_1, we also obtain a statement for all the remainder, using the change of the sign of x_j. Note that the change of sign of x_j does not affect the form (6.32). We shall introduce the coordinates $y_j = \pi/2 - \varphi_j$ to $T_0{}^m$. In these coordinates the mapping $X: T_0{}^m \to K$ takes the form

$$x_j = \cos y_j, \qquad j = 1, \ldots, m, \tag{6.33}$$

and is the homeomorphism between $T_0{}^m$ and K_1. However the inverse mapping has an unbounded derivative as $x_j \to 1$. We will assume

$$T_1^m = \{x \in \mathbb{R}^m : 0 < z_i < \pi^2/4, \forall i\}, \qquad y_j(x) = \sqrt{z_j}. \tag{6.34}$$

Since the function $\cos\sqrt{z}$ is analytic (this is obvious from its series expansion) and its derivative is nonzero when $0 \leqslant z < \pi^2$, the mapping

$$X_1 : z \to x, \qquad x_j = \cos\sqrt{z_j} \tag{6.35}$$

is an analytic diffeomorphism $T_1{}^m$ to K_1. Using $z = z(x)$ to denote the mapping which is inverse to X_1,

$$f(x)=f(X_1(z(x))=f_1(z(x)), \tag{6.36}$$

we obtain that by virtue of Lemma 6.5

$$\|f(x)\|_{C^l(K_1)}\leqq C_0C_1^l\|f_1(z)\|_{C^l(T_0^m)}. \tag{6.37}$$

Therefore, in order to prove (6.32), it is sufficient to estimate

$$f_1(z(x))=f(X(\sqrt{z_1},\ldots,\sqrt{z_m}))$$

in terms of $f(X(z_1, \ldots, z_m))$, i.e. to obtain the inequality

$$\|v(\sqrt{z_1},\ldots,\sqrt{z_m})\|_{C^l(T_1^m)}\leqq C_0'C_2^l\|v(z_1,\ldots,z_m)\|_{C^{2l}(T_0^m)} \tag{6.38}$$

for the functions $v(y)$ of the form

$$v(y)=f(\cos y_1,\ldots,\cos y_m).$$

In order to derive (6.38), we shall prove a lemma.

Lemma 6.7. Suppose

$$T_{00}=\{0<y_i<r_i, i=1,\ldots,m\},\ v(y)\in C^n(T_{00}),\ v(y)$$

$v(y)$ extends up to the function of the class C^n, which is even with respect to the variable y_j, $-r_i < y_j < r_j$. We will assume

$$u(x)=v(x_1,\ldots,x_{j-1},\sqrt{x_j},x_{j+1},\ldots,x_m).$$

Then

$$u(x)\in C^n(T_{10}),\ T_{10}=\{0<x_i<r_i(i\neq j), 0<x_j<r_j^2\}.$$

When $2k + |\alpha| < \eta$ the following inequality holds

$$\sup_{T_{10}} |\partial_j^k \partial^\alpha u|/(k!\alpha!) \leqq C_0 C_1^{2k+|\alpha|} \max_{s \leqq 2k, |\beta| \leqq \alpha} \sup_{y \in T_{00}} |\partial_j^s \partial^\beta v(y)|/(s!\beta!), \tag{6.39}$$

where $2k + \alpha < \eta$, ∂^α and ∂^β do not contain ∂_j (with respect to x and y respectively).

Proof. To simplify the notation we will assume j=1. We shall fix

$$x = y' = (y_2, \ldots, y_m) = (x_2, \ldots, x_m).$$

We shall fix $k > 1$, $k \in \mathbb{N}$. We shall expand $v(y_1, y')$ using Taylor's formula

$$v(y_1, y') = v_k(y_1, y') + v_{0k}(y_1, y'),$$
$$v_k(y_1, y') = \sum_{i=0}^{2k} (\partial_1^i v(0, y')) y_1^i / i!. \tag{6.40}$$

Since $v(y_1, y')$ is an even function with respect to y_1, this polynomial contains only even powers y_1. Substituting

$$y_1 = x_1^{1/2},\ x_j = y_j \qquad (j \neq 1),$$

we obtain:

$$u_k(x_1, x') = \sum_{i=0}^{k} \partial_1^{2i} v(0, x') x_1^i / (2i)!. \tag{6.41}$$

It is obvious that (6.39) holds for $u = u_k$ whilst C_0 and C_1 do not depend on k and v. We shall now estimate the derivatives of the residual term $v_{0k} = v - v_k$. Obviously, v_{0k} has a zero of order $2k$ when $y_1 = 0$. Therefore, when $l < 2k$

$$\partial_1^l v_{0k}(y) = y_1^{2k-l} \int_0^1 \int_0^{\theta_1} \cdots \int_0^{\theta_{2k-l-1}} \partial_1^{2k} v_{0k}(\theta_{2k-l} y_1, y')\, d\theta_{2k-l} \cdots d\theta_1. \tag{6.42}$$

We shall estimate

$$\partial_1^k \partial^\alpha u_{0k}(x)(y_1 = x_1^{1/2})$$

using (6.29). Obviously,

$$|\partial_1^\gamma x_1^{1/2}|/\gamma! \leqq |x_1|^{1/2-\gamma}.$$

It is also obvious that if $\gamma_1 + \ldots + \gamma_l = k$, then $1/2 - \gamma_1 + \ldots + 1/2 - \gamma_l = l/2 - k$. Therefore, assuming

$$\varphi_1(x) = \sqrt{x}, \quad f(x) = u_{0k}(x) = v_{0k}(y(x))$$

in (6.29), from (6.29) we obtain:

$$\frac{|\partial_1^k \partial^\alpha u_{0k}(x)|}{k!\, \alpha!} \leqq 2^{(m+1)(|\alpha|+k)} \sum_{\beta \leqq \alpha, l \leqq k} |\partial_1^l \partial^\beta v_{0k}(y)| \frac{|x_1|^{l/2-k}}{l!\, \beta!}, \tag{6.43}$$

where $\partial^\beta = \partial_{y'}^\beta$. By virtue of (6.42)

$$|\partial_1^l \partial^\beta v_{0k}(y)| \leqq |y_1|^{2k-l} \sup_{0 \leqq \theta \leqq 1} |\partial_1^{2k} \partial^\beta v_{0k}(\theta y_1, y')|/(2k-l)!$$

Using this inequality to estimate the right-hand side of (6.43), bearing in mind that

$$|y_1|^{2k-l} = |x_1|^{k-l/2}$$

and bearing in mind Remark 6.3, we obtain:

$$\sup_{x \in T_{10}} \left| \frac{\partial_1^k \partial^\alpha u_{0k}(x)}{k!\, \alpha!} \right| \leqq 2^{2(m+1)(|\alpha|+k)} \max_{\beta \leqq \alpha} \sup_{y \in T_{00}} \frac{|\partial_1^{2k} \partial^\beta v_{0k}(y)|}{\beta!\,(2k)!} \frac{(2k)!}{l!\,(2k-l)!}.$$

Hence follows (6.39) for $u = u_{0k}$. Since $u = u_k + u_{0k}$, then (6.39) is proved, and Lemma 6.7 is proved.

We shall now complete the proof of Lemma 6.6. For this it is sufficient to obtain estimate (6.38). It is obvious that

$$v(y)=f(\cos y_1,\dots,\cos y_m),\quad f\in C^{2l}(T_0^m)$$

satisfies the conditions of Lemma 6.6. Replacing $y_1=\sqrt{x_1}$, we obtain estimate (6.39) for $v^1=v(\sqrt{x},y')$. Successively applying Lemma 6.7 when $j = 2, \dots, m$ to the functions

$$v^1(x_1,y')=v(\sqrt{x_1},y_2,\dots,y_m),\ v^2=v^1(x_1,\sqrt{x_2},y_3,\dots,y_m)$$

etc., we obtain estimate (6.38) using (6.39). Lemma 6.6 is proved.

Proof of Theorem 1.1′ . Suppose $x_0 \in \partial\Omega$, $Y : K \to \mathcal{O}(x_0)$ is such a mapping, as in Condition 1.2. When deriving (6.15) and (6.17) according to Remarks 6.1 and 6.2 it is not necessary to consider Y an identical mapping. Therefore the estimates $\|y_N\|_s$, obtained in proving Theorem 6.1, hold for $y_N = y_N(x)$ and for $y_N = y_N(Y(x))$. (Note that these functions are analytic on $\overline{\Omega}$ and therefore $\chi X_h{}^* y_N$ are also analytic.) It is proved in Theorem 6.1 that series (6.18) converges in $W_2{}^s(T^m)$ when $s < s_0$. Using Sobolev's imbedding theorem, we obtain that this series converges in $C^l(T^m)$ when $l < s_0 - m/2$. According to Lemma 6.3, the series from $X_h{}^* y_N$ converges in $C^{l-m}(T^m)$ to $X^* w$. It follows from Lemma 6.6 that

$$w=w(Y(x))\in C^{l_1}(K_1)$$

when $l \leqslant (l - m)/2$. Using Lemma 6.5, we obtain that

$$w(x)\in C^{l_1}(Y(K_1)).$$

Since the compactum $K \subset \Omega \cap \Omega_1$ is covered by a finite number of sets of the form $Y(K_1)$, the statement of Theorem 1.1′ follows hence.

Remark 6.4. If the point $x_0 \in \partial\Omega$ is such that $\partial\Omega$ near x_0 does not have singularities, i.e. it is a smooth (analytic) surface, we can define the mapping $Y : K_1 \to \Omega$, such that $x_0 = Y(y_0)$, where y_0 lies at the centre of the bound of the cube K_1 with the equation $y_1 = 0$. It is obvious from the proof of Lemmas 6.3, 6.6 and 6.7, that in this case

$$u(x) \in C^{l_2}(\mathcal{O}(x_0))$$

when $2l_2 < s_0 - m/2 - 1$, if $\mathcal{O}(x_0)$ is small. Note that the derivatives $\partial^\alpha u$ with respect to the directions that are tangential to $\partial\Omega$ are continuous, if $|\alpha| < s_0 - m/2 - 1$.

§7. Proof of Theorems 1.2 and 1.2′ in the general case.

Theorem 1.2 is a corollary of the following theorem.

Theorem 7.1. Suppose the operator B satisfies Condition 1.1,

$$f \in \mathscr{D}_0,\ b \leqq b_0(B),\ A = B - bI,\ R_0 = R_0(A, f) > 0.$$

Suppose

$$g \in \mathscr{A}(\mathscr{I}_\infty^+),\ u(x) = g(A)f(x).$$

Then $\forall x_0 \in \Omega\ \forall \varepsilon > 0$ there exist a neighbourhood Ω_1 of the point x_0 and the constants q, C_1 and C_2, such that

$$\|\partial^\alpha u(x)|_{\Omega_1} \leqslant C_1 M_0 C_3^{|\alpha|} |\alpha|!\, g^{**}(|\alpha|/(r_0 - \varepsilon)), \tag{7.1}$$

where $g^{**}(s)$ is the majorant g, determined by (1.3′), $r_0 = 2R_0/\pi$.

To prove Theorem 7.1 and Theorem 1.2, we need certain lemmas on the effect of an analytic change of variable and of multiplication by an analytic function on the rate of increase of the norms of the derivatives.

Note, first of all, that with a change of variable the operators ∂_j become second-order differential operators with variable coefficients. Let us therefore consider the set $B_1, \ldots, B_m$ of first-order differential operators

$$B_j u(y) = \sum_{j=1}^{m} a_{ij}(y)\, \partial_i u(y), \qquad \partial_i = \partial/\partial y_i, \tag{7.2}$$

where $a_{ij}(y)$ are analytic functions in the compact domain $\bar{\Omega} \subset \mathbb{R}^m$. The following estimate therefore holds:

$$|\partial^\alpha a_{ij}(y)| \leqq C_1 R_1^{|\alpha|} |\alpha|!, \quad \forall \alpha \in \mathbb{Z}_+^m, \qquad y \in \bar{\Omega}. \tag{7.3}$$

Lemma 7.1. Suppose the function $a_{ij} = a$ satisfies (7.3). Suppose the function $f(y)$ is infinitely differentiable on $\bar{\Omega}$ and satisfies the condition

$$\|\partial^\alpha f\|_\Omega \leqq C_f R_f^{|\alpha| + l_0} (|\alpha| + l_0)!\, g(|\alpha| + l_0) \tag{7.4}$$

$\forall \alpha \in \mathbb{Z}_+^m$ where $g(t)$ is a function that monotonically increases when $t > 0$,

$$R_f \geqq 1,\ R_f > R_1,\ \|\ \|_\Omega$$

is defined in (1.5), and $l_0 \in \mathbb{Z}_+$. Then

$$\|\partial^\alpha (af)\|_\Omega \leqq C_f C_1 (1 - R_1/R_f)^{-1} g(|\alpha| + l_0) R_f^{|\alpha| + l_0} (|\alpha| + l_0)! \tag{7.5}$$

Proof. In the expansion $\partial^{\alpha}(af)$ using the Leibnitz formula the number of terms of the form $\partial^{\beta}a\partial^{\gamma}f$ with $|\beta| = k$ equals $|\alpha|!/(k!(|\alpha| - k)!)$. Therefore

$$\|\partial^{\alpha}(af)\|_{\Omega} \leqq \sum_{k=0}^{|\alpha|} \frac{|\alpha|!}{k!(|\alpha|-k)!} \max_{|\beta|=k} \sup_{y\in\Omega} |\partial^{\beta}a(y)| \max_{|\gamma|=|\alpha|-k} \left(\int_{\Omega} |\partial^{\gamma}f(y)|^2\,dy\right)^{1/2}. \tag{7.6}$$

Using estimates (7.3) and (7.4) and the monotony of g, from (7.6) we obtain:

$$\|\partial^{\alpha}(af)\|_{\Omega} \leqq C_f C_1 \sum_{k=0}^{|\alpha|} \frac{|\alpha|!}{(|\alpha|-k)!} R_1^k R_f^{|\alpha|-k+l_0} \times (|\alpha|-k+l_0)!\, g(|\alpha|-k+l_0). \tag{7.7}$$

Since

$$(|\alpha|+l_0-k)!/(|\alpha|-k)! \leqq (|\alpha|+l_0)!/|\alpha|!$$

and $g(|\alpha| + l_0 - k) < g(|\alpha| + l_0)$, then, making an upper estimate of the right-hand side of (7.7) we obtain

$$\|\partial^{\alpha}(af)\|_{\Omega} \leqq C_f C_1 (|\alpha|+l_0)!\, g(|\alpha|+l_0) R_f^{|\alpha|+l_0} \Sigma (R_1/R_f)^k$$

Hence we directly obtain (7.5).

Lemma 7.2. Suppose the operators B_j are determined using Formula (7.2), where a_{ij} satisfy condition (7.3). Suppose the function f satisfies condition (7.4), where $R_f > R_1$, $l_0 = 0$. Then $R_2 > 0$ exists, such that $\forall \alpha \in \mathbb{Z}_+^m\ \forall k \in \mathbb{Z}_+$

$$\left\|\partial^{\alpha} \prod_{l=1}^{k} B_{j_l} f\right\|_{\Omega} \leqq C_f R_f^{|\alpha|+k} R_2^k (|\alpha|+k)!\, g(|\alpha|+k). \tag{7.8}$$

Proof. We shall carry out induction with respect to k. When $k=0$ (7.8) directly follows from (7.4). Suppose (7.8) holds when $k < n - 1$. We shall prove (7.8) when $k=n$. Since

$$f_n=\prod_{l=1}^{n} B_{j_l}f=B_{j_n}f_{n-1},\quad f_{n-1}=\prod_{l=1}^{n-1} B_{j_l}f,$$

then

$$\|\partial^\alpha f_n\|_\Omega\leqq\sum_{i=1}^{m}\|\partial^\alpha a_{ij}\partial_i f_{n-1}\|_\Omega\qquad (j=j_n). \tag{7.9}$$

Estimate (7.8) for $\partial^\alpha f_{n-1}$ has the form (7.4), where $l_0 = n - 1$. Therefore $\partial_i f_{n-1}$ satisfies (7.4), where $l_0 = n$, and we have the constant $C_f R_2^{n-1} = C_{f1}$ instead of C_f. According to Lemma 7.1 $a_{ji}\partial_i f_{n-1}$ satisfies (7.5), i.e.

$$\|\partial^\alpha a_{ji}\partial_i f_{n-1}\|_\Omega\leqq C_f R_2^{n-1}C_1(1-R_1/R_f)^{-1}(|\alpha|+n)!\,g(|\alpha|+n). \tag{7.10}$$

Bearing in mind that there are m terms on the right-hand side of Formula (7.9), we obtain when $R_2 > C_1(1 - R_1/R_f)^{-1}m$ that (7.8) follows from (7.9) and (7.10) for $k=n$.

Lemma 7.3. Suppose

$$\Omega_0,\Omega_1\subset\mathbb{R}^m,\ X:\Omega_0\to\Omega_1,$$

X maps $\overline{\Omega}_0$ one-to-one into $\overline{\Omega}_1$, $\det X(y) \neq 0\ \forall\ y \in \Omega_0$, and X is analytic on $\overline{\Omega}_0$. Suppose $f \in C^\infty(\overline{\Omega}_0)$ satisfies estimate (7.4), where $\Omega = \Omega_0$, $l_0 = 0$. Then the constants C_1, C_2, R_1, depending only on X, exist, such that if $R' = \max(R_f, 2R_1)$, where R_f is a number from (7.4), then $\forall\,\beta \in \mathbb{Z}_0^m$

$$\|\partial_x^\beta X^{-1*}f\|_{\Omega_1}\leqq C_f C_1(C_2R')^{|\beta|}g(|\beta|)|\beta|!. \tag{7.11}$$

Here

$$(X^{-1*}f)(x)=f(X^{-1}(x)),\quad x\in\Omega_1,\quad g(t)$$

and C_f are the same as in (7.4).

Proof. We will assume $f_1(x) = f(X^{-1}(x))$. Obviously,

$$\partial f_1/\partial x_j(x) = \Sigma\, \partial f/\partial y_i(y)\, \partial y_i/\partial x_j(x), \quad y = X^{-1}(x). \tag{7.12}$$

We shall put $a_{ij}(y) = \partial y_i/\partial x_j(X(y))$. By virtue of (7.12) the differentiation operator with respect to x on $C^{\infty}(\Omega_1)$ is represented in the form

$$\partial/\partial x_j f_1 = X^{*-1} B_j X^* f_1, \qquad \forall\, f_1 \in C^{\infty}(\Omega_1), \tag{7.13}$$

where

$$X^*: f(x) \to f(x(y)), \qquad x = X(y),$$

and the operator B_j is determined by Formula (7,2). According to Formula (7.13)

$$\partial^{\beta} f_1 = X^{*-1} B^{\beta} f_1, \qquad f = X^* f_1. \tag{7.14}$$

Since

$$|\det\{\partial x_j/\partial y_i\}| \geqq C^{-1} > 0,$$

then

$$\|X^{*-1} v\|_{\Omega_1} \leqq C^{1/2} \|v\|_{\Omega_0}, \qquad \forall\, v \in C(\bar{\Omega}_0). \tag{7.15}$$

Therefore

$$\|\partial^{\beta} f_1\|_{\Omega_1} \leqq C^{1/2} \left\| \prod_{l=1}^{|\beta|} B_{j_l} f \right\|_{\Omega_0}. \tag{7.16}$$

Since X is analytic on $\overline{\Omega}_0$, and $\det X'(y) \neq 0$, then, according to the theorem on the implicit function, the function $y = y(x)$, which is defined by X^{-1}, is analytic on $\overline{\Omega}_1$. Therefore $a_{ij} = \partial y_i/\partial x_j(X(y))$ are analytic on Ω_0. Consequently, estimate (7.3) holds and Lemma 7.2 is applicable. Using (7.8), where $\alpha = 0$, $k = |\beta|$, from (7.16) we obtain inequality (7.11), where $C_1 = C^{1/2}$, $C_2 = R_2$.

Proof of Theorem 7.1. According to Lemma 2.1 $u = g(A)f = u_0 + w$, where w is represented in the form of series (2.5). It is sufficient to estimate w. If h is fairly small, then $K_h(x_0) \subset \Omega$ and $\chi X_h^* w$ is represented in the form of series (6.18). At the same time estimate (6.17) holds. According to this estimate inequality 2.37 holds for $y_N^1 = \chi X_h^* y_N$. By virtue of Lemma 2.1 estimate (2.5) holds, and according to Lemma 6.1 estimate (2.5) holds for y_N with $\| \ \| = \| \ \|_0$. According to Lemma 3.4, where $\omega = 0$

$$\|\chi X_h^* w\|_s \leqq C_1 C_2^s s^s g^{**}(s/(r_0-\varepsilon)), \tag{7.17}$$

By virtue of(2.15), (2.16) and (2.17), if $v \in (T^m)$, then

$$\int_{[-\pi,\pi]^m} |\partial^\beta v(y)|^2\,dy \leqq C\|v\|_\beta^2. \tag{7.18}$$

Note that the mapping $X = X_h$, determined by (6.6), where $\varphi = y$ is one-to-one when $|y| < \pi/6$ and maps the cube $[-\pi/6, \pi/6]^m = \overline{\Omega}_0$ into the cube $K_{h/2}(x_0) = \Omega_1$. Estimating $|\beta|^{|\beta|}$ in terms of $|\beta|!C^{|\beta|}$, from (7.17) and (7.18) we obtain:

$$\|\partial^\beta \chi X_h^* w\|_{\Omega_0} \leqq C C_3^s |\beta|!\, g^{**}(|\beta|/(r_0-\varepsilon)). \tag{7.19}$$

Since the function $1/\chi$ is analytic on $\overline{\Omega}_0$, the estimate for

$$\|\partial^{\beta} X_h^* w\|_{\Omega_0},$$

which is analogous to (7.19) follows from Lemma 7.1 and (7.19), i.e. we can assume $\chi = 1$ in (7.19).

We shall now use Lemma 7.3, where we will assume $f = X_h^* w$, $X = X_h$, Ω_0 and Ω_1 are defined above. By virtue of (7.19), where $\chi = 1$, f satisfies (7.4), where $g(t) = g^{**}(t/(r_0 - \varepsilon))$, $\Omega = \Omega_0$. Therefore, estimate (7.11) holds by virtue of Lemma 7.3. Since

$$X^{-1*} f = X_h^{-1*} X_h^* w = w,$$

(7.1) follows from (7.11), where $g(t) = g^{**}(t/(r_0 - \varepsilon))$, and Theorem 7.1 is proved.

Proof of Theorem 7.2. Suppose $Y : K \to \Omega_1$ is a mapping from condition 1.2, and the function w is defined by Formula (2.3). We will assume $w_1(x) = w(Y(x))$. According to Remark 6.2 estimate (7.17) holds for $w_2 = \chi X_h^* w_1$. From (6.32) and (6.22) we obtain the estimate

$$\|w_1\|_{C^l(K)} \leqq C_3 C_4^l \|X_h^* w_1\|_{C^{2l}(T^m)} \leqq C_5 C_6^l \|\chi X_h^* w_1\|_{C^{2l+m}(T^m)}. \tag{7.20}$$

If $\lambda > m/2$, then by virtue of Sobolev's embedding theorem, (2.15) and (2.17)

$$|\partial^{\alpha} w_2(x)| \leqq C_1 \|\partial^{\alpha} w_2\|_{\lambda} \leqq C_1 \|w_2\|_{\lambda + |\alpha|}. \tag{7.21}$$

Using (7.21) and bearing in mind (6.20), from (7.20) we obtain

$$\|w_1\|_{C^l(K)} \leqq C_5 C_6^l C_1 \max_{k \leqq 2l+m} (1/k! \|w_2\|_{\lambda+k}). \tag{7.22}$$

Using (7.17), when $k \leqslant p$, $p = \lambda + 2l + m$ we obtain

$$1/k!\|w_2\|_{\lambda+k} \leqq CC_2^{\lambda+k}(\lambda+k)^{\lambda+k}/k!\,g^{**}((\lambda+k)/(r_0-\varepsilon))$$
$$\leqq C'C_2^p g^{**}((\lambda+p)/(r_0-\varepsilon)) \max_{k\leqq p}(\lambda+k)^{\lambda+k}/k! \tag{7.23}$$
$$\leqq C''C_7^p g^{**}((\lambda+p)/(r_0-\varepsilon)).$$

From (7.22) and (7.23) we obtain:

$$\|w_1\|_{C^l(K)} \leqq C_9 C_8^{2l} g^{**}((2l+m+\lambda)/(r_0-\varepsilon)). \tag{7.24}$$

Since

$$w(y)=w_1(Y^{-1}(y)), \qquad y\in\Omega_1\cap Y(K),$$

and Y^{-1} is analytic, then using Lemma 6.5 we obtain inequality (1.6′) from (7.24).

§8. The membership of the solution of the equation $Bu = f$ in Nikol'skii's space.

It follows from Theorem 1.3 that if $b_0(B) > b = \rho^2$, then

$$B^{-1}f\in W_{2\,\mathrm{loc}}^{2R_0\rho/\pi-\varepsilon}(\Omega),\ \forall\,\varepsilon>0.$$

In this paragraph, when $\Omega = T^m$, we will consider the question of whether we can assume $\varepsilon = 0$. The example from paragraph 4 shows that this is impossible if, instead of the Sobolev space $W_2^s(T^m)$, we take the wider Besov space $B_{2,\Theta}^s(T^m)$. However, in this paragraph we will prove that, by replacing W_2^s with the Nikol'skii space $H_2^s(T^m) = B_{2,\infty}^s(T^m)$ we are able to prove the inclusion $B^{-1}f \in H_2^s(T^m)$ in the limiting case $s = 2R_0\rho/\pi$. At the same time it is required to impose a limit on the increase $\|\mathrm{ch}(t\sqrt{A})f\|$ when $t \to R_0(A, f)$.

Namely, we shall require that

$$\|\operatorname{ch} t\sqrt{A} f\|^2 \leqq CM_p(R_0 - t), \quad M_p(\tau) = \exp(\exp(\tau^{-p})), \tag{8.1}$$

where $p < 1$.

Theorem 8.1. Suppose B is an operator from Example 4.3.1, determined by Formula (4.3.1), where the elements of the matrices $a_{ij}(x)$ are trigonometric polynomials. Suppose $f \in A(T^m)$ is also a trigonometric polynomial. Suppose B is invertible, and the number $b = \rho^2 > 0$ is such that only a finite number of points of the spectrum of the operator B, $A = B - bI$, is contained on the semi-axis $]-\infty, b]$. Suppose condition (8.1) holds. Then

$$B^{-1} f \in H_2^{\sigma_0}(T^m), \qquad \sigma_0 = 2R_0(B, f)/\pi. \tag{8.2}$$

A theorem which is analogous to Theorem 8.1 holds for operators with polynomial coefficients in the domain $\Omega \subset \mathbb{R}^m$.

Theorem 8.2. Suppose B is an operator from Example 4.3.2 or 4.3.3, whilst the coefficients B and the weight G are algebraic polynomials. Suppose $f \in A(\overline{\Omega})$ is also an algebraic polynomial. Suppose condition (8.1) holds, the operator B is invertible, and the numbers b and σ_0 are the same as in Theorem 8.1. Then:

$$B^{-1} f \in H_{2\,\mathrm{loc}}^{\sigma_0}(\Omega). \tag{8.3}$$

Theorems 8.1 and 8.2 are a corollary of the following theorem on equations in a Hilbert space.

Theorem 8.3. Supopse B is a self-adjoint invertible operator in a Hilbert space, whilsst there is only a finite number of points of the spectrum of the operator B on the semiaxis $]-\infty, b]$. Suppose f satisfies condition (8.1), $\sigma_0 = 2R_0\rho/\pi$. Then the sequence of polynomials $P_n(B)$, $n \in \mathbb{N}$ exists, such that

$$\|P_n(B) f - B^{-1} f\| \leqq Cn^{-\sigma_0}, \qquad \forall n \in \mathbb{N}. \tag{8.4}$$

We shall prove Theorem 8.3 later, but shall now derive Theorem 8.1 and Theorem 8.2 from Theorem 8.3.

Proof of Theorem 8.1. Suppose the degrees of the polynomial-coefficients a_{ij} in (4.3.1) and the function f do not exceed the number q. Then $B^{j}f(x)$ is a polynomial of a degree no higher than $qj + q = q(j + 1)$, and $P_n(B)f(x) = G(x)$ is a trigonometric polynomial of a degree no higher than $q(n + 1) : G(x) = G_{q(n+1)}(x)$. The estimate of the rate of approximation using polynomials of the function $u(x) = B^{-1}f(x)$ to $L_2(T^{m})$ then follows from (8.4);

$$\|u(x)-G_{q(n+1)}(x)\|_0 \leqq Cn^{-\sigma_0} \leqq C_1(q(n+2))^{-\sigma_0}, \tag{8.5}$$

$\forall n \in \mathbb{N}$. Hence it follows that the polynomial $G_N(x)$ of a degree no higher than N, will be obtained for any sufficiently large $N \in \mathbb{N}$, such that

$$\|u(x)-G_N(x)\|_0 \leqq C_2 N^{-\sigma_0} \tag{8.6}$$

Indeed, for this it is sufficient to take $G_{q(n+1)}$, $q(n + 1) < N$, $q(n + 2) > N$ as G_N, and (8.6) holds by virtue of (8.5). (8.2) follows from estimate (8.6) and Bernshtein's multi-dimensional inverse theorem (see Nikol'skii [1]), and Theorem 8.1 is proved.

Proof of Theorem 8.2. As in the proof of Theorem 8.1, we obtain that (8.6) holds, where $G_N(x)$ are algebraic polynomials of a degree no higher than N. Suppose $x_0 \in \Omega$, and h is fairly small. Then the cube $K_h(x_0)$, determined by (6.5), belongs to Ω. From (8.6) and (6.9) we obtain

$$\|\chi X_h^* u-\chi X_h^* G_N\|_0 \leqq C_3 N^{-\sigma_0} \leqq C_4(N+m)^{-\sigma_0}. \tag{8.7}$$

Since, by virtue of (6.6), $X_h{}^*G_N$ is a trigonometric polynomial of the N-th degree, and χ, by virtue of (6.8), is a trigonometric polynomial of the m-th degree, it follows from (8.7) that $\chi X_h^* u \in H_2^{\sigma_0}(T^m)$. Hence it follows that

$$X_h^* u \in H_2^{\sigma_0}(\Omega_0), \qquad \Omega_0 = [-\pi/6, \pi/6]^m \tag{8.8}$$

since $\chi \neq 0$ on Ω_0. By virtue of the invariance of the Nikol'skii classes with respect to a smooth change of variable, from (8.8) we obtain:

$$u = X_h^{*-1} X_h^* u \in H_2^{\sigma_0}(\Omega_1), \quad \Omega_1 = X_h \Omega_0 = K_{h/2}(x_0).$$

Hence follows (8.3).

To prove Theorem 8.3, we need several lemmas which we shall prove below.

Lemma 8.1. Suppose $\mu(\lambda)$ is a nondecreasing function of bounded variation on the half-line $\mathbb{R}_+ = \{\lambda > 0\}$. Suppose

$$\int_0^\infty e^{2\sqrt{\lambda t}}\, d\mu(\lambda) \leq C M_p(R_0 - t), \tag{8.9}$$

where $M_p(\tau)$ is the same as in (8.1), $0 < p < 1$. Then $C > 0$ exists, such that

$$\int_0^\infty e^{2R_0\sqrt{\lambda}} e^{-C\sqrt{\lambda}/\ln^{q_0}\lambda}\, d\mu(\lambda) < \infty, \tag{8.10}$$

where $q_0 = 1/p$, R_0 is the same as in (8.1).

Proof. We will assume

$$\varphi(z) = \int_0^{R_0} e^{zt}[M_p(R_0 - t)]^{-1}\, dt. \tag{8.11}$$

Using condition (8.9) and Fubini's theorem, we obtain that

$$\int_0^\infty \varphi(2\sqrt{\lambda})\,d\mu(\lambda)=\int_0^{R_0}\int_0^\infty e^{2\sqrt{\lambda}t}[M_p(R_0-t)]^{-1}\,d(\mu(\lambda))\,dt\leqq\int_0^{R_0} C\,dt. \tag{8.12}$$

We shall make an upper estimate of $\varphi(z)$ for large z. Replacing t by $R_0 - t$ and using the definition of $M_p(\tau)$, from (8.11) we directly obtain that

$$\varphi(z)=e^{R_0z}\int_0^{R_0} e^{-zt}e^{-e^{t^{-p}}}dt=e^{R_0z}\int_0^{R_0} e^{-t(z+g(t))}\,dt, \tag{8.13}$$

where $g(t) = t^{-1}e^{t^{-p}}$. Suppose t_0 and t_1 are solutions of equations

$$g(t_0)=z,\quad g(t_1)=2z. \tag{8.14}$$

Since $g(t)$ decreases when $t > 0$, then $t_0 > t_1$. Since for large z t_0, $t_1 < R_0$, and

$$z \leqslant g(t) \leqslant 2z \text{ when } t_1 \leqslant t \leqslant t_0 \tag{8.15}$$

from (8.13) we obtain:

$$\varphi(z)\geqq e^{zR_0}\int_{t_1}^{t_0} e^{-3tz}\,dt\geqq e^{zR_0}(t_0-t_1)\,e^{-3t_0z}. \tag{8.16}$$

Obviously,

$$g(2(\ln z)^{-1/p})=\tfrac{1}{2}(\ln z)^{1/p}z^{p_1}\leqq z \qquad (p_1=2^{-p})$$

for large z. Therefore, by virtue of the decrease $g(t)$

$$t_0\leqq 2(\ln z)^{-1/p}=2\ln^{-q}z,(z\geqq z_1), \qquad q=1/p, \tag{8.17}$$

for fairly large z_1. Further, by virtue of (8.14) $t_0 - t_1 = -z/g'(t)$, for some $t \in [t_1, t_0]$. Since $-g'(t) = g(t)(t^{-1} + pt^{-1-p})$ then, bearing in mind (8.15),

$$t_0 - t_1 = (z/g(t))(t^{-1} + pt^{-1-p})^{-1}$$

$$\geqq \tfrac{1}{2}(t^{-1} + pt^{-1-p})^{-1} = t^{1+p}(2t^p + 2p).$$

Bearing in mind that t_0 and t_1 approach 0 as $z \to \infty$, and that $t_1 < t < t_0$, we obtain hence:

$$t_0 - t_1 \geqq C_1 t^{1+p} \geqq C_1 t_1^{1+p} \geqq C_2 t_1^2 \tag{8.18}$$

when $z > z_2$. From (8.14) and from the definition of $g(t)$ we obtain that $t_1 2z = e^{t_1^{-p}} > 1$. Therefore from (8.18) follows:

$$t_0 - t_1 \geqq C_2/(4z^2). \tag{8.19}$$

Using (8.19) and (8.17), we obtain from (8.16)

$$\varphi(z) \geqq C_3 e^{zR_0} z^{-2} e^{-6z \ln^{-q} z} \geqq C_4 e^{zR_0 - 7z \ln^{-q} z}, \tag{8.19'}$$

where $q = 1/p$, C_3, $C_4 > 0$, when $z > z_2$ for fairly large z_2. Since $\varphi(z) > C > 0$ when $2 < z < z_2$, and the right-hand side of (8.19') is bounded when $2 < z < z_2$, (8.19') holds for all $z > 2$ (with the other constant C_4). Substituting $z = 2\sqrt{\lambda}$ into (8.19'), we obtain:

$$\varphi(2\sqrt{\lambda}) \geqq C_4 \exp(2R_0\sqrt{\lambda} - 14\sqrt{\lambda}(\tfrac{1}{2}\ln\lambda + \ln 2)^{-q})$$

$$\geqq C_4 \exp(2R_0\sqrt{\lambda} - 14 \cdot 2^q \ln^{-q}\lambda\sqrt{\lambda}).$$

Hence and from (8.12) we obtain (8.10).

Lemma 8.2. Suppose $q > 0$. Then $C_0 > 0$ exist, such that when $z > 0$

$$\psi(z)=\prod_{j=9}^{\infty}\left(1+\frac{z^2}{(j\ln^q j)^2}\right)\geqq\tfrac{1}{2}e^{C_0 z/\ln^q(z+2)}. \tag{8.20}$$

Proof. The function

$$\ln(1+z^2/(j\ln^q j)^2)$$

decreases as j increases. Therefore

$$\begin{aligned}\ln\psi(z)&=\sum_{j=9}^{\infty}\ln\left(1+\frac{z^2}{(j\ln^q j)^2}\right)\geqq\int_9^{\infty}\ln\left(1+\frac{z^2}{t^2\ln^{2q}t}\right)dt\\&=z\int_{9/z}^{\infty}\ln\left(1+\frac{1}{x^2\ln^{2q}(zx)}\right)dx.\end{aligned} \tag{8.21}$$

Taking the integral not up to $+\infty$ but up to z in (8.21), and bearing in mind that when $x \leqslant z$ $(zx) \leqslant 2$ z, we obtain, making a lower estimate of the right-hand side of (8.21):

$$\ln\psi(z)\geqq z\int_{9/z}^{z}\ln\left(1+\frac{1}{4^q x^2\ln^{2q}z}\right)dx=\frac{z}{\ln^q z}\int_{9\ln^q z/z}^{z\ln^q z}\ln\left(1+\frac{1}{4^q t^2}\right)dt. \tag{8.22}$$

Obviously $9\ \ln^q z/z \leqslant C$ when $z \geqslant 1$, and $z\ln^q z \to \infty$ as $z \to \infty$, and therefore $z\ln^q z \geqslant C+1$ for fairly large z. Consequently, the integral on the right-hand side of (8.22) is bounded from below by the positive constant $C_0' > 0$, and from (8.22) we obtain (8.20) when $z \geqslant 2$. When $0 \leqslant z \leqslant 2$ the validity of (8.20) for some C_0 is obvious, since $\varphi(z) \geqslant 1$.

We shall use $\varkappa(j)$ to denote the integer-valued function of the natural argument $j \in \mathbb{N}$, $j \geqslant 9$:

$$\kappa(j)=\max\{\kappa\in\mathbb{N}: 2\kappa-1\leqq j\ln^q j\}, \tag{8.23}$$

where $q > 1$. Since when $t > 8$

$$(t\ln^q t)' = \ln^q t + (q-1)\ln^{q-1} t \geqq \ln t > 2, \tag{8.23'}$$

different $\varkappa(j)$ correspond to different integers $j > 9$, i.e. $\{\varkappa(j)\}$ is a subsequence of natural numbers whose terms are not repeated.

We will assume

$$\Phi(\lambda) = \prod_{k=0}^{\infty}\left(1+\frac{\sigma^2\lambda}{(2k-1)^2}\right)\prod_{j=9}^{\infty}\left(1+\frac{\sigma^2\lambda}{(2\kappa(j)-1)^2}\right)^{-1}. \tag{8.24}$$

Lemma 8.3. Suppose condition (8.10) holds, and suppose the number q in the definition (8.23) satisfies the condition $1 < q < q_0$, where q_0 is the same as in (8.10), $\sigma = 2R_0/\pi$. Suppose $Q(\lambda)$ is a polynomial. Then

$$\int_0^\infty \Phi^2(\lambda)Q^2(\lambda)\,d\mu(\lambda) < \infty. \tag{8.25}$$

Proof. Note that by virtue of (8.23)

$$2\kappa(j) - 1 \leqq j\ln^q j.$$

Therefore

$$\psi_1(z) = \prod_{j=9}^{\infty}\left(1+\frac{\sigma^2\lambda}{(\kappa(j)-1)^2}\right) \geqq \psi(\sigma\sqrt{\lambda}), \tag{8.26}$$

where $\psi(z)$ is determined by (8.20). Bearing in mind (3.11), from (8.24) and (8.26) we obtain:

$$\Phi(\lambda) \leqq \mathrm{ch}(R_0\sqrt{\lambda})/\psi(\sigma\sqrt{\lambda}). \tag{8.27}$$

Since $\mathrm{ch} z < e^z$ when $z > 0$, and bearing in mind estimate (8.20), from (8.27) we obtain

$$\int_2^\infty \Phi^2(\lambda)Q^2\, d\mu(\lambda) \leqq \int_2^\infty e^{2R_0\sqrt{\lambda}} e^{-C_1\sqrt{\lambda}/\ln^q(\sigma^2\lambda)} Q^2(\lambda)\, d\mu(\lambda).$$

Hence by virtue of (8.10) we obtain

$$\int_2^\infty \Phi^2(\lambda)Q^2(\lambda)\, d\mu(\lambda) \leqq C \sup_{\lambda \geqq 2} (\lambda^{2\nu} e^{C_0\sqrt{\lambda}/\ln^{q_0}\lambda - C_1\sqrt{\lambda}/\ln^q(\sigma^2\lambda)}), \tag{8.28}$$

where ν is the degree of the polynomial Q. Since $q < q_0$, then

$$2\nu \ln \lambda + C_0\sqrt{\lambda}/\ln^{q_0}\lambda - C_1\sqrt{\lambda}/\ln^q(\sigma^2\lambda + 2) < 0$$

for fairly large λ. Therefore the right-hand side of (8.28) is finite. Hence follows (8.25).

Proof of Theorem 8.3. Suppose the polynomials Q and Q_1 are the same as in the proof of Theorem 1.3. Then

$$u = B^{-1}f = Q(B)f + u_2, \qquad u_2 = B^{-1}Q_1(B)f.$$

If P is some polynomial, then since $Q_1(\lambda) = 0$ at points of the increase $(E_\lambda f, f)$ when $\lambda < b$, then

$$\begin{aligned} \|(P(B) - B^{-1})Q_1(B)f\|^2 &= \int_{b_0}^\infty |P(\lambda) - \lambda^{-1}|^2 |Q_1(\lambda)|^2\, d(E_\lambda f, f) \\ &= \int_b^\infty |P(\lambda) - \lambda^{-1}|^2 |Q_1(\lambda)|^2\, d(E_\lambda f, f). \end{aligned} \tag{8.29}$$

Note that by virtue of (8.1)

$$\int_b^\infty \mathrm{ch}^2(t\sqrt{\lambda - b})\, d(E_\lambda f, f) \leqq \int_{b_0}^\infty \mathrm{ch}^2(t\sqrt{\lambda - b})\, d(E_\lambda f, f) \leqq CM_p(R_0 - t). \tag{8.30}$$

Thus, putting

$$f_1 = Q_1(B)f, \qquad \mu(\lambda) = (E_{\lambda - b} f, f),$$

from (8.29) we obtain:

$$\|(P(B)f - B^{-1})f_1\|^2 = \int_0^\infty |P(\lambda+b) - (\lambda+b)^{-1}|^2 |Q(\lambda)|^2 \, d\mu(\lambda). \tag{8.31}$$

Bearing in mind that $\mu(\lambda)$ satisfies (8.9) by virtue of (8.30), we obtain hence:

$$\|(P(B) - B^{-1})f_1\|^2 \leq \left[\sup_{\lambda \geq 0} \frac{|P(\lambda+b) - (\lambda+b)^{-1}|}{\Phi(\lambda)} \right]^2 \int_0^\infty \Phi^2(\lambda) |Q(\lambda)|^2 \, d\mu(\lambda), \tag{8.32}$$

where $\Phi(\lambda)$ is determined by Formula (8.24) and satisfies (8.25) by virtue of Lemma 8.3. We will assume

$$S_n'(z) = \prod_{j=1}^{n} \left(1 - \frac{i\sigma z}{(2k(j)-1)}\right)^2, \tag{8.33}$$

where $k(j)$ is the j-th term of the sequence of natural numbers, equal to

$$\mathbb{N} \setminus \{\kappa(j),\ j \in \mathbb{N},\ j \geq 9\},$$

where $\varkappa(j)$ is determined by (8.23). Suppose $T_n(z)$ is determined by Formula (2.5.17), where $S_n(z) = S_n'(z)$ is determined by Formula (8.33), which obviously has the form (2.5.16). Suppose the polynomial $P_n(\lambda)$, $\lambda = x^2$ is determined by Formula (2.5.18), where $R=1$. Note that since, according to (8.33), the numbers $\beta(j)$ in (2.5.16) are identical with $(2k(j) - 1)/\sigma$, the function $\Psi_\infty(z)$, which is defined by (2.5.11), is identical with the function $\Phi(z)$, which is defined by Formula (8.24). Therefore inequality (2.5.19), where $\rho^2 = b$, takes the form

$$|(b+\lambda)^{-1} - P_n(\lambda)| \leq b^{-1} \Phi(\lambda) / T_{n+1}(i\sqrt{b}) \tag{8.34}$$

when $\lambda > 0$. Assuming $P(\lambda + b) = P_n(\lambda)$ in (8.32), and using (8.34) and (8.25), we obtain:

$$\|(P_n(B-bI)-B^{-1})f_1\| \leqq C/T_{n+1}(i\sqrt{b}). \tag{8.35}$$

We shall make a lower estimate of $T_{n+1}(i\sqrt{b})$, in order to obtain an inequality of the form (8.4) from (8.35). Note that, according to (2.5.17),

$$T_n(ir) \geqq S_n'(ir)/2$$

when $r \in \mathbb{R}$. Therefore

$$T_n(ir) \geqq \tfrac{1}{2} \prod_{j=1}^{n} \left(1+\frac{r\sigma}{2k(j)-1}\right)^2. \tag{8.36}$$

Suppose $l = l(n)$ terms of the sequence $\varkappa(j)$, $j = 9, \ldots, l+8$ are included between the numbers 1 and $k(n)$. It is then obvious that since $k(n) > n$, then when $r > 0$

$$\begin{aligned}\prod_{j=1}^{n} \left(1+\frac{r\sigma}{2k(j)-1}\right)^2 &= \prod_{j=1}^{k(n)} \left(1+\frac{r\sigma}{2j-1}\right)^2 \prod_{j=9}^{l+8} \left(1+\frac{r\sigma}{2\kappa(j)-1}\right)^{-2} \\ &\geqq \prod_{j=1}^{n} \left(1+\frac{r\sigma}{2j-1}\right)^2 \prod_{j=9}^{\infty} \left(1+\frac{r\sigma}{2\kappa(j)-1}\right)^{-2}.\end{aligned} \tag{8.37}$$

Note that by virtue of (8.23′) the interval between the numbers $(j-1)\ln^q(j-1)$ and $j\ln^q j$ when $j > 9$ has a length of larger than 2. Therefore the number $2\varkappa(j)-1$, which is determined by (8.23), is included in this interval, and consequently

$$2\kappa(j)-1 \geqq (j-1)\ln^q(j-1).$$

Therefore, since $q > 1$,

$$\prod_{j=9}^{\infty}\left(1+\frac{r\sigma}{2\kappa(j)-1}\right)^2 \leqq \prod_{j=9}^{\infty}\left(1+\frac{r\sigma}{(j-1)\ln^q(j-1)}\right)^2 < +\infty. \tag{8.38}$$

From (8.36), (8.37) and (8.38) we obtain

$$T_n(ir) \geqq C_1 S_n(ir\sigma), \qquad C_1 > 0, \tag{8.39}$$

where $S_n(z)$ is determined by Formula (3.1.7). Using (3.2.32) and (3.2.38′), where $j=0$, we obtain that

$$S_n(ir\sigma) \geqq C_2 n^{r\sigma}.$$

Hence and from (8.39) we obtain that

$$T_n(ir) \geqq C_3 n^{r\sigma}.$$

Therefore from (8.35) follows the inequality

$$\|P_n(B-bI)f_1 - B^{-1}f_1\| \leqq C_4 n^{-r\sigma} \leqq C_5 (n+\nu)^{-r\sigma}, \tag{8.40}$$

where ν is the degree of the polynomial $Q_1(f_1 = Q_1 f)$. Bearing in mind that $B^{-1}f = Q(B)f + B^{-1}f_1$ and denoting $P_n(B - bI)Q_1(B)$ by $P_{n+\nu}(B)$, from estimate (8.40) we obtain estiamte (8.4), where $\sigma_0 = r\sigma$. Since

$$r = \sqrt{b} = \rho, \; \sigma = 2R_0/\pi,$$

hence follows the statement of Theorem 8.1.

Remark 8.1. By virtue of Proposition 4.1.2 and inequality (4.1.10), in condition (8.1) we can take B instead of A.

The following lemma is useful to verify estimate (8.1).

Lemma 8.4. Suppose B is a self-adjoint operator in a Hilbert space, $u(t)$, $C < t < R_0$ is a soluton of Cauchy's problem (4.5.1), satisfying all the conditions of Lemma 4.5.1. Suppose

$$\|u(t)\|^2 \leqq C_1 M_p(R_0 - t), \tag{8.41}$$

where M_p is the same as in (8.1). Then estimate (8.1) holds.

Proof. The estimate

$$\|\operatorname{ch} t\sqrt{A} f\|^2 \leqq C_2 \|\operatorname{ch} t\sqrt{B} f\|^2.$$

follows from inequality (4.8.10'). Therefore, using Lemma 4.5.1, from (8.41) we obtain estimate (8.1).

Remark 8.2. Since the function $M_p(\tau)$ in (8.1) and (8.41) approaches ∞ extremely rapidly as $\tau \to 0$, condition (8.41) permits an extremely strong singularity when $t = R_0$ of the solution of problem (4.5.1).

We shall now give an example of the operator B and the function f, for which this (in fact much stronger) estimate holds.

Suppose the operator B is determined on $A(T^1)$ using Formula (4.6.1). According to Theorem 4.6.1, if $f \in A(T^1)$ is an entire function, then $R_0(B, f) > \pi/2$. (Note that the example constructed in §5 of this chapter shows that $R_0(B, f) = \pi/2$, since if $R_0(B, f) > \pi/2$, then $\kappa_0 = 1 + \varepsilon$, $\varepsilon < 0$, in estimate (5.22), which contradicts Theorem 5.2). Below we shall make an upper estimate of the increase $\|\operatorname{ch}(t\sqrt{B}) f\|$ as $t \to R_0(B, f)$ for the operator B.

Lemma 8.5. Suppose the operator B is determined on $A(T^1)$ by Formula (4.6.1), where $a(x)$ is a trigonometric polynomial. Suppose the function $f(x) \in A(T^1)$ is also a trigonometric polynomial. Suppose $u(t)$ is a solution of problem (4.5.1) with the above f

and *B*. Then $u(t)$ is analytic with respect to t when $-\pi/2 < t < \pi/2$, whilst

$$\|u(t)\|^2 \leq \exp[C(\pi/2-|t|)^{-\nu}]. \tag{8.41'}$$

Proof. Cauchy's problem (4.5.1), where B is determined by Formula (4.6.1), takes the form (4.6.2). As shown in Proposition 4.6.2, the solution $v(t) = \mathrm{ch}(t\sqrt{B})f$ is analytic with respect to t when $|t| < \pi$ in the intervals

$$|\kappa-\pi/2|<\delta, \quad |\kappa+\pi/2|<\delta, \ \delta>0$$

It is thus sufficient to estimate $v(t)$ when $x \in]-\pi/2, \pi/2[$. The estimate in the interval $]\pi/2, 3\pi/2[$ reduces to an estimate in $]-\pi/2, \pi/2[$, since problem (4.6.2) preserves this form when x is changed to $x + \pi$.

As in the proof of Proposition 4.6.3, we shall change the variable x to the variable z in the interval $]-\pi/2, \pi/2[$:

$$x(z)=2\,\mathrm{arctg}\,\mathrm{th}(z/2),\ z=\xi+i\eta,\ \xi\in\mathbb{R},\ \eta\in]-\pi/2,\pi/2[. \tag{8.42}$$

The function $u(x(\xi + i\eta)) = v(\xi, \eta)$ satisfies Eq.(4.6.16), having the form

$$\partial_t^2 v-\partial_\eta^2 v=-i\,\mathrm{th}(\xi+i\eta)\,\partial_\eta v+a(x(\xi+i\eta))v, \tag{8.43}$$

with the initial conditions

$$v|_{t=0}=f(x(\xi+i\eta)), \qquad \partial_t v|_{t=0}=0. \tag{8.44}$$

Here $\xi \in \mathbb{R}$ is the parameter. In Proposition 4.6.3 it is proved that $v(t, \xi, \eta)$ for each $\xi \in \mathbb{R}$ is analytic with respect to ξ, η in the domain

$$\Omega_{\pi/2} = \{t \geqq 0\} \cap \{t+\eta < \pi/2\} \cap \{t-\eta < \pi/2\}.$$

Since all the estimates are henceforth uniform with respect to $\xi \in \mathbb{R}$, we shall not mention in the notation the dependence of v on ξ, and shall write $v(t, \eta)$ instead of $v(t, \xi, \eta)$.

We shall use $\Omega(T, \tau)$ to denote a trapezoid with the bases $t = 0$, $t = \tau$ where $0 < \tau < T < \pi/2$ and with the lateral sides $(t - T) = \pm \eta$, which lies in $\Omega_{\pi/2}$. Suppose Γ_0, Γ_τ, Γ_+ and Γ_- are the sides of this trapezoid, Γ_0 and Γ_τ are segments of the lines $t = 0$ and $t = \tau$ respectively, and Γ_+ and Γ_- are segments of the lines $(t - T) = \eta$ and $(t - T) = -\eta$ respectively. We shall multiply (8.4.3) by $2\partial_t \bar{v}$ and shall integrate it over $\Omega = \Omega(T, \tau)$:

$$\begin{aligned} \mathcal{J} &= 2\,\mathrm{Re} \int_\Omega [\partial_t^2 v\, \partial_t \bar{v} - \partial_\eta^2 v\, \partial_t \bar{v}]\, dt\, d\eta \\ &= 2\,\mathrm{Re} \int_\Omega [-ith(\xi + i\eta)\, \partial_\eta v\, \partial_t \bar{v} + a(x(\xi + i\eta)) v\, \partial_t \bar{v}]\, dt\, d\eta. \end{aligned} \tag{8.45}$$

It is easy to see that

$$2\,\mathrm{Re}[\partial_t^2 v\, \partial_t \bar{v} - \partial_\eta^2 v\, \partial_t \bar{v}] = \partial_t |\partial_t v|^2 - 2\,\mathrm{Re}\, \partial_\eta(\partial_\eta v\, \partial_t \bar{v}) + \partial_t(\partial_\eta |v|^2).$$

Using this identity and Green's formula, we shall transform the left-hand side of (8.45) to the form

$$\begin{aligned} \mathcal{J} = &- \int_{\Gamma_0} (|\partial_t v|^2 + |\partial_\eta v|^2)\, d\eta + \int_{\Gamma_\tau} (|\partial_t v|^2 + |\partial_\eta v|^2)\, d\eta \\ &+ \int_{\Gamma_+} (|\partial_t v|^2 + |\partial_\eta v|^2 + 2\,\mathrm{Re}(\partial_\eta v\, \partial_t \bar{v}))\, d\eta \\ &+ \int_{\Gamma_+} (|\partial_t v|^2 + |\partial_\eta v|^2 - 2\,\mathrm{Re}(\partial_\eta v\, \partial_t \bar{v}))\, d\eta. \end{aligned} \tag{8.45'}$$

Bearing in mind that the integrals over Γ_+ and Γ_- are nonnegative, and putting

$$G(\tau)=\int_{\Gamma_\tau}(|\partial_t v|^2+|\partial_\eta v|^2)\,d\eta=\int_{\tau-T}^{T-\tau}(|\partial_t v|^2+|\partial_\eta v|^2)\,d\eta, \tag{8.46}$$

from (8.45) we obtain:

$$G(\tau)-G(0)\leq\int_{\Omega}[|\mathrm{th}(\xi+i\eta)|(|\partial_\eta v|^2+|\partial_t v|^2) \\ +|a(x(\xi+i\eta))|(|v|^2+|\partial_t v|^2)]\,dt\,d\eta. \tag{8.47}$$

Note that since $a(x)$ is a trigonometric polynomial, a and f are polynomials of $\sin x$ and $\cos x$. Since $x(z)$ is determined by Formula (8.42), then

$$\sin x(z)=\frac{2\mathrm{th}(z/2)}{1+\mathrm{th}^2(z/2)}=\mathrm{th}\,z,\quad \cos x(z)=\frac{1-\mathrm{th}^2(z/2)}{1+\mathrm{th}^2(z/2)}=\frac{1}{\mathrm{ch}\,z}.$$

Since

$$|\cos(is+\theta)|\geq|\cos\theta|\mathrm{ch}\,s$$

when $s,\theta\in\mathbb{R}$, then when $|\eta|<\pi/2$ the following estimate holds:

$$|\mathrm{th}(\xi+i\eta)|\leq C(\pi/2-|\eta|)^{-1},\ |f(x(\xi+i\eta))|+|\partial_\eta f(x(\xi+i\eta))| \\ +|a(x(\xi+i\eta))|\leq C_1(\pi/2-|\eta|)^{-\nu}, \tag{8.48}$$

where C does not depend on $\xi\in\mathbb{R}$.

We shall write the dual integral on the right-hand side of (8.47) in the form of a repeated integral, and using inequalities (8.48) we will obtain, bearing in mind that $\tau<T<\pi/2$:

$$G(\tau)-G(0)\leqq C_2\int_0^\tau\int_{t-T}^{T-t}[(\pi/2-|\eta|)^{-1}(|\partial_\eta v|^2+|\partial_t v|^2)$$

$$+(\pi/2-|\eta|)^{-\nu}(|\partial_t v|^2+|v|^2)]\,d\eta\,dt \tag{8.49}$$

$$\leqq C_3(\pi/2-|T|)^{-\nu}\int_0^\tau(G(t)+\int_{t-T}^{T-t}|v|^2\,d\eta)\,dt.$$

Noting that

$$|v(\theta,\eta)|=\left|\int_0^\theta \partial_t v(t,\eta)\,dt+v(0,\eta)\right|\leqq\left(\int_0^\theta|\partial_t v(t,\eta)|^2\,dt\right)^{1/2}t^{1/2}+|v(0,\eta)|, \tag{8.50}$$

we obtain when $t<\tau<T$, bearing in mind that

$$v(0,\eta)=f(x(\xi+i\eta)),$$

and estimate (8.48) for f:

$$\int_{t-T}^{T-t}|v|^2\,d\eta\leqq\pi\int_{t-T}^{T-t}\int_0^t|\partial_t v(\theta,\eta)|^2\,d\theta\,d\eta$$

$$+\int_{-T}^{T}|v(0,\eta)|^2\,d\eta\leqq\pi\int_0^t G(\theta)\,d\theta+C_4|\pi/2-T|^{-2\nu}.$$

Hence and from (8.49) we obtain:

$$G(\tau)-G(0)\leqq C_3(\pi/2-T)^{-\nu}\int_0^\tau[G(t)+\pi\int_0^\tau G(\theta)\,dt]\,dt\quad+\pi C_4|\pi/2-T|^{-2\nu}. \tag{8.51}$$

By virtue of (8.44), (8.46) and (8.48) $|G(0)|<C_4(\pi/2-T)^{-2\nu}$,

$$G(\tau)\leqq C_5(\pi/2-T)^{-\nu}\int_0^\tau G(t)\,dt+C_6(\pi/2-T)^{-2\nu}$$

Using Gronwall's inequality, hence we obtain:

$$G(\tau) \leq C_6(\pi/2 - T)^{-2\nu} \exp(\tau C_5(\pi/2 - T)^{-\nu}) \tag{8.52}$$

when $0 \leqslant \tau \leqslant T < \pi/2$, where C_5 and C_6 do not depend on τ, T or on $\xi \in \mathbb{R}$.

Note, now, that the dual integral on the right-hand side of (8.45), after noting it in the form of the repeated one, is estimated from above by the integral of $G(t)$ and by $|a|$. The left-hand side of (8.45) is written in the form (8.45'). The integrals over Γ_0 and Γ_τ in (8.45') are estimated using (8.52). The integral over Γ_- is nonnegative. Therefore from (8.45) and (8.45') follows the estimate

$$\int_{\Gamma+} [|\partial_t v|^2 + |\partial_\eta v|^2 + 2\,\mathrm{Re}(\partial_\eta v\, \partial_t \bar{v})|]\, d\eta$$
$$\leq C_7 \exp[2\tau C_5(\pi/2 - T)^{-\nu}]. \tag{8.53}$$

Since $\eta = t - T$ on Γ_+, the left-hand side of this inequality equals

$$\int_{t=0}^{\tau} \left| \frac{d}{dt}(v(t, t-T)) \right|^2 dt.$$

Therefore from (8.53) when $\tau = T$ follows the estimate

$$|v(T, 0)| \leq |v(0, T)| + C_7 \exp[2TC_5(\pi/2 - T)^{-\nu}]. \tag{8.54}$$

Bearing in mind that $v(0, T) = f(x(\xi + iT))$, and using (8.48), from (8.54) we obtain the estimate

$$|v(T, \xi, 0)| \leq C_8 \exp[C_9(\pi/2 - T)^{-\nu}], \tag{8.55}$$

where $\xi \in \mathbb{R}$, $0 \leqslant T < \pi/2$. Since

$$v(t, \xi, 0) = u(t, 2\,\mathrm{arctg\,th}\,\xi)$$

from estimate (8.55) follows the estimate

$$|u(t,x)| \leqq C_8 \exp[C_9(\pi/2-t)^{\nu}] \tag{8.56}$$

when $x \in\ ]-\pi/2, \pi/2[$. Estimate (8.41′) is thus proved.

Remark 8.3. Estimate (8.41), and consequently also (8.1) obviously follows from estimate (8.41′). Consider Eq.(4.3). According to (4.4) $b_1(B_\beta) > b_0(B_\beta) > (\beta - 1/2)$. Therefore, by virtue of Theorem 8.1,

$$B^{-1}f \in H_2^{\beta-1/2}(T^1).$$

Moreover, by virtue of Lemma 4.4,

$$B^{-1}f \notin B_{2,\theta}^{\beta-1/2}(T^1)$$

when $1 < \theta < \infty$.

Remark 8.4. For fixed value of s the Sobolev spaces $W_2^s = B_{2,2}^s$, Besov spaces $B_{r,\Theta}^s$, $1 < \Theta < \infty$ and Nikol'skii spaces $H_2^s = B_{r,\infty}^s$ consist of functions of different smoothness. The widest of these spaces is the space H_2^s. A question arises in this connection: is it possible to refine the statement of Theorem 8.1, i.e. to take the space $B_{2,\Theta}^s$, $\Theta < \infty$ instead of $H_2^s = B_{2,\infty}^s$? Remark 8.3 shows that this is impossible, i.e. $\Theta = \infty$ is the unimprovable value of the parameter Θ and H_2^s is the natural limiting class of smoothness of the solutions of the degenerating equations.

CHAPTER 6. CONSTRUCTION OF FUNCTIONS OF NON-SELFADJOINT OPERATORS

In this chapter we shall construct polynomial representations of the form (4.1.1) of the function $g(A)f$ of the operator A, when A is a nonselfadjoint operator. We shall consider two cases: firstly, differential operators corresponding to first-order systems, and secondly, second-order differential operators. The coefficients of the operator and the function f are assumed to be analytic. We shall first construct representations of the form (4.1.1) for abstract operators in the Banach space E, and shall then obtain applications to differential operators.

§1. The resolvent and functions of operators in a Banach space.

In this paragraph we briefly discuss the necessary known results on expressing functions of operators in a Banach space in terms of the contour integral of the resolvent. See Dunford and Schwartz [1] and Kato [1] for a more detailed discussion.

Suppose $\mathcal{D} \subset E$, where E is a Banach space, and $\mathcal{D}$ is a linear space which is dense in E. Consider the equation

$$Au - \zeta u = f, \qquad \zeta \in \mathbb{C}. \tag{1.1}$$

If the constant $C = C(\zeta)$ exists, such that

$$\|u\| \leqq C\|(A - \zeta I)u\|, \qquad \forall u \in \mathcal{D}, \tag{1.2}$$

the operator $(A - \zeta I)^{-1}$ is determined on the subspace $E_0 = [(A - \zeta I)\mathcal{D}]$ where [ ] is a closure in E. When $E_0 = E$, we say that ζ belongs to the resolvent set $\rho_0(A)$ of the operator A, and $(A - \zeta I)^{-1}$ is the resolvent of the operator A when $\zeta = \zeta_0$.

For brevity, we shall henceforth write $(A - \zeta)^{-1}$ instead of $(A - \zeta I)^{-1}$

We shall denote the set of $\zeta \in \mathbb{C}$, for which estimate (1.2) holds with fixed C, by $\rho(C)$ (it is not assumed that $\zeta \in \rho_0(A)$, i.e. that $[(A - \zeta)\mathcal{D}] = E$).

Proposition 1.1. Suppose $\zeta_0 \in \rho(C) \cap \rho_0(A)$. Then the circle

$$K=\{\zeta:|\zeta-\zeta_0|<C^{-1}\}$$

lies in $\rho_0(A)$ and $(A - \zeta)^{-1}f$ is analytic with respect to ζ in this circle.

Proof. Consider Eq.(1.1), where $\zeta = \zeta_0 + \delta$. Since $\zeta_0 \in \rho_0(A)$, this equation can be rewritten in the form

$$u-\delta(A-\zeta_0)^{-1}u=(A-\zeta_0)^{-1}f. \tag{1.3}$$

We shall see the solution of this equation in the form of series

$$u=\sum_{k=0}^{\infty}\delta^k(A-\zeta_0)^{-k}(A-\zeta_0)^{-1}f. \tag{1.4}$$

Since $\zeta_0 \in \rho(C)$, this series converges when $|\delta| < C^{-1}$. It is easy to see that u, which is determined by this series, is the solution of (1.3). By virtue of (1.3) $u \in (A - \zeta_0)^{-1}E$. Consequently, it follows from (1.3) that the sequence $u_j \in \mathcal{D}$ exists, such that

$$u_j\to u, \qquad (A-\zeta_0)u_j\to\delta u+f.$$

Therefore $(A - \zeta_0)u_j - \delta u \to f$. Since $\delta u_j \to \delta u$, then

$$(A-\zeta_0-\delta)u_j\to f,\ u_j\in\mathcal{D}.$$

Consequently, $f \in [(A-\zeta_0-\delta)\mathscr{D}]$. Since f is an arbitrary vector E, $[(A-\zeta_0-\delta)\mathscr{D}]=E$, and therefore $\zeta_0+\delta \in \rho_0(A)$. Since the function u, determined by (1.4), obviously analytically depends on δ, the proposition is proved.

Proposition 1.2. If

$$\zeta_0, \zeta_1 \in \rho(C), \zeta_0 \in \rho_0(A)$$

and the continuous curve $\zeta(t)$, $\zeta(t) \in \rho(C)$ exists, when

$$t \in [0, 1], \zeta(0)=\zeta_0, \zeta(1)=\zeta_1,$$

then $\zeta_1 \in \rho_0(A)$. At the same time

$$(A-\zeta_1)^{-1}E=(A-\zeta_0)^{-1}E.$$

Proof. Covering the curve $\zeta(t)$ with circles of radius $1/C$, and using Proposition 1.1, we obtain a statement of proposition 1.2.

Proposition 1.3. Suppose

$$\rho(A)=\bigcup_C\{\zeta \in \rho(C), C>0\}$$

Suppose there is a point from $\rho_0(A)$ in each connected component $\rho(A)$. Then $\rho(A) = \rho_0(A)$. If ζ_1, $\zeta_2 \in \rho_0(A)$, then

$$(A-\zeta_0)^{-1}E=(A-\zeta_1)^{-1}E. \tag{1.5}$$

Proof. $\rho(A) = \rho_0(A)$ by virtue of Proposition 1.2. If ζ_0, $\zeta_1 \in \rho_0(A)$, we shall consider the equation $Au - \zeta_1 u = f$. We can rewrite this equation in the form

$$Au-\zeta_0 u+(\zeta_0-\zeta_1)u=f$$

On one hand,

$$u=(A-\zeta_1)^{-1}f\in(A-\zeta_1)^{-1}E$$

On the other,

$$u=(A-\zeta_0)^{-1}((\zeta-\zeta_0)u+f)\in(A-\zeta_0)^{-1}E.$$

Therefore $(A - \zeta_0)^{-1}E = (A - \zeta_1)^{-1}E$.

We shall henceforth put $\mathcal{D}(A) = (A - \zeta_0)^{-1}E$, where $\zeta_0 \in \rho_0(A)$.

The operator A is defined on $\mathcal{D}(A)$ using the formula $Au = Au - \zeta_0 u + \zeta_0 u$. (This extension of the operator A agrees with its closure.)

We shall denote the set

$$(A-\zeta_0 I)^{-n}E,\ \zeta_0\in\rho_0(A),\ n\in\mathbb{N}$$

by $\mathcal{D}(A^n)$. As in (1.5), it is proved that

$$(A-\zeta_0 I)^{-1}\mathcal{D}(A)=(A-\zeta_1 I)^{-1}\mathcal{D}(A),$$

and, by induction,

$$(A-\zeta_0 I)^{-1}\mathcal{D}(A^k)=(A-\zeta_1 I)^{-1}\mathcal{D}(A^k)$$

when $\zeta_0, \zeta_1 \in \rho_0(A)$. Hence it follows that $\mathcal{D}(A^n)$ does not depend on $\zeta_0 \in \rho_0(A)$.

The following Hilbert identity holds for $\zeta_1, \zeta_2 \in \rho_0(A)$

$$(A-\zeta_1)^{-1}-(A-\zeta_2)^{-1}=(\zeta_1-\zeta_2)(A-\zeta_1)^{-1}(A-\zeta_2)^{-1}. \tag{1.6}$$

Let us now consider the operators A with the domain of definition $\mathcal{D}(A)$, satisfying the following condition.

Condition 1.1. $\omega_0 > 0$ exists, such that each number $\zeta \in \mathbb{C}$, for which $|\mathrm{Im}\zeta| > \omega_0$, belongs to the resolvent set $\rho_0(A)$ of the operator A. At the same time $\forall\, \varepsilon > 0$ the constant $C = C(\varepsilon)$ exists such that

$$\|(A-\zeta I)^{-1}\| < C \text{ when } |\mathrm{Im}\,\zeta| > \omega_0 + \varepsilon \tag{1.7}$$

We shall now define the functions of operators which satisfy condition 1.1.

Suppose

$$g \in \mathcal{A}(\mathcal{J}_\beta),\ \beta > \omega_0,\ \mathcal{J}_\beta=\{\zeta:\mathrm{Im}\,\zeta|<\beta\}.$$

(See Ch.3 for the definition of the class $\mathcal{A}(\mathcal{J}_\beta)$.)

We will assume, when $f \in \mathcal{D}(A^2)$,

$$g(A)f = -\frac{1}{2\pi i}\int_{L_\omega} g(\zeta)(A-\zeta)^{-1}f\, d\zeta, \tag{1.8}$$

where $\omega_0 < \omega < \beta$,

$$L_\omega=L_\omega^+\cup L_\omega^-,\quad L_\omega^\pm=\{\zeta:\zeta=\pm i\omega+s,\ s\in\mathbb{R}\}. \tag{1.9}$$

We understand the convergence of integral (1.8) in the sense of the principal value (i.e. the integral is taken over $L_\omega \cap \{|\mathrm{Re}\,\zeta| < N\}$ and $N\to\infty$). Integral (1.8) exists in the above sense. Indeed, suppose $|\mathrm{Im}\,\zeta_0| > \omega$. Then, according to (1.6),

$$(A-\zeta)^{-1}(A-\zeta_0)^{-1}(A-\zeta_0)f=(\zeta-\zeta_0)^{-1}((A-\zeta)^{-1}f_1-(A-\zeta_0)^{-1}f_1), \tag{1.10}$$

where $f_1 = (A - \zeta_0)f$. Therefore the integral on the right-hand side of (1.8) equals

$$-\frac{1}{2\pi i}\int_{L_\omega}\frac{g(\zeta)}{(\zeta-\zeta_0)}(A-\zeta)^{-1}f_1\,d\zeta+\frac{1}{2\pi i}\int_{L_\omega}\frac{g(\zeta)}{(\zeta-\zeta_0)}\,d\zeta f. \tag{1.11}$$

Since the function $g(\zeta)/(\zeta - \zeta_0)$ is holomorphic in the strip $\mathcal{J}_\omega$, the integral over a rectangle with sides formed by the lines L_ω and those perpendicular to them, $|\mathrm{Re}\,\zeta| = N$, equals zero. The function $g(\zeta)/(\zeta - \zeta_0)$ in the segments of the lines $|\mathrm{Re}\,\zeta| = N$, which are included in $\mathcal{J}_\omega$, approaches 0 as $N\to\infty$. Therefore the second integral in (1.11), understood in the sense of the principal value, converges and equals zero. Consequently, Formula (1.8) takes the form

$$g(A)f=-\frac{1}{2\pi i}\int_{L_\omega}\frac{g(\zeta)}{\zeta-\zeta_0}(A-\zeta)^{-1}(A-\zeta_0)f\,d\zeta. \tag{1.12}$$

Integral (1.12) has the form (1.8), where the function $g_1(\zeta) = g(\zeta)/(\zeta - \zeta_0)$ is substituted instead of $g(\zeta)$. Transforming this integral, as done above, to the form (1.11) and (1.12) with new g and f, we obtain that (1.8) is represented in the following form when $f \in \mathcal{D}(A^2)$:

$$g(A)f=-\frac{1}{2\pi i}\int_{L_\omega}\frac{g(\zeta)}{(\zeta-\zeta_0)^2}(A-\zeta)^{-1}f_2\,d\zeta, \qquad f_2=(A-\zeta_0)^2 f. \tag{1.13}$$

This integral obviously converges. Thus, the integrals (1.8) and (1.12) converge in the sense of the principal value.

It follows from Formulae (1.12) and (1.13) that

$$(A-\zeta_0)^{-1}[g(A)f]=[(A-\zeta_0)^{-1}g(A)]f=g(A)[(A-\zeta_0)^{-1}f]. \tag{1.14}$$

Note that

$$\text{if} \quad g(\zeta)\equiv 1, \quad \text{then} \quad g(A)f=f. \tag{1.15}$$

Indeed, if we take the point ζ_0 inside the contour L_ω, i.e. $\omega_0 < |\mathrm{Im}\,\zeta_0| < \omega$, the integral on the right-hand side of (1.8), as previously, equals (1.11). The second term in (1.11) equals $fg(\zeta_0)$. Bearing in mind that the integral over L_ω of $g(\zeta)(\zeta_0 - \zeta)^2$ equals $g'(\zeta_0)$, in the same way as Formula (1.13) we obtain formula

$$g(A)f=-\frac{1}{2\pi i}\int_{L_\omega} g(\zeta)\frac{(A-\zeta I)^{-1}}{(\zeta-\zeta_0)^2}f_2\,d\zeta+fg(\zeta_0)+(A-\zeta_0 I)g'(\zeta)f. \tag{1.16}$$

Since the integral does not change when ω increases by virtue of the holomorphicity of $(A - \zeta)^{-1}f_2$, $g \equiv 1$ and $(\zeta - \zeta_0)^{-2}$ in the domain $|\mathrm{Im}\zeta| > |\mathrm{Im}\zeta_0|$, then, directing ω to $+\infty$, and using estimate (1.7), we obtain (1.15) from (1.16).

Note that if $\zeta_1, \ldots, \zeta_k \in \rho_0(A)$, the following operator is defined:

$$Q^{-1}(A)f=(A-\zeta_1 I)^{-1}\ldots(A-\zeta_k I)^{-1}. \tag{1.17}$$

Proposition 1.4. If A satisfies Condition 1.1,

$$f\in\mathscr{D}(A^{l+2}),\ g_1\in\mathscr{A}(\mathscr{I}_\beta),\ g_2(\zeta)=P(\zeta)/Q(\zeta)$$

is a rational function, P and Q are polynomials $Q(\zeta) \neq 0$ in $\mathcal{J}_\omega$, and the degrees Q and P are connected by the inequality

$$\deg P\leq\deg Q+l, \tag{1.18}$$

the function $[g_1g_2(A)]f$, defined by (1.8), where $g = g_1g_2$, agrees with

$$g_1(A)[P(A)Q^{-1}(A)f]$$

and with

$$[P(A)Q^{-1}(A)][g_1(A)f]$$

where $Q^{-1}(A)$ is determined by Formula (1.17).

Proof. When P=1 the statement of the proposition directly follows from (1.14). If $l > 0$, then by virtue of (1.14)

$$Q^{-1}g_1(A)f = Q^{-1}g_1(A)(A-\zeta_0)^{-l}(A-\zeta_0)^l f = Q_1^{-1}g_1(A)f_1,$$

where

$$\deg Q_1 = \deg Q + l,\ f_1 \in \mathscr{D}(A_2).$$

Therefore, from the very beginning we can take $l = 0$, $Q_1 = Q$. We shall put $\deg Q = n$. Consider the case when

$$P(A) = (A-\zeta_0)^k,\ k \leq n,\ Q(A) = (A-\zeta_0)^n.$$

In this case the statement of the proposition follows from (1.14). Since

$$g_1(\zeta)Q^{-1}(\zeta) = g_1(\zeta)(A-\zeta_0)^n Q^{-1}(\zeta)(A-\zeta_0)^{-n},$$

then, assuming $g_{10} = g_1Q^{-1}(A - \zeta_0)^n$, and noting that g_{10} also belongs to $\mathcal{A}(\mathcal{J}_{\beta'})$, $\beta' > \omega$, we obtain that the statement of the proposition also holds for Q^{-1} of general form. Expanding $P(A)$ in powers of $A - \zeta_0$ using Taylor's formula, we obtain a statement of the proposition for $P(A)$ of general form.

Let us now consider the operator A, which satisfies the following condition.

Condition 1.2. $\omega_0 > 0$ exists, such that for each $\zeta \in \mathbb{C}$, satisfying the condition $|\mathrm{Im}\zeta| > \omega_0$, the number ζ^2 belongs to the resolvent set $\rho_0(A)$ of the operator A. At the same time for any $\varepsilon > 0$ the constant $C = C(\varepsilon)$ exists such that

$$\|(A-\zeta^2 I)^{-1}\| < C \text{ when } |\mathrm{Im}\,\zeta| > \omega_0 + \varepsilon \tag{1.19}$$

We shall introduce functions of the operator A, which satisfies condition 1.2.

We shall use $L^2{}_\omega$ to denote the image of the contour L_ω determined by (1.9), under the mapping $\zeta \to \zeta^2$.

$$L^2_\omega = \{\zeta \in \mathbb{C} : \mathrm{Re}\,\zeta = -\omega^2 + (\mathrm{Im}\,\zeta)^2/(4\omega^2)\}. \tag{1.20}$$

It is easy to see that $L^2{}_\omega$ is the boundary of the set

$$\mathcal{J}^+_\omega = \{\zeta \in \mathbb{C} : \zeta = z^2, z \in \mathcal{J}_\omega\}. \tag{1.21}$$

If $g \in A(\mathcal{J}_\beta{}^+)$ (see Chapter 3), $\beta > \omega_0$, $f \in \mathcal{D}(A^2)$, we will assume

$$g(A)f = -\frac{1}{2\pi i}\int_{L^2_\omega} g(z)(A-zI)^{-1} f\, dz, \tag{1.22}$$

where $\omega_0 < \omega < \beta$, and the integral is understood in the sense of the principal value. We will find another form of noting $g(A)f$, obtained using the substitution $z = \zeta^2$, more convenient:

$$g(A)f = -\frac{1}{2\pi i}\int_{L_\omega} g(\zeta^2)(A-\zeta^2 I)^{-1} f\zeta\, d\zeta. \tag{1.23}$$

Similarly to Proposition 1.4, we obtain:

Proposition 1.5. Suppose A satisfies condition 1.2,

$$f \in \mathscr{D}(A^{l+2}),\ g_1 \in \mathscr{A}(\mathscr{I}_\beta^+),\ \beta > \omega > \omega_0,\ g_2(\zeta^2) = P(\zeta^2)/Q(\zeta^2),$$

where P and Q are polynomials that satisfy (1.18). Suppose $Q(\zeta^2) \neq 0$ in $\mathcal{J}_\omega$. Then the function $g_1 g_2(A)f$, defined by (1.23), agrees with

$$P(A)Q^{-1}(A)[g_1(A)f]$$

and with

$$g_1(A)[P(A)Q^{-1}(A)f],$$

where $Q^{-1}(A)$ is determined by Formula (1.17).

§2. Construction of functions of self-determined operators.

Suppose E is a Banach space, $\mathcal{D}_0 \subset E$, $\mathcal{D}_0$ is a linear space which is everywhere dense in E. Suppose the operator A is determined on $\mathcal{D}_0$, whilst

$$A\mathscr{D}_0 \subset \mathscr{D}_0. \tag{2.1}$$

Definition 2.1. The operator A, which is determined on $\mathcal{D}_0$, is called a first-order self-determined operator if (2.1) holds and the closure of this operator satisfies condition 1.1. In addition, $\forall f \in \mathcal{D}_0, R > 0$ exists, such that

$$M_R(A, f) = \sum_{j=0} \|A^j f\| R^j/j! < \infty. \tag{2.2}$$

We will assume for $f \in \mathcal{D}_0$

$$R_{01}(A,f)=\sup\{R\geqq 0: M_R^1(A,f)<\infty\}. \tag{2.3}$$

The term "self-determined" was chosen because, as is obvious from subsequent results, the value of the functions $g(A)f$ of the operator A in $f \in \mathcal{D}_0$ is uniquely determined by A^jf, $j \in \mathbb{Z}_+$. This situation is fairly unusual. For example, if A is a differential operator, then $(A - \zeta)^{-1}f$ is the solution of the equation $Au - \zeta u = f$. As a rule, for u to be uniquely determined, it is required to impose some boundary conditions. Therefore u depends not only on A and f, but also on the boundary conditions. Self-determined operators are such that we are not required to introduce boundary conditions in order to isolate the unique solution u. (Examples of these differential operators will be given in §§3 and 4 of this chapter.)

Remark 2.1. It follows from (2.1) that $(A - \zeta)^j\mathcal{D} \subset \mathcal{D}$ $\forall$ $j \in \mathbb{N}$, $\zeta \in \mathbb{C}$. Therefore

$$\mathcal{D}_0 \subset \bigcap_{k\in\mathbb{N}} \mathcal{D}(A^k)=\mathcal{D}_\infty(A). \tag{2.4}$$

Lemma 2.1. If $\zeta \in \rho_0(A)$, then

$$R_{01}(A,f)=R_{01}(A,(A-\zeta_0)^l f)$$

for any $l \in \mathbb{N}$.

Proof. Suppose $0 < R < R_1 < R_2 < R_{01}(A, f)$. By virtue of (2.3)

$$\|A^jf\|\leqq CR_2^{-j}j!.$$

Therefore

$$\|A^j(A-\zeta_0)^l f\| \leqq C_1 R_2^{-j-l}(j+l)! \leqq C_2 R_1^{-j} j!.$$

Consequently,

$$M_R^1(A, (A-\zeta_0)^l f) < +\infty$$

when $R < R_1$. Since R is as close to $R_{01}(A, f)$ as desired, then

$$R_{01}(A, (A-\zeta_0)^l f) \geqq R_{01}(A, f).$$

On the other hand, suppose $0 < R < R_{01}(A, (A-\zeta_0)^l f)$. Since

$$\|A^j f\| = \|(A-\zeta_0)^{-l} A^j (A-\zeta_0)^l f\| \leqq C\|A^j(A-\zeta_0)^l f\|,$$

then $M_R{}^1(A, f) < \infty$. Since R is as close to $R_{01}(A, (A-\zeta_0)^l f)$, as desired, the lemma is proved.

We will assume when $j < n < +\infty$

$$\Phi(j, \zeta, N) = \prod_{k=j}^{N}\left(1 + \frac{4\zeta^2}{\pi^2(2k-1)^2}\right). \tag{2.5}$$

Obviously, since the product converges for all $\zeta \in \mathbb{C}$ and vanishes when

$$\zeta = \pm\pi i(2k-1)/2, \; k \geqq j,$$

then

$$1/\Phi(j, R\zeta, N) \in A(\mathcal{J}_\beta) \text{ when } \pi(2j-1)/(2R) > \beta. \tag{2.6}$$

We will assume $\Phi(j, \zeta) = \Phi(j, \zeta, \infty)$.

Lemma 2.2. Suppose A is a self-determined first-order operator, $f \in \mathcal{D}_0$, $0 < R < R_{01}(A, f)$. Then $\forall j \in \mathbb{N}$ and the following limit exists:

$$\lim_{N\to\infty} \Phi(j, RA, N)f \equiv \Phi(j, RA)f. \tag{2.7}$$

Proof. It is obvious from Formula (2.5) that the coefficients of the polynomial $\Phi(j, \zeta, N)$ are nonnegative and monotonically increase when N increases and when j decreases. The function $\Phi(0, \zeta, \infty)$ is identical with chζ. Therefore for the Taylor coefficient a_{jNn} of ζ^{2n} of the function $\Phi(j, \zeta, N)$, the following estimate holds:

$$0 \leqq a_{jNn} \leqq 1/(2n)!. \tag{2.8}$$

Suppose

$$R < R_{01}(A, f),\ l \in \mathbb{N}.$$

As proved in Lemma 1.1

$$\|(A-\zeta)^l \Phi(j, RA, N) f\| \leqq C, \qquad \forall N \in \mathbb{N}, \quad N \geqq j, \tag{2.9}$$

where C does not depend on N.

We shall now note that when $j \leqslant n_1 < N_2$

$$\begin{aligned}\Phi(j, RA, N_2)f - \Phi(j, RA, N_1)f \\ = [I - \Phi^{-1}(N_1+1, RA, N_2)]\Phi(j, RA, N_2)\, f.\end{aligned} \tag{2.10}$$

The function

$$g(\zeta, N_1, N_2) = 1 - \Phi^{-1}(N_1+1, R\zeta, N_2) \in \mathscr{A}(\mathscr{I}_\beta), \qquad \beta > \omega$$

if N_1 is sufficiently large. Assuming $f_0 = \Phi(j, RA, N_2)(A-\zeta_0)^2 f$, from (1.13) we obtain:

$$\|g(A, N_1, N_2)f_0\| \leqq \frac{C}{2\pi} \int_{L_\omega} \frac{|g(\zeta, N_1, N_2)|}{|(\zeta-\zeta_0)^2|} |d\zeta| \, \|f_0\|. \tag{2.11}$$

We shall prove that for any $\varepsilon > 0$ $\exists$ N_2, such that $\|g(A, N_1, N_2)f_0\| < \varepsilon$ when N_1, $N_2 > N$. Indeed, since by virtue of (2.9) $\|b_0\| < C$, $g(\zeta, N_1, n_2)$ is uniformly bounded with respect to N_1 and N_2, there exist $s > 0$ such that for all N_1, $N_2 > N$

$$\frac{C}{2\pi} \int_{L_\omega \cap |\mathrm{Re}\zeta| > s} \frac{|g(\zeta, N_1, N_2)|}{|\zeta-\zeta_0|^2} |d\zeta| \, \|f_0\| \leqq \varepsilon/2. \tag{2.12}$$

Note, now, that by virtue of the convergence of the infinite product $\mathrm{ch} z = \Phi(0, \zeta, \infty)$, the function $\Phi(N, R\zeta, \infty) \to 1$ as $N \to \infty$ uniformly in any set bounded in $\mathbb{C}$. Since

$$\Phi(N_1+1, R\zeta, N_2) = \Phi(N_1+1, R\zeta, \infty)/\Phi(N_2+1, R\zeta, \infty),$$

then $\Phi(N_1+1, R\zeta, N_2) \to 1$ as N_1, $N_2 \to \infty$ in $L_\omega \cap \{|\mathrm{Re}\,\zeta| < s\}$. Bearing in mind (2.12), we obtain that $\|g(A, N_1, N_2)f_0\| < \varepsilon$ when N_1, $N_2 > N_3$, and N_3 is sufficiently large. Since ε is arbitrary, it follows hence, by virtue of (2.10), that the sequence $\Phi(j, RA, N)f$ is fundamental, and it has a limit. Formula (2.7) is thus proved.

Lemma 2.3. Suppose A is a self-determined first-order operator, $f \in \mathcal{D}_0$. Then for any R, $0 < R < R_{01}(a, f)$ and for any ω, $0 < \omega_0 < \omega$ and any $l \in \mathbb{Z}_+$ the constant C exists such that for any function $g \in A(\mathcal{J}_\beta)$, $\beta > \omega$, and any polynomial P

$$\|A^l(g(A)f - P(A)f)\| \leqq C\mu_{\omega,R}(P, g), \tag{2.13}$$

where $\mu_{\omega,R}$ is determined by Formula (3.32).

Proof. Since by virtue of Proposition 1.4 $A^l g(A)f = g(A)A^l f$ and according to Proposition 1.1

$$R_{01}(A,f)=R_{01}(A,A^l f),$$

it is sufficient to prove (2.13) for $l = 0$. According to Proposition 1.4

$$(g(A)-P(A))f=$$

$$\frac{-1}{2\pi i}\int_{L_\omega}\frac{g(\zeta)-P(\zeta)}{(\zeta-\zeta_0)^{j+2}}\Phi^{-1}(j,R\zeta,N)(A-\zeta I)^{-1}f_1\,d\zeta, \tag{2.14}$$

where

$$f_1=\Phi(j,RA,N)(A-\zeta_0)^{j+2}f, \tag{2.15}$$

and the number j satisfies the condition $\pi(2j-1)/(2R) > \omega$.

Let N approach ∞ in (2.15) and (2.14). According to Lemma 2.1

$$R<R_{01}(A,(A-\zeta_0)^{j+2}f).$$

Therefore Lemma 2.2 is applicable to the vector $(A-\zeta_0)^{j+2}f = f_2$. According to this lemma

$$\Phi(j,RA,N)f_2\to\Phi(j,RA)f_2.$$

By virtue of estimate (1.7) it follows hence that

$$(A-\zeta I)^{-1}\Phi(j,RA,N)f_2\to(A-\zeta I)^{-1}\Phi(j,RA)f_2 \tag{2.16}$$

uniformly with respect to $\zeta \in L_\omega$. At the same time the functions $(g(\zeta) - P(\zeta)) \times \Phi^{-1}(j, R\zeta, N)$ for large N are uniformly bounded in L_ω and converge uniformly on any bounded subset L_ω to $\Phi^{-1}(j, R\zeta)$. Since the function $|\zeta - \zeta_0|^{(j+2)}$ is integrable over L_ω, it follows hence and from (2.16) that a passage to the limit as $N \to \infty$ is possible in (2.14). As a result we obtain:

$$(g(A)-P(A))f = \frac{-1}{2\pi i}\int_{L_\omega} \frac{g(\zeta)-P(\zeta)}{(\zeta-\zeta_0)^{j+2}}\Phi^{-1}(j, R\zeta)(A-\zeta I)^{-1}f_1\,d\zeta, \tag{2.17}$$

where f_1 is determined by Formula (2.15), in which $N=\infty$. Bearing in mind (1.7), from (2.17) we obtain

$$\|g(A)f - P(A)f\| \leqq \frac{C}{2\pi}\int_{L_\omega}\frac{|d\zeta|}{|\zeta-\zeta_0|^2}\sup_{\zeta\in L_\omega}\frac{|g(\zeta)-P(\zeta)|}{|\zeta-\zeta_0|^j\Phi(j, R\zeta)}. \tag{2.18}$$

Note that

$$\Phi(j, R\zeta) = \Phi(0, R\zeta, \infty)/\Phi(0, R\zeta, j-1).$$

Since $\Phi(0, R\zeta, j-1)$ has the degree j, then $|\Phi(0, R\zeta, j-1)|/|\zeta - \zeta_0|^j$ is bounded and therefore

$$\sup_{\zeta\in L'_\omega}\frac{|g(\zeta)-P(\zeta)|}{|\zeta-\zeta_0|^j|\Phi(j, R\zeta)|} \leqq C_1 \sup_{L'_\omega}\frac{|g(\zeta)-P(\zeta)|}{|\mathrm{ch}(R\zeta)|}, \tag{2.19}$$

where $L'_\omega = L_\omega \cap \{|\mathrm{Re}\,\zeta| > 1\}$. Note that

$$\mathrm{ch}(s - i\tau) = \cos(is+\tau) = \cos(is)\cos\tau - \sin(is)\sin\tau.$$

Therefore when $\tau = \pm\omega \in \mathbb{R}$, $s \in \mathbb{R}$

$$|\mathrm{ch}(Rs - i\tau)|^2 = \mathrm{ch}^2(Rs)\cos^2\tau + \mathrm{sh}^2(Rs)\sin^2\tau.$$

Since when

$$s \geqq 1,\ \mathrm{sh}^2(Rs) \geqq C_2^2\,\mathrm{ch}^2(Rs),\ C_2 > 0,$$

then

$$|\mathrm{ch}(R(s-i\tau))| \geqq C_2\,\mathrm{ch}(Rs), \qquad s+i\tau \in L'_\omega. \tag{2.20}$$

It is obvious that

$$|\zeta-\zeta_0|^j\,|\Phi(j, R\zeta)| \geqq C_3 \geqq C_4\,\mathrm{ch}(R(\mathrm{Re}\,\zeta)),\ \zeta \in L_\omega \backslash L'_\omega, \tag{2.21}$$

where C_3, $C_4 > 0$. From (2.19), (2.20) and (2.21) we obtain:

$$\sup_{\zeta \in L_\omega} \frac{|g(\zeta)-P(\zeta)|}{|\zeta-\zeta_0|^j\,|\Phi(j, R\zeta)|} \leq C_5 \sup_{\zeta \in L_\omega} \frac{|g(\zeta)-P(\zeta)|}{\mathrm{ch}(R(\mathrm{Re}\,\zeta))}. \tag{2.22}$$

From (2.18) and (2.22) we obtain (2.13), and the lemma is proved.

Theorem 2.1. Suppose A is a self-determined first-order operator. Suppose $g \in A(\mathcal{J}_\beta)$, $\beta > \omega_0$, ω_0 is the same as in (1.7). Suppose

$$f \in \mathscr{D}_0,\ 0 < R < R_{02}(A, f),\ r = 2R/\pi,$$

and the polynomial $P_{2n-1} = \Pi_r^{2n-1} g$. Then Formula (4.1.1) holds, whilst $\forall\, \omega > \omega_0$, $\omega < \beta$, $\forall\, l \in \mathbb{Z}_+$, C exists such that

$$\|A^l(g(A)f - P_{2n-1}(A)f)\| \leqq C n^{-r(\beta-\omega)}, \qquad \forall\, n \in \mathbb{N}. \tag{2.23}$$

Proof. Inequality (2.23) follows from (2.13) and from estimate (3.3.3) for $\mu_{\omega,R}(g, P_{2n-1})$. Formula (4.1.1) follows from (2.23).

Definition 2.2. The operator A, which is determined in $\mathcal{D}_0$, is called a self-determined second-order operator if (2.1) holds and the closure of the operator A satisfies condition 1.2. In addition, $\forall f \in \mathcal{D}_0$ and $R > 0$ exists, such that

$$M_R^2(A, f) \equiv \sum_{j=0}^{\infty} \|A^j f\| R^{2j}/(2j)! < \infty. \tag{2.24}$$

We will assume when $f \in \mathcal{D}_0$

$$R_{02}(A, f) = \sup\{R \geqq 0 : M_R^2(A, f) < \infty\}. \tag{2.25}$$

Lemma 2.4. Suppose A is a self-determined second-order operator, $f \in \mathcal{D}_0$. Then for any R, $0 < R < R_{02}(A, f)$ for any $l \in \mathbb{Z}_+$, and for any ω, $\omega_0 < \omega$, where ω_0 is the same as in (1.19), the constant C exists such that for any function $g^+ \in A(\mathcal{J}_\beta^{\,+})$, $\beta > \omega$ and any polynomial P

$$\|A^l(g(A)f - P(A)f)\| \leqq C\mu_{\omega,R}^+(P, g^+), \tag{2.26}$$

where $\mu_{\omega,R}^+$ is determined by Formula (3.6.2).

Proof. Lemma 2.4 is proved in a similar way to Lemma 2.3. We shall indicate only the differences. Formula (1.23) is used instead of Formula (1.8), Proposition 1.5 instead of Proposition 1.4, and estimate (1.19) instead of estimate (1.7). Note that the finiteness of

$$\|\Phi(j, R\sqrt{A}, \infty)f\|,$$

follows from (2.24), and since the function $g^+(\zeta^2)$ is substituted in Formula (1.23) (instead of $g(\zeta)$) into (1.8), then

$$\Phi^{-1}(j, R\sqrt{\zeta^2}, N) = \Phi^{-1}(j, R\zeta, N).$$

enters formulae of the form (1.23), which correspond to Formulae (2.14) and (2.17) of the form (1.8). Therefore all the calculations in proving Lemma 2.4 are carried out in the same way as in proving Lemma 2.3. The finiteness of the integral of $|\zeta||\zeta^2 - \zeta_0^2|^{-2}$ in L_ω is used instead of that of the integral of $|\zeta - \zeta_0|^{-2}$ in L_ω.

A theorem similar to Theorem 2.1 holds.

Theorem 2.2. Suppose A is a self-determined second-order operator. Suppose

$$g^+ \in \mathscr{A}(\mathscr{J}_\beta^+),\ \beta > \omega_0,$$

ω_0 is the same as in (1.19). Suppose

$$f \in \mathscr{D}_0,\ 0 < R < R_{02}(A, f),\ r = 2R/\pi,$$

and the polynomial $P_n = \Pi_r^{+n} g^+$. Then Formula (4.1.1) holds, whilst $\forall \omega > \omega_0$, $\omega < \beta$, $\forall l \in \mathbb{Z}_+$, C exists such that

$$\|A^l(g^+(A)f - P_n(A)f)\| \leq Cn^{-r(\beta-\omega)}, \qquad \forall n \in \mathbb{N}. \tag{2.27}$$

Proof. Inequality (2.27) follows from (2.26) and from estimate (3.6.7) for $\mu^+_{\omega,R}(P_n, g^+)$. (4.1.1) follows from (2.27).

Lemma 2.5. Suppose $f \in D_\infty(A)$, whilst $R_{02}(A, f) > 0$. Then

$$\forall \zeta \in \mathbb{C},\ R_{02}(A - \zeta I, f) = R_{02}(A, f).$$

Proof. We shall prove that $R_{02}(A - \zeta I, f) > R_0(A, f)$. Suppose $R < R_{02}(A, f)$. We will assume $|\zeta| = b$. Obviously

$$\|(A - \zeta I)^j f\| \leq \sum_{i=0}^{j} \|A^i f\| b^{j-i} j!/(i!(j-i)!).$$

Therefore

$$\sum_{j=0}^{\infty}\|(A-\zeta I)^j f\|\frac{R^{2j}}{(2j)!}\leq\sum_{j=0}^{\infty}\sum_{i=0}^{j}\|A^i f\|\frac{b^{j-i}R^{2j}j!}{i!(j-i)!(2j)!}$$
$$=\sum_{i=0}^{\infty}\|A^i f\|\frac{R^{2i}}{(2i)!}\sum_{j=i}^{\infty}\frac{b^{j-i}R^{2(j-i)}j!(2i)!}{i!(j-i)!(2j)!}.$$

Since $(2i)!/(2j)! < i!/j!$ when $i < j$, hence we obtain, bearing in mind (2.24):

$$M_R^2(A-\zeta I,f)\leq\sum_{i=0}^{\infty}\|A^i f\|\frac{R^{2i}}{(2i)!}e^{bR^2}=e^{bR^2}M_R^2(A,f).$$

Therefore $R_{02}(A - \zeta I, f) > R_{02}(A, f)$. Putting $A - \zeta I = A_1$, $A = A_1 + \zeta I$ and applying the inequality obtained to A_1, we obtain that $R_{02}(A, f) > R_{02}(A - \zeta I, f)$. Hence we obtain a statement of the lemma.

Lemma 2.6. Suppose $f \in \mathcal{D}_\infty(A)$, whilst $R_{01}(A, f) > 0$. Then

$$\forall\zeta\in\mathbb{C},\ R_{01}(A-\zeta I,f)=R_{01}(A,f).$$

The proof of Lemma 2.6 is analogous to that of Lemma 2.5.

§3. First-order self-determined differential operators

Suppose $\Omega \subset \mathbb{R}^m$ is a bounded domain, and F is an operator defined on $\mathcal{D}_0 = (\mathcal{A}(\overline{\Omega}))^\varkappa$ by the formula

$$Fu=\sum_{j=1}^{m}a_j\partial_j u+a_0 u, \tag{3.1}$$

where a_j, $j = 0, \ldots, m$ is a $\varkappa \times \varkappa$-th order matrix, whose elements belong to $\mathcal{A}(\overline{\Omega})$, and the matrix a_j when $j = 0$ is Hermitian. It is assumed that the boundary $\partial\Omega$ of the domain Ω is piecewise-smooth, and the following condition of degeneracy holds on it:

$$\sum_{j=1}^{m} n_j(x)a_j(x)=0 \qquad \text{when } x\in\partial\Omega. \tag{3.2}$$

We will assume

$$b_1(x)=-\tfrac{1}{2}\sum_{j=1}^{m}(\partial_j a_j)(x)+\tfrac{1}{2}(a_0(x)+a_0^*(x)), \tag{3.3}$$

and we will assume $b_{1+}(x)$ is the maximum eigenvalue of the matrix $b_1(x)$, $b_{1-}(x)$ is the minimum eigenvalue,

$$b_+=\sup_{x\in\Omega} b_{1+}(x), \quad b_-=\inf_{x\in\Omega} b_{1-}(x). \tag{3.3'}$$

Lemma 3.1. Suppose $\lambda = \xi + i\eta$, $\xi, \eta \in \mathbb{R}$. Then $\forall u \in \mathcal{D}_0$

$$(\xi+b_+)\|u\|^2\geqq \operatorname{Re}((F+\lambda I)u,u)\geqq(\xi+b_-)\|u\|^2. \tag{3.4}$$

Proof. Obviously,

$$\operatorname{Re}(Fu,u)=\int_\Omega\left[\operatorname{Re}\sum_{j=1}^{m}\langle a_j\partial_j u,u\rangle+\operatorname{Re}\langle a_0u,u\rangle\right]dx, \tag{3.5}$$

where $\langle\,,\rangle$ is a standard scalar product in $\mathbb{C}^{\varkappa}$. Obviously, by virtue of the Hermitian character of a_j

$$2\operatorname{Re}\langle a_j\partial_j u,u\rangle=\partial_j\langle a_j u,u\rangle-\langle(\partial_j a_j)u,u\rangle. \tag{3.6}$$

Substituting (3.6) into (3.5), using Green's formula and bearing in mind that the boundary terms equal zero by virtue of Formula (3.2), we obtain:

$$\operatorname{Re}(Fu,u)=\int_\Omega\langle b_1(x)u,u\rangle\,dx. \tag{3.7}$$

Hence it follows that

$$b_-(u,u) \leqq \mathrm{Re}(Fu,u) \leqq b_+(u,u). \tag{3.8}$$

(3.4) obviously follows from (3.8).

Let us now consider the differential operator

$$A = i(F - b_0 I), \quad i = \sqrt{-1}, \quad b_0 = (b_+ + b_-)/2. \tag{3.9}$$

It is easy to see that by virtue of (3.4), if

$$\zeta \in \mathbb{C},\ \mathrm{Im}\,\zeta > \omega_0, \quad \omega_0 = (b_+ - b_-)/2,$$

then

$$\mathrm{Im}((A - \zeta I)\,u, u) \geqq (|\mathrm{Im}\,\zeta| - \omega_0)\,\|u\|^2. \tag{3.10}$$

Therefore when $|\mathrm{Im}\zeta| > \omega_0$

$$\|u\| \leqq (|\mathrm{Im}\,\zeta| - \omega_0)^{-1} \|(A - \zeta I)u\| \qquad \forall\, u \in \mathscr{D}_0. \tag{3.11}$$

Lemma 3.2. The set $|\mathrm{Im}\zeta| > \omega_0$ belongs to the resolvent set $\rho_0(A)$ of the operator A, determined by (3.9).

Proof. The set $\rho(A)$ of those ζ for which an estimate of the form (1.2) holds, contains the set $|\mathrm{Im}\zeta| > b$, by virtue of (3.11). This set consists of two connected components: $\rho_+ = \{\mathrm{Im}\zeta > \omega_0\}$ and $\rho_- = \{\mathrm{Im}\zeta < -\omega_0\}$. By virtue of Proposition 1.3, it is sufficient to prove that the point ζ_0 exists in each of these components, such that

$$(A - \zeta_0 I)H \text{ is dense in } H. \tag{3.12}$$

We will assume $\zeta_0 = iM$, where $|M|$ is fairly great. Obviously,

$$A-\zeta_0 M=i(F-b_0 I+MI)=iF_1,$$

where $F_1 = F - b_0 I + MI$. To prove (3.12), it is sufficient to prove that $F_1\mathcal{D}_0$ is dense in H. Suppose F_1^* is a differential operator, which is formally adjoint to F_1 (see Sect.4, Chapter 4). Then $F_1^*F_1 = B$ is a self-adjoint second-order differential operator, which is bounded from below. Since for large M, by virtue of (3.4),

$$\|F_1^* u\| \geqq \|u\|, \quad \|F_1 u\| \geqq \|u\|, \tag{3.13}$$

the operator B satisfies the condition $(Bu, u) > \|u\|^2$ in $\mathcal{D}_0$. Since $\mathcal{D}_0$ satisfies conditions (4.2.2)-(4.2.5) (see Example 3.2, Chapter 4), Theorem 4.2.3 is applicable. According to section 1) of this theorem $\forall f \in \mathcal{D}_0$ the sequence of polynomials P_n is obtained, such that

$$\|(B-bI)^l(B^{-1}f-P_n(B-bI)f)\| \to 0 \qquad (l \in \mathbb{Z}_+)$$

as $n \to \infty$ $\forall f \in \mathcal{D}_0$. Hence it follows that

$$\|B(B^{-1}f-P_n(B-bI)f\| = \|f-BP_n(B-bI)f\| \to 0 \tag{3.14}$$

as $n \to \infty$. We will assume $P_n(B - bI)f = u_n$. Since $B\mathcal{D}_0 \subset \mathcal{D}_0$, then $u_n \in \mathcal{D}_0$. Thus, it follows from (3.14) that $\forall f \in \mathcal{D}_0$ and the sequence $u_n \in \mathcal{D}_0$ is obtained, such that

$$Bu_n \to f \text{ as } n \to \infty \tag{3.15}$$

We shall now prove (3.12). For this it is sufficient to prove that $F_1\mathcal{D}_0$ is dense in H. Suppose $f \in \mathcal{D}_0$, $f_1 = F_1^*f$. According to (3.15)

$$F_1^* F_1 u_n - F_1^* f \to 0.$$

Since $F_1 u_n$ and f belong to $\mathcal{D}_0$, and $\|u\| \leqslant \|F_1^* u\|$ in $\mathcal{D}_0$, it follows that

$$F_1 u_n - f \to 0, \; u_n \in \mathcal{D}_0.$$

Since $f \in \mathcal{D}_0$ is arbitrary, and $\mathcal{D}_0$ is dense in H, it follows that $F_1 \mathcal{D}_0$ is dense in H, i.e. (3.12) is proved. It follows from (3.12) that $\zeta_0 \in \rho_0(A)$ and, by virtue of Proposition 1.3, both the domain ρ_+ and the domain ρ_- lie in $\rho_0(A)$.

Theorem 3.1. Suppose the operator A is determined by Formula (3.9), where F is determined by (3.1) and condition (3.2) holds. Suppose $\mathcal{D} = \mathcal{D}_0 = (A(\overline{\Omega}))^{\varkappa}$. Then A is a first-order self-determined operator. At the same time ω_0 in (1.7) is the same as in (3.10).

Proof. The existence of a resolvent follows from Lemma 3.2, and estimate (1.7) follows from (3.10). The finiteness of $M_k^1(A, f)$ for small R follows from Theorem 1.1.1, where $p = 1$, $K = \overline{\Omega}$.

We can thus calculate the functions in an operator of the form (3.9) and, consequently, of the form (3.1), using Formulae (4.1.1), where $P_n = \Pi_r^n g$, with the corresponding value of the parameter r.

Remark 3.1. If

$$A = \zeta_1 A_1 + \zeta_2 I, \; \zeta_1, \zeta_2 \in \mathbb{C},$$

the polynomial in A is obviously a polynomial in A_1. Note that the spectrum $\sigma(A)$ of the operator A (a complement with respect to the resolvent set) is connected to the spectrum $\sigma(A_1)$ of the operator A_1 by the relation $\sigma(A) = \zeta_1 \sigma(A_1) + \zeta_2$. In a number of cases $A = \zeta_1 A_1$ is a self-determined operator, at the same time as A_1 is not it.

Remark 3.2. According to Lemma 2.6, if A_1 is a first-order self-determined operator, then $A = A - \zeta_2 I$ is also a first-order self-determined operator, but possibly with another value of the constant ω_0 in (1.7). Since, as is obvious from (2.23), the rate of convergence in (4.1.1) essentially depends on ω_0, it is feasible to choose ζ_2, such that we minimise ω_0. For this it is required, roughly, to take ζ_2, such that the spectrum of the operator A is included in a strip of minimal width, arranged symmetrically with regard to the real axis (on which lie the nodes of the interpolation polynomials P_{2n-1}). The number b_0 in (3.9) was also chosen with this aim.

We shall now construct, on the basis of Theorem 2.1, polynomial representations of solutions of differential equations containing the operator F, which is determined by (3.1).

Consider the equation

$$Fu - \lambda u = f. \tag{3.16}$$

Theorem 3.2. Suppose F is determined by (3.1), whilst (3.2) holds. Suppose either $\operatorname{Re}\lambda > b_+$, or $\operatorname{Re}\lambda < b_-$, where b_+ and b_- are the same as in (3.3′). Suppose A is determined by Formula (3.9),

$$\zeta_0 = i\lambda - i(b_+ + b_-)/2, \quad g(\zeta) = (\zeta - \zeta_0)^{-1}.$$

Suppose

$$0 < R < R_{01}(A, f), \quad r = 2R/\pi, \quad P_{2n-1} = \Pi_r^{2n-1} g.$$

Suppose u is the solution of (3.16). Then $P_{2n-1}f \to u$ $(n \to \infty)$, whilst $\forall l \in \mathbb{Z}_+$ and $\varepsilon > 0$ C exists such that

$$\|A^l(u - iP_{2n-1}f)\| \leqq Cn^{-\sigma} \qquad \forall n \in \mathbb{N}, \tag{3.17}$$

where

$$\sigma = r(|\mathrm{Re}\,\lambda - (b_- + b_+)/2| - (b_+ - b_-)/2 - \varepsilon). \tag{3.18}$$

Proof. By virtue of (3.9) Eq.(3.16) takes the form $-iAu + b_0u - \lambda u = f$, or

$$Au - i(\lambda - b_0)u = if. \tag{3.19}$$

The solution of this equation is given by the formula $u = (a - \zeta_0 I)^{-1}if$. Obviously the function

$$g(\zeta) = (\zeta - \zeta_0)^{-1} \in \mathscr{A}(\mathscr{I}_\beta), \qquad \forall \beta < |\mathrm{Im}\,\zeta_0| = |\mathrm{Re}\,\lambda - (b_+ + b_-)/2| = \beta_0.$$

By virtue of Theorem 3.1, estimate (1.7), in which $\omega_0 = (b_+ - b_-)/2$, holds. Using Theorem 2.1, where $\beta = \beta_0 - \varepsilon/2$, $\omega = \omega_0 + \varepsilon/2$, we obtain estimate (3.17) from (2.23).

Let us now consider Cauchy's problem

$$\partial_t u(t) = Fu(t) + f_0, \quad f_0 \in \mathscr{D}_0, \tag{3.20}$$

$$u(0) = f_1, \quad f_1 \in \mathscr{D}_0. \tag{3.21}$$

Theorem 3.3. Suppose F is determined by (3.1), whilst (3.2) holds. Suppose A is determined by the formula

$$A=iF, \quad g_0(\zeta)=(e^{-i\zeta t}-1)/(i\zeta), \quad g_1(\zeta)=e^{-i\zeta t},\ t\in\mathbb{R}.$$

Suppose

$$f\in\mathscr{D}_0, \qquad 0<R<\min(R_{01}(A,f_0),\ R_{01}(A,f_1)),\ r=2R/\pi.$$

Suppose

$$P^0_{2n-1}=\Pi_r^{2n-1}g_0, \quad P^1_{2n-1}=\Pi_r^{2n-1}g_1$$

Suppose $u(t)$, $t \in \mathbb{R}$ is the solution of (3.20), (3.21). Then

$$P^0_n f_0+P^1_n f_1\to u\ (n\to\infty),$$

whilst

$$\|A^l(u(t)-P^0_{2n-1}(A)f_0-P^1_{2n-1}(A)f_1)\|\leqq C\,e^{-\sigma n}, \tag{3.22}$$

where $\sigma > 0$, and C does not depend on n.

Proof. Since $A=iB$ differs from A in (3.9) only by the terms ib_0I, A is a first-order self-determined operator. The soultion of problem (3.20), (3.21) is given by the formula

$$u(t)=g_0(A)f_0+g_1(A)f_1. \tag{3.23}$$

It is obvious that $g_0(A)f_0 + g_1(A)f_1$ and that the following estimate holds for the majorants $g_0^*(s)$ and $g_1^*(s)$ (see (3.5.0)) of the functions g_0 and g_1:

$$g_0^*(s)\leqq C\,e^{|t|s}, \quad g_1^*(s)\leqq C\,e^{|t|s}. \tag{3.24}$$

Using Theorem 3.5.1 and Theorem 3.5.3, we obtain from (3.5.1) and (3.5.14) an estimate for $g = g_0$ and $g = g_1$:

$$\mu_{\omega,R}(P_{2n-1}, g) = Cn^{r\omega}[(1+\varepsilon)(1-e^{-|t|/r})]^{2n}. \tag{3.25}$$

Using Lemma 2.3, from (3.23) and (3.13) we obtain

$$\|u - P^0_{2n-1} f_0 - P^1_{2n-1} f_1\| \leqq C(\mu_{\omega,R}(P^0_{2n-1}, g_0) + \mu_{\omega,R}(P^1_{2n-1}, g_1)).$$

Using (3.25), where $\varepsilon < e^{-t/2}$, we obtain (3.22).

At the end we will obtain the estimate of the numbers $R_{01}(A, f)$, where $A = iF$, F is determined by (3.1). For this it is useful to consider Cauchy's problem (3.20), (3.21), where

$$f_0 = 0,\ f_1 = f \in \mathscr{D}_0 = (\mathscr{A}(\bar{\Omega}))^{\kappa}.$$

According to the Cauchy-Kowalewska theorem, there exists a solution of this theorem that is analytic with respect to x and t when

$$x \in \bar{\Omega},\ t \in \mathbb{C}, \quad |t| < t_0$$

This solution decomposes into the series

$$u(t) = \sum_{k=0}^{\infty} \frac{(-iA)^k}{k!} f. \tag{3.26}$$

We shall use $R^K_{01}(A, f, p)$ to denote the radius of the convergence of this series in $L_p(\Omega)$. If $p=2$, we will assume, for brevity,

$$R^K_{01}(A, f, p) = R^K_{01}(A, f).$$

Lemma 3.3. Suppose $u(t, x)$ is the solution of (3.21), (3.22), where $f_0 = 0$, $f_1 = f$, $u(t, x)$ is analytic with respect to

$$(t, x) \in \{|t| < t_0\} \times \{x \in \bar{\Omega}\}, \quad F = L_p(\Omega).$$

Then $R_{01}(a, f) > t_0$.

Proof. Obviously, by virtue of Eq.(3.20)

$$\partial_t^k u(t, x)|_{t=0} = (-iA)^k f(x), \qquad k \in \mathbb{Z}_+. \tag{3.27}$$

According to Cauchy's formula, when $0 < \rho < t_0$

$$\frac{1}{k!}\partial_t^k u(t, x)|_{t=0} = \frac{1}{2\pi i}\int_{|t|=\rho} \frac{u(t, x)}{(\tau - t)^{k+1}} d\tau. \tag{3.28}$$

From (3.27) and (3.28) we obtain:

$$\|A^k f\|_{L_p}/k! \leqq \rho^{-k} \sup_{|t|=\rho} \|u(t, x)\|_{L_p}. \tag{3.29}$$

Therefore, if $e = L_p(\Omega)$, the value $M_R^1(A, f)$, determined by (2.2), where $\| \ \|$ is a norm in L_p, is finite when $R < \rho$. Since ρ is as close to t_0 as desired, the statement of the lemma follows.

Lemma 3.4. Suppose $E = L_p(\Omega)$. Then

$$R_{01}(A, f) \geqq R_{01}^K(A, f).$$

Proof. It is obvious that $u(t, x)$ is analytic with respect to t when $0 < t < R_{01}^K(A, f)$ Using Lemma 3.3, we obtain the statement of Lemma 3.4.

Example 3.1. Suppose in $T^1 = \mathbb{R}/(2\pi\mathbb{Z})$

$$Fu = \sin x\, \partial_x u + a(x)u \qquad (\partial_x = \partial/\partial x), \tag{3.30}$$

and $a(x) \in A(T^1)$ is an entire function.

Lemma 3.5. If $f \in A(T^1)$ is an entire function, $E = L_p(T^1)$, $p > 1$, then $R_{01}(F, f) > \pi/2$.

Proof. Problem (3.20), (3.21) takes the form

$$\partial_t u - \sin x\, \partial_x u = a(x)u, \qquad u|_{t=0} = f(x). \tag{3.31}$$

Carrying out the change

$$x = x(z) = 2 \operatorname{arctg} \operatorname{th} z/2 + \pi/2,$$

when $0 < x < \pi$, and noting that $\partial_x = \operatorname{ch} z\, \partial_z$, $\sin x(z) = 1/\operatorname{ch} z$, we obtain that (3.31) takes the form

$$\partial_t v - \partial_z v = a(x(z))v, \qquad v|_{t=0} = f(x(z)), \tag{3.32}$$

where

$$z \in \mathscr{I}_{\pi/2} = \{z \in \mathbb{C} : |\operatorname{Im} z| < \pi/2\}, \quad v(t, z) = u(t, x(z)).$$

The function $x(z)$ is analytic for these z (see the proof of Proposition 4.6.3.) The solution of (3.32) is given by formula

$$v(t, z) = f(x(z+t)) \exp \int_0^t a(x(z+t-s))\, ds. \tag{3.33}$$

Note that for each $z \in \mathbb{R}$ the function $x(z + \zeta)$ is holomorphic with respect to ζ when $|\zeta| < \pi/2$. Since $f(x)$ and $a(x)$ are entire functions, and for s, which belongs to the segment $[0, t]$, $|s| < |t|$, then the expression on the right-hand side of (3.33) is holomorphic with respect to t when $|t| < \pi/2$.
Therefore $v(t, x)$ is also holomorphic with respect to t and with respect to x when $|t| < \pi/2$, $x \in]0, \pi[$.

Since Eq.(3.31) does not change form when x is replaced by $\pi + x$, and t is replaced by $-t$, $u(t, x)$ is a solution of (3.31) which is holomorphic with respect to t and with respect to x when $x \in\,]0, \pi[\, \cup\,]\pi, 2\pi[$.

We shall estimate the radius of the convergence of series (3.26) in the small neighbourhood of the point $x=0$ using the majorant method. Suppose $K = \{x : |x| \leqslant \delta\}$, δ is sufficiently small. The following estimates hold for derivatives of $a(x)$ and $f(x)$ when $x \in K$:

$$|\partial_x^j a(x)| \leqq C_0 C_1^k k!, \quad |\partial_x^j f(x)| \leqq C_0 C_1^k k!, \tag{3.34}$$

$$|\partial_x \sin x| \leqslant 1 \text{ when } j \geqslant 1, \ |\sin x| \leqslant \delta \text{ when } x \in K.$$

Therefore

$$|\partial_x^j f(x)|, |\partial_x^j a(x)| \leqq \partial_x^j \frac{C_0}{1-C_1 x}\bigg|_{x=0}, \tag{3.35}$$

$$|\partial_x^j \sin x| \leqq \partial_x^j \frac{\delta + x}{1-x}\bigg|_{x=0}. \tag{3.36}$$

It is easy to derive from (3.35) and (3.36) that

$$\sup_{x \in K} |A^j f(x)| \leqq \left[\left(\frac{\delta + x}{1-x}\right)\partial_x + \frac{C_0}{1-C_1 x}\right]^j \frac{C_0}{1-C_1 x}\bigg|_{x=0}. \tag{3.37}$$

Consider the subsidiary problem

$$\partial_t v = \frac{\delta + x}{1-x}\partial_x v + \frac{C_0}{1-C_1 x} v, \quad v|_{t=0} = \frac{C_0}{1-C_1 x}. \tag{3.38}$$

We shall make the following change for x in the neighbourhood of zero $|x| < \delta/2$

$$z = (1+\delta)\ln(\delta + x) - x, \tag{3.39}$$

where z is in the neighbourhood of $-\infty$, $z < -M$. Obviously,

$$x=e^{z/(1+\delta)}\cdot e^{x/(1+\delta)}-\delta. \tag{3.40}$$

If Re$z < -M$, where M is sufficiently large the derivative of the right-hand side of (3.40) with respect to x is small. Therefore, by virtue of the theorem on the implicit function, x is an analytic function of $e^{z/(1+\delta)}$ for small $|e^{z/(1+\delta)}|$. Consequently, (3.40) determines the analytic function $x = x(z)$, which is holomorphic in the domain Re$z < -M$ and takes values in the small neighbourhood of zero. A formula which is analogous to Formula (3.33) (with other f and a, which are analytic in the neighbourhood of zero), holds for the function $v(t, x(z))$ - the solution of (3.38). If M is sufficiently great and Re$z < -2M$, then $z + \zeta \in \{\text{Re}z < -M\}$ when $|\zeta| < M$. Therefore $v(t, x(z))$ is holomorphic with respect to t when $|t| < M$, Re$z < -M$. Consequently, $v(t, x)$ is holomorphic with respect to t when $|t| < M$, $|x| < \delta_1$.

Taking $M > \pi$ and using the fact that the right-hand side of (3.37) equals - by virtue of (3.38) - the derivatives with respect to t from $v(t, x)$ when $t=0$, we obtain that

$$|A^j f(x)| \leqq C(\pi-\varepsilon)^{-j}/j!.$$

Hence it follows that the function $u(t, x)$ - the solution of (3.31) - is holomorphic with respect to t when $|x| < \delta_1$, $|t| \leqslant \pi/2$.

Since the point $x=\pi$ becomes the point $x=0$ under the translation $x \to x - \pi$, and (3.31) does not change its form when t is replaced by $-t$ and x is replaced by $x-\pi$, this holds at the point $x=\pi$. Therefore $u(t, x)$ is analytic with respect to t when $|t| < \pi/2$.

Using Lemma 3.3, where $t_0 = \pi/2$, we obtain a statement of Lemma 3.5.

§4. Second-order self-determined differential operators.

Let us first consider operators in the Hilbert space H of the form

$$Au = B_2u + B_1u. \tag{4.1}$$

It is assumed that $B_2\mathcal{D}_0 \subset \mathcal{D}_0$, $B_1\mathcal{D}_0 \subset \mathcal{D}_0$ and the following conditions hold in the set $\mathcal{D}_0$:

$$(B_2u, u) \geqq 0, \qquad \forall u \in \mathcal{D}_0, \tag{4.2}$$

$$(B_2u, v) = (u, B_2v), \qquad \forall u, v \in \mathcal{D}_0, \tag{4.3}$$

$$(B_1u, B_1u) \leqq C(B_2u, u), \qquad \forall u \in \mathcal{D}_0. \tag{4.4}$$

Lemma 4.1. Suppose

$$\zeta = \xi + i\eta, \quad \xi, \eta \in \mathbb{R}, \quad |\eta| > \sqrt{C/2} + \varepsilon, \quad \varepsilon > 0.$$

Then the constant $C_1 = C_1(\varepsilon)$ exists, such that $\forall\, u \in \mathcal{D}_0$

$$\|u\| \leqq C_1 \|Au - \zeta^2 u\| / \eta^2. \tag{4.5}$$

Proof. Suppose

$$Au - \zeta^2 u = f, \quad u \in \mathcal{D}_0, \quad f \in Au - \zeta^2 u.$$

Then

$$(f, u) = ((A - \zeta^2)u, u) = (B_2u, u) + (B_1u, u) - (\xi^2 - \eta^2 + 2\xi\eta i)(u, u).$$

Taking the real and imaginary part of the second equation, we obtain

$$(B_2u, u) - (\xi^2 - \eta^2)\|u\|^2 \leqq \|f\|\,\|u\| + \|B_1u\|\,\|u\|, \tag{4.6}$$

$$2|\xi||\eta|\|u\|^2 \leqq \|f\|\|u\| + \|B_1u\|\|u\|. \tag{4.7}$$

Using (4.4) and (4.7), from (4.6) we obtain:

$$C^{-1}\|B_1u\|^2 + \eta^2\|u\|^2 \leqq \|f\|\|u\| + \|B_1u\|\|u\| + (\|f\| + \|B_1u\|)^2/(4\eta^2).$$

We shall rewrite this inequality in the form

$$\begin{aligned}&[C^{-1} - 1/(4\eta^2)][\|B_1u\| - (2C^{-1} - 2/(4\eta^2))^{-1}\|u\|]^2\\ &\quad + [\eta^2 - (4C^{-1} - 1/\eta^2)^{-1}]\|u\|^2\\ &\leqq \|f\|\|u\| + (\|f\|^2 + 2\|f\|\|B_1u\|)/(4\eta^2).\end{aligned} \tag{4.8}$$

If $\eta^2 > C/2$, the coefficients for the quadratic terms on the left-hand side of (4.8) are positive, whence follows the boundedness of $\|u\|$ and $\|B_1u\|$ for bounded $\|f\|$. Hence follows inequality (4.5).

Lemma 4.2. Suppose conditions (4.1)-(4.4) hold. Suppose B_3 is a self-adjoint extension of the Friedrichs operator B_2, $\mathcal{D} = \mathcal{D}(B_3)$. Suppose $M > C + 1/C$, where C is a constant from (4.4). Then $\forall f \in H$ and the solution $u \in \mathcal{D}(B_3)$ of the following equation exists:

$$B_3u + B_1u + Mu = f. \tag{4.9}$$

This solution is given by the Formula

$$u = \sum_{k=0}^{\infty} (-1)^k[(B_3 + M)^{-1}B_1]^k(B_3 + M)^{-1}f. \tag{4.10}$$

Proof. By virtue of (4.3) the operator B_2 is symmetric, and by virtue of (4.4) it is nonnegative. Therefore the Friedrichs extension B_3 of the operator B_2 with the domain of definition $\mathcal{D}(B_3)$ exists. As is well known, $\mathcal{D}(B_3)$ is contained in a Hilbert space obtained by supplementing $\mathcal{D}_0$ using the norm produced by the scalar product $(B_2u, v) + (u, v)$. We will assume

$$\|u\|_1^2 = \|B_1u\|^2 + \|u\|^2. \tag{4.11}$$

By virtue of (4.4) $\mathcal{D}(B_3) \subset H_1$, where H_1 is the closure of $\mathcal{D}_0$ using the norm $\| \ \|_1$. Obviously, B_1 is continuous from H_1 to H. Since $M > 0$, the operator $(B_3 + M)$ is invertible. We shall rewrite (4.9) in the form

$$u + (B_3 + M)^{-1}B_1u = (B_3 + M)^{-1}f. \tag{4.12}$$

We shall seek the solution of (4.12) in the form of series (4.10). We shall prove that this series converges to H_1. We shall estimate the norm of the operator $(B_3 + M)^{-1}B_1$. Consider the equation

$$B_3v + Mv = B_1u, \tag{4.13}$$

where $u \in H_1$. Multiplying (4.13) scalarly by v and using (4.4), we obtain:

$$C^{-1}\|B_1v\|^2 + M\|v\|^2 \leq \|B_1u\| \, \|v\| \leq C^{-1}/2\|B_1u\|^2 + C/2\|v\|^2.$$

Hence, by virtue of (4.11), we obtain:

$$\|B_1v\|^2 + (CM - C^2/2)\|v\|^2 \leq 1/2\|u\|_1^2.$$

Since $M > C + 1/C$, then $CM - C^2/2 > C^2/2 + 1 > 1$. Therefore

$$\|v\|_1 < 1/\sqrt{2}\|u\|_1.$$

Consequently,

$$\|(B_3+M)^{-1}B_1\| \leqq 1/\sqrt{2} < 1,$$

in H_1, and series (4.10) converges to H_1. The function u, which is determined by (4.10), is obviously the solution of (4.12). It is obvious from (4.12) that $u \in \mathcal{D}(B_3)$, and therefore u is a solution of (4.9).

Lemma 4.3. Suppose the operator B, which is defined on $\mathcal{D}_0$, satisfies conditions (4.2.2)-(4.2.5). Then B is a second-order self-determined operator (in the space $E = H_G$).

Proof. Condition (2.24) of Definition 2.2 holds by virtue of (4.2.5), and condition (2.1) holds by virtue of (4.2.2). It remained to verify condition 1.2. Note that the set $\rho(B)$ (for $\zeta \in \rho(B)$ $\|u\| <$ $C\,\|(B - \zeta)u\|$) consists of all the points which do not lie on the real semiaxis $\zeta > b$. By virtue of Proposition 1.3, in order to verify Condition 1.2 it is sufficient to prove that the set $(B - bI + I)\mathcal{D}_0$ is dense in H. The operator B has a Friedrichs extension B' with the domain of definition $\mathcal{D}(B')$. The equation

$$B'u - bu + u = f \tag{4.14}$$

has a unique solution in $\mathcal{D}(B')$. To prove the density of $(B' - bI + I)\mathcal{D}_0$ in H, it is sufficient to prove that $\forall f \in H$ and the sequence $\{u_n\} \subset \mathcal{D}_0$ exists such that

$$(B' - bI + I)u_n \to f. \tag{4.15}$$

Since $\mathcal{D}_0$ is everywhere dense in H, it is sufficient to prove (4.15) for $f \in \mathcal{D}_0$. We will assume $A' = B' - bI$ and shall use Theorem 4.1.1, where $g(\lambda) = (\lambda + 1)^{-1}$, $A = A'$. It follows from Formula (1.17) that a sequence of polynomials will be obtained, such that $\forall j \in \mathbb{Z}_+$

$$A'^j P_n(A')f \to A'^j(A'+I)^{-1}f. \tag{4.16}$$

Hence it follows that, taking $u_n = P_n(A')f$, we have:

$$(A'+I)u_n \to (A'+I)(A'+I)^{-1}f = f. \tag{4.17}$$

Since $A'\mathcal{D}_0 \subset \mathcal{D}_0$, then $u_n = P_n(A')f \in \mathcal{D}_0$. Since $A' + I = B' - bI + I$, (4.15) follows from (4.17).

Theorem 4.1. Suppose the operator B_2 satisfies conditions (4.2.2)-(4.2.5), where $H_G = H$. Suppose the operator B_1 is defined on $\mathcal{D}_0$, $B_1\mathcal{D}_0 \subset \mathcal{D}_0$ and condition (4.4) holds. Suppose, in addition, $R_{02}(A, f) > 0$, $\forall f \in \mathcal{D}_0$ for the operator A, determined by (4.1). Then A is a second-order self-determined operator (in the space $E=H$).

Proof. It follows from Lemma 4.1 that the set $\rho(A)$ contains the supplement Ω_ω to the set $\mathcal{J}^1_\omega$, if ω is fairly great. Since Ω_ω is connected, then to satisfy condition 1.2 it is sufficient to prove that $(A - \zeta_0^2)\mathcal{D}_0$ is everywhere dense in $\mathcal{D}_0$ for some $\zeta_0^2 \in \Omega$. We shall take as ζ_0^2 the number $-M$, $\zeta_0^2 = -M$, where $M > 0$ is sufficiently large. It is required to prove that the equation

$$B_2u + B_1u + Mu = f \tag{4.18}$$

has the solution $u \in \mathcal{D}_0$ for the set of functions f, which is everywhere dense in H. Since $\mathcal{D}_0$ is dense in H, it is sufficient to prove that $\forall f \in \mathcal{D}_0$ and the sequence u_n exists, such that

$$B_2u_n + B_1u_n + Mu_n \to f, \qquad u_n \in \mathcal{D}_0. \tag{4.19}$$

Consider Eq.(4.9), which differs from (4.18) only in that the solution belonging to $\mathcal{D}(B_3)$ is sought in (4.9). According to Lemma 4.2 $\forall f \in H$ and the solution $u \in \mathcal{D}(B_3) \subset H_1$ of Eq.(4.9) exists. According to Lemma 4.3, since the operator $(B_2 + MI)$ satisfies all its conditions, the sequence $u_n \in \mathcal{D}_0$ exists,

$$(B_2+MI)u_n=(B_3+MI)u_n\to(B_3+MI)u. \tag{4.20}$$

Note that, by virtue of (4.4) and (4.11), it follows from (4.20) that $u_n \to u$ in H_1. Since B_1 is continuous from H_1 to H, and $B_3 = B_2$ on $\mathcal{D}_0$, then

$$(B_2+MI)u_n+B_1u_n\to(B_3+MI)u+B_1u=f,$$

and (4.19) is proved. Thus $\zeta_0^2 = -M$ belongs to $\rho_0(A)$. Consequently, A satisfies condition 1.2. Condition (2.24) holds for any

$$f\in\mathcal{D}_0=(\mathcal{A}(\bar{\Omega}))^{\kappa}$$

by virtue of the requirements of the theorem. Consequently, all the requirements of Definition 2.2 hold for A, and the theorem is proved.

Let us now consider examples of differential operators. Suppose the operator B is determined in the bounded domain $\Omega \subset \mathbb{R}^m$ by Formula (4.3.1), whilst conditions (4.3.3) hold, and the condition of degeneracy (4.3.5) holds on $\partial\Omega$. Suppose the operator b_1 is determined in $\mathcal{D}_0 =)\mathcal{A}(\bar{\Omega}))^{\varkappa}$ by the formula

$$B_1u=\sum_{j=1}^{m} a_j(x)\,\partial_j u+a_0(x)u. \tag{4.22}$$

It is assumed that the elements of the matrix a_j, $j = 0, \ldots, m$ belong to $A(\overline{\Omega})$, and that the following condition of subordination holds:

$$b^1(\zeta) \equiv \left\langle \sum_{j=1}^{m} a_j(x)\zeta_j, \sum_{j=1}^{m} a_j(x)\zeta_j \right\rangle \leqq C_0 a(\zeta)$$
$$\forall \zeta = (\zeta_1, \ldots, \zeta_m) \in \mathbb{C}^{\kappa m} \qquad (\zeta_j \in \mathbb{C}^{\kappa}). \tag{4.23}$$

Here $a(\zeta)$ is the same as in (4.3.3).

Theorem 4.2. Suppose the operator B and B_1 satisfy the conditions $\mathcal{D}_0 = (A(\overline{\Omega}))^{\varkappa}$, formulated above. Then the operator $A = B + B_1$ is a second-order self-determined operator.

Proof. We will assume $B_2 = B + MI$, where $M > 0$ is a fairly large number, $B_1' = B_1 - MI$. It follows from the calculations in Formulae (4.3.4) and (4.3.6) that

$$(B_2 v, v) = \int_{\Omega} a(\partial v)\, dx + \int_{\Omega} \langle a_{00} v, v \rangle\, dx + M\|v\|^2. \tag{4.24}$$

It follows from (4.22) that

$$(B_1' v, B_1' v) \leqq 2 \int_{\Omega} b^1(\partial v)\, dx + 2 \int_{\Omega} \langle (M - a_0)v, (M - a_0)v \rangle\, dx. \tag{4.25}$$

It follows from (4.24) and (4.25) that if M is so great that

$$((M + a_{00})v, v) \geqq \langle v, v \rangle, \quad \forall v \in \mathbb{C}^{\kappa},$$

inequality (4.4) holds, where $B_1 = B_1'$ with some constant C. Conditions (4.2.2)–(4.2.5) hold for B_2 by virtue of Theorem 4.3.2. Thus, all the conditions of Theorem 4.1 hold. Using this theorem, we obtain a statement of Theorem 4.2.

We shall now consider differential equations which contain the operator $B + B_1$. Consider in $\Omega \subset \mathbb{R}^m$ the equation

$$Bu + B_1 u - \zeta_0^2 u = f, \qquad f \in (\mathscr{A}(\bar{\Omega}))^\kappa, \tag{4.26}$$

where B is determined by Formula (4.3.1), and B_1 is determined by Formula (4.22).

Theorem 4.3. Suppose conditions (4.3.3), (4.3.5) and (4.23) hold for the operators B and B_1. Then $\omega_0 > 0$ will be obtained, such that when $|\mathrm{Im}\,\zeta_0| > \omega_0$ Eq.(4.2.6) has the solution u. Suppose

$$0 < R < R_{02}(A, f), \quad A = B + B_1, \quad r = 2R/\pi.$$

Suppose $g^+(\zeta) = (\zeta - \zeta_0^2)^{-1}$, and the polynomial $P_n = \Pi_r^{+n} g^+$. Then when $|\mathrm{Im}\,\zeta| > \omega_0$, $P_n(A)f \to u$ $(n \to \infty)$ At the same time for any $l \in \mathbb{Z}_+$ and for any β, ω, $|\mathrm{Im}\,\zeta_0| > \beta > \omega > \omega_0$ the constant C exists, such that

$$\|A^l(u - P_n(A)f)\| \leqq Cn^{-r(\beta-\omega)}. \tag{4.27}$$

Proof. The operator A is a second-order self-determined operator, by virtue of Theorem 4.2. Therefore Formula (1.19) holds for some ω_0. When $|\mathrm{Im}\,\zeta_0| > \omega_0$, we derive the solvabilty of (4.26) from (1.19). Since

$$g^+(\lambda) = (\lambda - \zeta_0^2)^{-1} \in \mathscr{A}(\mathscr{I}_\beta^+)$$

when $\beta < |\mathrm{Im}\,\zeta_0|$, Theorem 2.2 is applicable. We obtain (4.27) from estimate (2.27), and Theorem 4.3 is proved.

We shall now give the results of polynomial representations of solutions of Cauchy's problems.

Theorem 4.4. Suppose $\mathcal{D}_0 = (A(T^{m}))^{\varkappa}$, and the operator A is determined by Formula (4.1), where $B_2 = B$ and B_1 are the same as in Theorem 4.3. Suppose f_0, $f_1 \in \mathcal{D}_0$, $H = L_2(\Omega)$. Then Cauchy's problem (11.3.5), (1.3.6), where $L = A$, has the solution $u(t)$ when $t > 0$. This solution is given by Formula (1.3.7), where the functions $g_{11}(\lambda)$ and $g_{10}(\lambda)$ are determined by (1.3.8), and the function of the operator is understood in the sense of (1.22), where ω is fairly great. Polynomial representations of the type (4.2.18), where $A = B - bI$, hold for $u(t)$ for fixed t. At the same time

$$P_n^1 = \Pi_r^{+n} g_{11}^+,\ P_n^0 = \Pi_r^{+n} g_{10},\ r = 2R/\pi,$$

$$0 < R < R_0 = \min(R_{02}(A, f_0), R_{02}(A, f_1)).$$

The following estimate holds:

$$\|A^l(u(t) - P_n^0(A) f_0 - P_n^1(A) f_1)\| \leq C e^{-\sigma_1 \ln^2 n}, \tag{4.28}$$

where $\sigma_1 = (1 - \varepsilon) r^2/(4\Theta)$, when $\operatorname{Re} t > 0$, $|t|^2/\operatorname{Re} t < \Theta$, and ε is arbitrarily small.

Proof. The functions g_{11} and $g_{10}(= g^+)$ belong to

$$\mathcal{A}(\mathcal{J}_\infty^+), \quad \mathcal{D}_0 \subset \bigcap \mathcal{D}(A^k).$$

The operator A is a second-order self-determined operator, therefore it satisfies condition 1.2 and the function $g(A)$ is defined by Formula (1.23). Direct verification shows that Formula (1.3.7) determines the solution $u(t)$ of problem (1.3.5), (1.3.6). We obtain estimate (4.28) from estimate (2.26) and estimates (3.6.9), (5.1.11) and (3.5.5), where $p = 2$, $t = \Theta$.

Remark 4.1. The form of estimates (4.28) and the analogous estimate (4.2.10), obtained in the case of the self-adjoint operator A is completely identical. This is connected with the fact that the term $n^{r\omega}$ in estimate (3.6.9), which gives an increasing factor when $\omega \neq 0$, i.e. in the nonselfadjoint case, plays a subordinate role with respect to the principal cofactor of the form $e^{-\sigma_1 \ln^2 n}$, and does not substantially affect σ_1.

Theorem 4.5. Suppose $\mathfrak{D}_0 = (A(T^m))^{\varkappa}$, and the operator A is determined by Formula (4.1), where $B_2 = B$ and B_1 are the same as in Theorem 4.2. Suppose $f_0, f_1, f_2 \in \mathfrak{D}_0$, $H = L_2(\Omega)$. Then Cauchy's problem (1.3.10), (1.3.11), where $L = A$, has a solution which is given by Formula (1.3.12), where the functions $g_{2j}(\lambda)$ are determined by Formula (1.3.13), and the functions of the operator are understood in the sense of Formula (1.22), where ω is fairly great. Polynomial representations of the type (4.2.18), where $B - bI = A$, hold for $u(t)$ for fixed t. At the same time

$$P_n^0 = \Pi_r^{+n} g_{20}, \quad P_n^1 = \Pi_r^{+n} g_{21}, \quad P_n^2 = \Pi_r^{+n} g_{22}, \qquad r = 2R/\pi,$$
$$0 < R < \min(R_{02}(A, f_0), R_{02}(A, f_1), R_{02}(A, f_2)).$$

The following estimate holds:

$$\| A^l [(u(t) - P_n^0(A) f_0 - P_n^1(A) f_1 - P_n^2(A) f_2] \| \leqq C e^{-\sigma_2 n}, \tag{4.29}$$

where $\sigma_2 > 0$.

Proof. Since the functions

$$g = g_{2i} \in \mathscr{A}(\mathscr{I}_\infty^+), \ \mathscr{D}_0 \in \cap \mathscr{D}(A^k),$$

the integral (1.22) is determined. Direct verification shows that it gives a solution of problem (1.3.10), (1.3.11). Estimate (4.29) follows from estimate (2.26) and estimates (3.6.9), (5.1.12) and (3.5.14).

§5. The smoothness of functions of differential operators.

In this paragraph, as in Chapter 5, we will obtain estimates of the smoothness of functions of differential operators from §§3 and 4.

Theorem 5.1. Suppose the operator A is determined by Formula (3.9), and suppose $g \in \mathcal{A}(\mathcal{J}_\beta)$, where $\beta > \omega_0$, ω_0, is a number from inequality (3.10). Suppose $f \in \mathcal{D}_0$, $R_0 = R_{01}(A, f)$. Then

$$u \in W^s_{2\,\mathrm{loc}}(\Omega), \quad \forall s: 0 \leqq s < r(\beta - \omega_0), \qquad (r = 2R_0/\pi). \tag{5.1}$$

If $\partial\Omega$ in Ω_1 satisfies condition 5.1.2, the compactum $K \subset \overline{\Omega} \cap \Omega$, then $u \in C^l(K)$ when $2l < r(\beta - \omega_0) - 3m/2$.

Proof. Theorem 5.1 is proved in a similar way to Theorem 5.6.1. The single difference is that in estimate (2.23), in the non-selfadjoint case $\omega \neq 0$, unlike the estimate in the selfadjoint case. The estimate $\|y_N\|$ in the proof of Theorem 5.6.1, which has the form $\|y_N\| \leqslant CM_0 2^{-Nr}\rho$, is derived from estimate (5.2.36), which is derived from estimate (5.2.5):

$$\begin{aligned} \|y_N\| = \|u_{N+1} - u_N\| &\leqq (\mu^+_{0,R}(P_n, g^+) \\ &+ \mu^+_{0,R}(R_{2n}, g^+)) \|\mathrm{ch}\, R\sqrt{A} f\|. \end{aligned} \tag{5.2}$$

The corresponding estimate for y_N, based on Formula (2.13), has the form

$$\|u_{N+1} - u_N\| \leqq C_1(\mu_{\omega,R}(P_{2n+1}, g) + \mu_{\omega,R}(P_{4n+2}, g)). \tag{5.3}$$

Bearing in mind (3.6.6), we see that these estimates differ only in that $\mu_{\omega,R}$ is substituted into (5.2) instead of $\mu^+_{0,R}$ Therefore, according to (3.3.3), we obtain the estimate $\|y_N\| \leqslant Cn^{-r(\rho-\omega)}$, where $\rho = \beta - \varepsilon$, and ε is small, instead of the estimate $\|y_N\| \leqslant Cn^{-r\rho}$ when $\omega \neq 0$. These estimates have an identical form, with the single difference that in the second estimate $\rho_1 = \rho - \omega$, $\omega \neq 0$. Replacing ρ in the proof of Theorem 6.1 by $\rho_1 = \beta - \omega - \varepsilon$, we obtain the statement (5.1), in the same way as (5.6.4).

When condition 5.1.2 holds, the estimate of the smoothness in $C^l(\overline{\Omega} \cap \Omega_1)$ is found in the same way as in the proof of Theorem 5.1.1′ in §6 of Chapter 5. At the same time, as has already been mentioned, β is replaced by $\beta - \omega$.

Theorem 5.2. Suppose the operator A is determined by Formula (4.1), where $B_2 = B$ and B_1 satisfy conditions (4.3.3), (4.3.5) and (4.23). Suppose ω_0 is such that condition (1.19) holds. Suppose

$$g^+ \in \mathscr{A}(\mathscr{I}_\beta^+),\ \beta > \omega_0.$$

Suppose

$$f \in \mathscr{D}_0,\ R_0 = \mathrm{R}_{02}(A, f),\ u = g^+(A)f.$$

Then

$$u \in W_{2\,\mathrm{loc}}^s(\Omega), \qquad \forall\, s: 0 < s < r(\beta - \omega_0). \tag{5.4}$$

If $\partial\Omega$ satisfies condition 5.1.2 in Ω_1, then for any compactum K $\Omega_1 \cap \overline{\Omega}$, $u \in C^l(K)$ when $2l < r(\beta - \omega_0)$.

Proof. Statement (5.4) is proved in exactly the same way as (5.6.4). The following estimate, based on inequality (2.26), is used instead of estimate (5.2) to derive estimate (5.2.36):

$$\|y_N\| = \|u_{N+1} - u_N\| \leqq C_1[\mu_{\omega,R}^+(P_n, g^+) + \mu_{\omega,R}^+(P_{2n}, g^+)]. \tag{5.5}$$

Using (3.6.7), where $\omega > 0$, we obtain the estimate $\|y_N\| < C_3 n^{-r(\rho-\omega)}$ instead of the estimate $\|y_N\| < C_2 n^{-r\rho}$. Hence, replacing ρ by $\rho_1 = \rho - \omega$, we prove (5.4) in the same way as (5.6.4) when proving Theorem 6.1.

We obtain the estimates in C^l in the same way as when proving Theorem 5.1.1′.

Taking $g(\xi) = (\zeta - \zeta_0)^{-1}$ and $g^+(\zeta) = (\zeta - \zeta_0{}^2)^{-1}$ as the corollary from Theorems 5.1 and 5.2, we obtain theorems on the smoothness of solutions of differential equations.

Theorem 5.3. Suppose the conditions of Theorem 5.2 hold, and $u(x)$ is a solution of the equation $Fu - \lambda u = f$, $A = i(F - b_0 I)$ where b_0 is the same as in (3.9). Then

$$u \in W^s_{2\,\mathrm{loc}}(\Omega), \qquad \forall s < s_0, \quad s_0 = r(|\mathrm{Re}\,\lambda - b_0| - (b_+ - b_-)/2), \tag{5.6}$$

where $r = 2R_0(A, f)/\pi$, and b_+ and b_- are the same as in (3.3′). If $\partial\Omega$ satisfies condition 5.1.2 in Ω_1, and the compactum $K \subset \overline{\Omega} \cap \Omega_1$, then $u \in C^l(K)$ when $2l < s_0 + 3m/2$.

Remark 5.1. Estimates of the smoothness of solutions of first-order systems on a torus are obtained by Moser [1] using other methods and in other terms.

Theorem 5.4. Suppose the operators B and B_1 are the same as in Theorem 4.3. Suppose u is a solution of the equation

$$Bu + B_1 u - \zeta_0^2 u = f, \quad f \in \mathscr{D}_0 = (\mathscr{A}(\overline{\Omega}))^\kappa, \quad |\mathrm{Im}\,\zeta_0| > \omega_0.$$

Then

$$u \in W^s_{2\,\mathrm{loc}}(\Omega), \qquad \forall s < s_0, \quad s_0 = r(|\mathrm{Im}\,\zeta_0| - \omega_0), \tag{5.7}$$

where $r = 2R_0(B + B_1, f)/\pi$. If $\partial\Omega$ satisfies condition 5.1.2 in Ω_1, then $u \in C^l(K)$ when $2l < s_0 - 3m/2$, $K \subset \Omega_1 \cap \overline{\Omega}$.

Remark 5.2. The estimates of smoothness given in (5,6) and (5.7) cannot be improved. This is proved by the following examples.

Example 5.1. Consider the equation

$$Fu \equiv \sin x\, \partial_x u + \cos x u + \beta u = \sin x, \tag{5.8}$$

$\beta > 1/2$, which is another form of noting the equation (5.4.11). The operator F, which acts on the circle T^1, (or on the segment $[0, 2\pi]$), obviously has the form (3.1). The function $b_1(x)$, which is defined by Formula (3.3), equals $b_1(x) = 1/2 \cos x + \beta$. Therefore the numbers b_- and b_+, which are determined by (3.3'), equal:

$$b_+ = \tfrac{1}{2} + \beta, \quad b_- = \beta - \tfrac{1}{2}. \tag{5.9}$$

The operator A, determined by (3.9), equals

$$Au = i(\sin x\, \partial_x u + \cos x u). \tag{5.10}$$

According to Lemma 3.5 $R_{01}(A, f) \geq \pi/2$. Therefore (5.6) for the above example takes the form ($\lambda=0$)

$$u \in W^s_{2\,\mathrm{loc}}(]0, 2\pi[), \qquad \forall s < |\beta| - \tfrac{1}{2}. \tag{5.11}$$

It follows from Lemmas 5.4.3 and 5.4.4 that

$$u \notin W_2^{\beta - 1/2}(]0, 2\pi[).$$

Therefore it is impossible to replace the sign of strict inequality in (5.6) with the sign of non-strict inequality, and it is impossible to replace $r(\beta - \omega_0)$ in (5.1) with $r(\beta - \omega_0) + \varepsilon$, $\varepsilon > 0$.

Example 5.2. Suppose the operator A is determined in $\mathcal{A}([0, 2\pi])$ by Formula (5.10). We will assume $B = A^2$.

Proposition 5.1. The operator B, which is determined in $\mathcal{D}_0 = A([0, 2\pi])$, is a second-order self-determined operator, whilst we can assume $\omega_0 = 1/2$ in (1.19), where $A = B$.

Proof. It is obvious that A^2 can be written in the form $A^2u = A^*Au + (A - A^*)Au$. Since $A - A^*$ is an operator of zero order (i.e. it is bounded), then A^2u can be represented in the form $Bu + B_1u$, and by virtue of Theorem 4.2 A^2 is a second-order self-determined operator.

We shall now prove that we can assume $\omega_0 = 1/2$ in (1.19).

The operator $A^2 - \zeta^2 I$ can be represented in the form $(A - \zeta I)(A + \zeta I)$. Since (3.10) holds, where, according to (5.9), $\omega_0 = 1/2$, then when $|\operatorname{Im}\zeta| > \frac{1}{2}$

$$\begin{aligned} \|(A-\zeta I)(A+\zeta I)u\| &\geqq (|\operatorname{Im}\zeta| - \tfrac{1}{2})\|(A+\zeta I)u\| \\ &\geqq (|\operatorname{Im}\zeta| - \tfrac{1}{2})^2\|u\|. \end{aligned} \tag{5.12}$$

Hence it follows that we can take $\omega_0 = 1/2$.

We shall now use Theorem 5.4 to estimate the smoothness of the solution of the equation

$$A^2u - \zeta^2 u = f, \tag{5.13}$$

where f is an entire function. According to Lemma 3.5 $R_{01}(A, f) > \pi/2$. The finiteness of $M_R^2(A^2, f)$, determined by (2.24), obviously follows from the finiteness of $M_R^1(A, f)$, determined by (2.2). Therefore, $R_{02}(A^2, f) > \pi/2$. Taking $\zeta = i\beta$, $\beta > 0$ in (5.13), from (5.7) we obtain

$$u \in W^s_{2\,\mathrm{loc}}(\Omega), \qquad \forall s < \beta - \tfrac{1}{2}, \quad (\Omega =]0, 2\pi[). \tag{5.14}$$

We shall take the function $i(A - \zeta I)\sin x$ as f. Then the solution of (5.13) is obviously identical with that of the equation $Au + \zeta u = i\sin x$, i.e. (since $\zeta = i\beta$, and A is determined by (5.10)) with the solution of Eq.(5.8). Since this solution does not belong to

$$W_{2\,\mathrm{loc}}^{\beta-1/2}(]0, 2\pi[),$$

it is impossible to replace the sign of strict inequality in (5.14) with the sign of non-strict inequality. Therefore it is also impossible to do so in (5.7), and it is also impossible to replace $r(\beta - \omega_0)$ in (5.4) with $r(\beta - \omega_0) + \varepsilon$.

Let us now consider the question of the smoothness with respect to x of the functions $g(A)f(x)$ and $g^{+}(A)f(x)$ for

$$g \in \mathscr{A}(\mathscr{I}_\infty),\ g^{+} \in \mathscr{A}(\mathscr{I}_\infty^{+}).$$

Theorem 5.5. Suppose A is determined by Formula (3.9), and suppose $g \in A(\mathscr{I}_\infty)$. Suppose ω_0 is a number from (3.10), $R_0 = R_{01}(A, f)$. Then for any compactum $K_0 \subset \Omega$, $\forall \varepsilon > 0$ C exists such that for $u = g(A)f$ the following estimate holds:

$$\|\partial^\alpha u\|_K \leqq C R_1^{|\alpha|}\,|\alpha|!\,g^{**}(\omega_0 + |\alpha|/(r-\varepsilon)), \tag{5.15}$$

where $\|\ \|_K$ is determined in (5.1.5), and g^{**} is the majorant of the function g. If $\partial\Omega$ in Ω_1 satisfies condition 5.1.2, then for any compactum $K \subset \overline{\Omega} \cap \Omega_1$

$$\|\partial^\alpha u\|_{C(K)} \leqq C R_2^{|\alpha|}|\alpha|!\,g^{**}(\omega_0 + (2|\alpha| + 3m/2)/(r-\varepsilon)). \tag{5.15'}$$

Proof. Theorem 5.5 is proved in a similar way to Theorem 5.7.1. The function u is represented by the formula $u = u_0 + w$, where $u_0 = g(\lambda_1)f$, w is determined by Formula (5.2.3), where by virtue of (2.13)

$$y_N = P_{2n+1}(A)f - P_{4n+1}(A)f$$

satisfy estimate (5.3), and not (5.2). As is obvious from the proof of Lemma 5.3.1 for

$$\mu_{\omega,R}(P_{2n+1}, g) + \mu_{\omega,R}(P_{4n+1}, g),$$

there is an estimate which is similar to (5.3.7). Estimate (5.6.17) holds as previously. Therefore the statement of Lemma 5.3.4 holds and we have the estimate (5.3.14) (in which there is a constant which is proportional to $M_R^{\ 1}$, as $C(\varepsilon)$). Unlike the case considered in Theorem 5.7.1 in (5.3.14) $\omega > 0$. Formally the case $\omega > 0$ reduces to the case $\omega = 0$, if we set $g^{**}(\omega + s) = g^{**}_{\omega}(s)$. Therefore estimate (5.7.17) holds when we replace g^{**} by g^{**}_{ω}. Consequently, estimates (5.7.1) and (5.1.6) hold, where g^{**} is replaced by g^{**}_{ω}. Bearing in mind that ω is arbitrarily close to ω_0, and the number ε in (5.1.6) is arbitrary, we obtain that estimate (5.15) holds.

Estimate (5.15′) is proved in the same way as estimate (5.1.6′). The proof is carried out in the same way as that of Theorem 5.1.2′ in §7 of Chapter 5, by replacing g^{**} with g^{**}_{ω}.

Theorem 5.6. Suppose the operator A, and the number ω_0 and the function f are the same as in Theorem 5.2. Suppose

$$g^+ \in \mathcal{A}(\mathcal{J}^+_\infty), \; r = 2R_{02}(A, f)/\pi.$$

Suppose $u = g^+(A)f$. Then the nunbers R_1 and C exist for any compactum $K \subset \Omega$, $\forall \varepsilon > 0$, such that

$$\|\partial^\alpha u\|_K \leq CR_1^{|\alpha|}|\alpha|!g^{**}(\omega_0 + |\alpha|/(r-\varepsilon)). \tag{5.16}$$

If $\partial\Omega$ in Ω_1 satisfies condition 5.1.2, the following estimate holds for the compactum $K \subset \Omega_1 \cap \Omega$:

$$\|\partial^\alpha u\|_{C(K)} \leqq CR_1^{|\alpha|}|\alpha|!g^*(\omega_0+(2|\alpha|+3m/2)/r-\varepsilon)). \tag{5.16'}$$

The proof of Theorem 5.6 is similar to that of Theorem 5.7.1, bearing in mind the remarks made when proving Theorem 5.5.

Remark 5.3. Replacing s by $\omega_0 + s$ for the functions $g^{**}(s) = e^{\sigma s^2+s}$ and $g^{**}(s) = e^{\sigma s+s}$ does not contribute significant changes to the increase in $g(s)$, since the change in the constant R_1 makes changes of that type. Therefore the following theorems hold.

Theorem 5.7. Suppose the operator A, the functions f_0 and f_1 and the number R_0 are the same as in Theorem 4.4. Suppose $u(t)$ is the solution of Cauchy's problem (1.3.5), (1.3.6), where $L=A$. Then estimate (5.1.10) holds for $u(t)$.

Theorem 5.8. Suppose the operator A and the functions f_0, f_1 and f_2 are the same as in Theorem 4.5, and $u(t, x)$ is the solution of Cauchy's problem (1.3.10), (1.3.11), where $L=A$. Suppose $\partial\Omega$ satisfies condition 5.1.2 in Ω_1. Then $u(t,x)$ is analytic with respect to x and t in $\mathbb{R} \times K$ for any compactum $K \subset \Omega$ or $K \subset \overline{\Omega} \cap \Omega_1$.

Theorem 5.9. Suppose the operator F and the functions f_0 and f_1 are the same as in Theorem 3.3. Suppose $\partial\Omega$ satisfies condition 5.1.2 in Ω_1. Then the solution $u(t, x)$ of problem (3.20), (3.21) is analytic with respect to x and t in $\mathbb{R} \times K$ for any compactum $K \subset \Omega$ or $K \subset \overline{\Omega} \cap \Omega_1$.

Proof. The analyticity with respect to x follows from estimate (5.16'), since estimate (3.24) holds for the majorant $g^*(s)$.

§6. Example of a self-determined operator in the space L_p $p \neq 2$

Consider in $A(T^1)$, $T^1 = \mathbb{R}/(2\pi\mathbb{Z})$ the operator A, determined by Formula (5.10). We shall use $\| \ \|_p$, $p > 1$ to denote the norm in the space $L_p(T^1)$

$$\|u\|_p^p = \int_0^{2\pi} |u|^p \, dx. \tag{6.1}$$

Proposition 6.1. Suppose $|\mathrm{Im}\,\zeta| > 1 - 1/p$. Then

$$\|Au - \zeta u\|_p \geqq (|\mathrm{Im}\,\zeta| - 1 + 1/p)\,\|u\|_p. \tag{6.2}$$

Proof. We shall multiply $Au - \zeta u$ by $|u|^{p-2}\bar{u}$

$$(Au - \zeta u)\bar{u}|u|^{p-2} = i[\sin x\,\partial u|u|^{p-2}\bar{u} + \cos x|u|^p + i\zeta|u|^p].$$

The imaginary part of this expression equals:

$$\begin{aligned} &\tfrac{1}{2}\sin x(\bar{u}\,\partial u + u\,\partial\bar{u})|u|^{p-2} + |u|^p(\cos x - \mathrm{Im}\,\zeta) \\ &= \tfrac{1}{2}\sin x|u|^{p-2}\,\partial|u|^2 + |u|^p(\cos x - \mathrm{Im}\,\zeta) \\ &= 1/p \sin x\,\partial(|u|^2)^{p/2}) + |u|^p(\cos x - \mathrm{Im}\,\zeta). \end{aligned}$$

Integrating this expression over T^1, we obtain

$$\begin{aligned} &\int_{T^1} \mathrm{Im}[(Au - \zeta u)\bar{u}|u|^{p-2}]\,dx \\ &\quad = \int_{T^1} |u|^p(\cos x(1 - 1/p) - \mathrm{Im}\,\zeta)\,dx. \end{aligned} \tag{6.3}$$

Hence we obtain, using Hölder's inequality:

$$\begin{aligned} (|\mathrm{Im}\,\zeta| - (1 - 1/p))\,\|u\|_p^p &\leqq \int_{T^1} |Au - \zeta u|\,|u|^{p-1}\,dx \\ &\leqq \|Au - \zeta u\|_p\,\|u\|_p^{p-1}. \end{aligned} \tag{6.4}$$

(6.2) follows directly from (6.4).

Proposition 6.2. The operator A, which is determined by (5.10) in $\mathcal{D}_0 = A(T^1)$, is a first-order self-determined operator in $E = L_p(T^1)$ whilst the constant ω_0 in inequality (1.7) equals $1 - 1/p$.

Proof. The satisfaction of (2.2) for small R follows from Theorem 1.1.1. We shall indicate $\zeta_0 \in \rho_0(A)$. If $\zeta_0 = iM$, where $M > 0$ is fairly great, it follows from Theorems 3.1 and 3.2 that the equation $Au - \zeta_0 u = f$ is solvable in $H = L_2(T^1)$ when $f \in \mathcal{D}_0$. At the same time $P_n(A)f_0 \to u$ in L_2. Note that if f is an entire function, then according to Theorem 4.5.4 and Remark 4.5.1 $R_{02}(A^2, f) > C > 0$, where C does not depend on f. Hence it follows that $R_{01}(A, f) > C - \varepsilon > 0$, also. Thus, if M is fairly great, the number $r(M - \omega_0)$ will be larger than 3. Therefore the following estimate holds:

$$\|u - P_n(A)f\|_{L_2} \leq Cn^{-3} \tag{6.5}$$

for any entire function f. The convergence of $u_n = P_n(A)f$, $n = 2^N$ to u in $W_2^2(T^1)$ follows from estimate (6.5), as is obvious from the proof of Theorem 5.2.1. According to Sobolev's imbedding theorem, $u_n \to u$ in $C^1(T^1)$. Therefore $Au_n - \zeta_0 u_n \to f$ in $C(\Omega)$, and consequently, also in $L_p(\Omega)$ for any p. The operator A is thus a first-order self-determined operator in $E = L_p(T^1)$. The satisfaction of estimate (1.7) with $\omega_0 = 1 - 1/p$ follows from (6.2).

Proposition 6.3. Suppose $f \in \mathcal{A}(T^1)$ is a trigonometric polynomial. Suppose

$$|\mathrm{Im}\,\zeta_0| > 1 - 1/p, \quad u = (A - \zeta_0 I)^{-1} f.$$

Then $u \in W_p^s(T^1)$ for any s, which satisfies the inequality

$$s < (|\mathrm{Im}\,\zeta| - 1 + 1/p). \tag{6.6}$$

Proof. The function $(\zeta - \zeta_0)^{-1}$ belongs to $\mathcal{A}(\mathcal{I}_\beta)$ when $\beta < |\mathrm{Im}\,\zeta_0|$. Using estimate (2.23), we obtain - like estimate (5.8.6) when proving Theorem 5.8.1 - the estimate

$$\|u(x) - G_N(x)\|_p \leq CN^{-\sigma_0}, \qquad N \in \mathbb{N}, \tag{6.7}$$

where G_N is a trigonometric polynomial of a degree no higher than N, $\forall\sigma_0$: $\sigma_0 < 2R_{01}(A, f)(\beta - \omega_0)/\pi$, where $\omega_0 = 1 - 1/p$, $\beta < |\mathrm{Im}\,\zeta_0|$. According to Bernshtein's inverse theorem, it follows from estimate (6.7) that

$$u\in H_2^{\sigma_0}(T^1)\subset W_2^{\sigma_0-\varepsilon}(T^1), \qquad \forall\varepsilon>0. \tag{6.8}$$

Since for the operator A in (5.10) and for entire f $R_{01}(A, f) > \pi/2$, (see Lemma 3.5), and the number β is as close to $|\mathrm{Im}\,\zeta|$, as desired, σ_0 is as close to $|\mathrm{Im}\,\zeta| - \omega_0$ as desired. Since the number ε in (6.8) is as small as desired, the statement of the proposition follows from (6.8).

Remark 6.1. In inequality (6.6) it is impossible to replace $|\mathrm{Im}\,\zeta| - 1 + 1/p$ by $|\mathrm{Im}\,\zeta| - 1 + 1/p + \varepsilon$, $\varepsilon > 0$. Indeed, if $f(x) = i\sin x$, $\zeta_0 = i\beta$, then $u = (A - \zeta_0 I)^{-1}f$ is identical with the solution of Eq.(5.8) or (5.4.11). According to Lemmas 5.4.3 and 5.4.4, u belongs to $B^s_{p,p}(T^1)$ when $s = \beta - 1 + 1/p$, and consequently does not belong to $W_p^s(T^1)$ when $s > \beta - 1 + 1/p$.

Remark 6.2. Like the solution of Eq.(5.4.11) with real $\beta > 0$, which was investigated in §5.4, we investigate the case of complex β with $\mathrm{Re}\,\beta > 0$. This solution does not belong to

$$B^s_{p,p}(T^1), \; s>\beta-1+1/p.$$

either. In particular, when $\beta = 1 - 1/p$, $u \notin L_p(T^1)$. Consequently, the whole strip $|\mathrm{Im}\,\zeta| > 1 - 1/p$ belongs to the spectrum of the operator A, which acts in $L_p(T^1)$, i.e. the spectrum A is identical with this strip.

§7. The Polynomial solvability of the equation $Au=f$ when the spectrum of the operator A does not encompass zero.

Suppose A is a first-order self-determined operator. It is obvious that the spectrum $\sigma(A)$ of the operator A is arranged in the strip $\mathcal{J}_{\omega 0}$ (ω_0 is the same as in (1.7)). According to Theorem 2.1, if $|\mathrm{Im}\,\zeta_0| > \omega_0$, the following polynomial representation holds for $u = (A - \zeta_0)^{-1}f$, $f(\mathcal{D}_0)$:

$$u = \lim_{n\to\infty} P_n(A)f. \tag{7.1}$$

In this paragraph we shall consider the case when we have the points ζ_0 of the resolvent set $\rho_0(A)$ of the operator A, which lie in the domain $|\mathrm{Im}\,\zeta_0| < \omega_0$. A question arises: does a representation of the form (7.1) exist in this case?

Note that the case $\zeta_0 \neq 0$ reduces to the case $\zeta_0 = 0$. Indeed, by virtue of Lemma 2.6 the operator $A_1 = A - \zeta_0 I$ is also a first-order self-determined operator, and the polynomial $P_n(A)$ is some polynomial $P_n(\zeta_0 I + A_1)$ of A_1. Therefore the existence of a representation of the vector $u = A^1 f$ in the form (7.1), where A is replaced by A_1, is equivalent to the existence of the representation $u = (A - \zeta_0)^{-1}f$ in the form (7.1).

Therefore we shall henceforth consider the self-determined operator A, whose spectrum does not contain zero. We shall call the equation

$$Au = f,\ f \in \mathcal{D}_0 \tag{7.2}$$

polynomially solvable (in E) if a sequence of polynomials exists, such that Formula (7.1) holds.

Note that, without additional limitations on the spectrum A, Eq.(7.1) cannot be polynomially solvable even if zero does not belong to the spectrum A. This is indicated by the following simple example.

Example 7.1. Consider the operator A of multiplication by the function e^{ix} in the one-dimensional torus T^1. Obviously, this operator is bounded, and is consequently a first-order self-determined operator. Consider the corresponding equation (7.2), where f is a function that equals unity. Thus (7.2) will take the form

$$e^{ix}u(x)=1 \tag{7.3}$$

and, obviously, $u(x) = e^{-ix}$. We shall now prove that Eq.(7.3) is not polynomially solvable in $L_2(T^1)$. Indeed, let us assume the contrary. Then Formula (7.1) holds, where $A = e^{ix}$, $f = 1$, $u = e^{-ix}$, and by virtue of (7.1)

$$e^{-ix}=\lim_{n\to\infty} P_n(e^{ix})1. \tag{7.4}$$

From (7.4) follows:

$$\int_0^{2\pi} e^{-ix}e^{ix}\,dx=\lim_{n\to\infty}\int_0^{2\pi} P_n(e^{ix})\,e^{ix}\,dx.$$

It is obvious that the integral on the left-hand side equals 2π, and on the right-hand side all the integrals equal zero, since they are taken from the sum of terms of the form $C_k e^{ikx}$, $k = 1, 2, \ldots$. We thus obtain a contradiction, which proves the polynomial solvability of Eq.(7.3).

Example 7.1 shows that, to guarantee the polynomial solvability of Eq.(7.2), it is required to impose some additional conditions on A,.

Henceforth we shall consider first-order self-determined operators that are invertible in E, i.e. such that $0 \notin \sigma(A)$. We shall says that $\sigma(A)$ does not encompass zero if the continuous curve $\zeta(\Theta)$, $\Theta \in [0, 1]$ exists, such that

$$\zeta(0)=0,\ |\operatorname{Im}\zeta(1)|=\omega_0,\quad \zeta(\theta)\notin\sigma(A),\qquad \forall\,\theta\in[0,1], \tag{7.5}$$

where ω_0 is the same as in (1.7).

Theorem 7.1. If A is a first-order self-determined operator. whilst $\sigma(A)$ does not encompass zero, and $f \in \mathcal{D}_0$ Eq.(7.2) is polynomially solvable.

We shall indicate only the basic stages of the proof of Theorem 7.1.

Since the spectrum $\sigma(A)$ of the operator A does not encompass zero, $\sigma(A)$ can be bounded by the contour Γ from two smooth connected curves Γ_+ and Γ_-. One of them, Γ_+, is identical with the straight line $\{\operatorname{Im}\zeta=\omega\}$, $|\omega|>\omega_0$ (we will assume, to be specific, that $\omega > 0$). The second, Γ_-, is identical with the line $\{\operatorname{Im}\zeta=-\omega\}$ for large $|\zeta|$. There are no points of the spectrum A outside the domain $\Omega\subset\mathbb{C}$, which is encompassed by Γ. Therefore, Formula (1.8), where L_ω is replaced by Γ, obviously holds for $g(A) = A^{-1}$.

The following formula is proved in a similar way to Formula (2.13):

$$\|(g(A)-P(A))f\|\leq C\sup_{\lambda\in\Gamma}\frac{|g(\lambda)-P(\lambda)|}{\operatorname{ch}(R|\operatorname{Re}\lambda|)}, \tag{7.6}$$

if $0 < R < R_{01}(A.\ f)$ and $g(\lambda)$ is a function that is analytic and bounded in $\overline{\Omega}$.

Assuming $g(\lambda) = \lambda^{-1}$ in (7.6), we obtain that, for (7.1) to be valid, it is sufficient to obtain the sequence of polynomials $P_n(\lambda)$, such that

$$\sup_{\lambda\in\Gamma}\frac{|\lambda^{-1}-P_n(\lambda)|}{\exp(R|\operatorname{Re}\lambda|)}\to 0\qquad (n\to\infty). \tag{7.7}$$

We shall show that Formula (7.7) is a special case of Formula (7.1). Consider the domain $\Omega\subset\mathbb{C}$, which is bounded by the curve Γ. We shall use $C_b(\overline{\Omega})$ to denote the space of functions that are continuous and bounded in $\overline{\Omega}$, and the norm of the function f from $C_b(\Omega)$ is determined like $\sup|f(\lambda)|$. It is obvious that $\forall \lambda_0 \in \overline{\Omega}$ is a delta-function and

$$\delta(\lambda-\lambda_0), (\delta(\lambda-\lambda_0), f)=f(\lambda_0)$$

is a continuous functional on $C_b(\overline{\Omega})$.

Consider the function $f_0(\lambda) = e^{-R|\mathrm{Re}\lambda|}$. Obviously, $f_0(\lambda) \in C_b(\overline{\Omega})$. We shall use A to denote the operator of multiplication by $i\lambda$, $Af(\lambda) = i\lambda f(\lambda)$. It is easy to see that

$$\|A^j f_0(\lambda)\| \leqq C j^j R^{-j} e^{-j} \leqq C_1 R_1^{-j} j!,$$

where $R_1 = R - \varepsilon$, $\varepsilon > 0$. We will assume that Formula (7.1) holds for this operator A and for $f = f_0$. Note that the convergence of (7.1) to $C_b(\Omega)$ is equivalent to the uniform with respect to λ_0 convergence

$$(\delta(\lambda-\lambda_0), u)=\lim_{n\to\infty} (\delta(\lambda-\lambda_0), P_n(A)f). \tag{7.9}$$

Assuming $f = f_0$, we hence obtain:

$$P_n(i\lambda_0)\, e^{-R|\mathrm{Re}\,\lambda_0|} \to (i\lambda_0)^{-1} e^{-R|\mathrm{Re}\,\lambda_0|}, \qquad \lambda_0 \in \Gamma,$$

uniformly with respect to $\lambda_0 \in \Gamma$, i.e.

$$|\lambda^{-1} - iP_n(i\lambda)|\, e^{-R|\mathrm{Re}\,\lambda|} \leqq \varepsilon_n, \qquad \varepsilon_n \to 0, \quad (n\to\infty).$$

Assuming $P_n^1(\lambda) = iP_n(i\lambda)$, we obtain hence (7.7). The existence of the sequence P_n, such that Formula (7.9) holds, follows from

Babin's results [4]. (In this book it is sufficient to consider the case when A is an operator of multiplication by $i\lambda$). The above book by Babin proves the existence of the polynomials P_n using reduction to the problem of approximation using polynomials on a compactum, and this problem has a solution by virtue of Mergelyan's theorem (see, for example, Markushevich [1]).

Remark 7.1. General questions of the theory of approximation of functions using polynomials with a weight in unbounded sets in $\mathbb{C}$ have been considered by Dzhrbashyan, Mergelyan and Shaginyan (see Mergelyan [2]).

A. V. BABIN

CHAPTER 7. ITERATIONS OF NONLINEAR DIFFERENTIAL OPERATORS

In this chapter we examine nonlinear operators. We examine the problem of obtaining an expression for $F^{-1}f$ in terms of the iteration $F^j h$, $j \in \mathbb{N}$, which is analogous to the linear problems considered above. To obtain these expressions we linearise non-linear operators by a change of variables in Banach space and construct complex powers F^z of the nonlinear operator F.

§1. Introductory remarks.

Here we shall briefly discuss the contents of this chapter. We consider nonlinear differential operators on a torus $T^m = \mathbb{R}^m/(2\pi\mathbb{Z})^m$ of the form

$$Fu = G(\partial^{(\nu)}u), \tag{1.1}$$

where $\partial^{(\nu)}u$ is a set of all partial derivatives from u of an order from 0 to ν, G is a function of its arguments that is analytic in the neighbourhood of zero, and $G(0) = 0$. We use A to denote the linear differential operator - the differential of the operator F at zero,

$$Av = \sum_{|\beta| \leq \nu} G_\beta(0)\, \partial^\beta v, \quad (G_\beta = \partial G/\partial(\partial^\beta v)). \tag{1.2}$$

It is assumed that A is elliptic and invertible.

We shall use E to denote Sobolev's complex Banach space

$$W_p^l(T^m),\ l > m/p.$$

Below we shall prove (see §4) that the product of functions in E has the following property:

$$\|A^k(A^{-k}uA^{-k}v\|_E \leqq CT^k\|u\|_E\|v\|_E, \tag{1.3}$$

where the constant T depends only on A, p and l. We shall formulate a local theorem on the reduction of A to linear form.

Theorem 1.1. Suppose $\|A^{-1}\| < 1$ (the norm in E), and $T < 1$ in (1.3). Then the operator ε,

$$\mathscr{E}:\mathscr{O}\to E,\quad \mathscr{E}(0)=0,\quad \mathscr{E}'(0)=I,$$

and its inverse operator ε^{-1}, which are analytic in some neighbourhood $\mathscr{O}$ of zero in E exist, such that the representation of A in the following form holds in $A^{-1}(\mathscr{O})$

$$F=\mathscr{E}\circ A\circ\mathscr{E}^{-1}. \tag{1.4}$$

Remark 1.1. Theorem 1.1 is a generalisation of Poincaré's theorem on the reduction of the analytic mapping F in the neighbourhood of zero in $\mathbb{C}^N$ to linear form A by means of Formula (1.4). Poincaré's theorem, unlike Theorem 1.1, contains conditions on the spectrum of the operator A (the condition of nonresonance of eigenvalues). (See Arnol'd [1]). The larger N, the more this condition is limiting. Theorem 1.1 does not require that there should be no resonance in the spectrum A. Instead, it is required that there be estimate (1.3) with $T < 1$. We need to note that the conditions $\|A^{-1}\| < 1$ and $T < 1$ are not very limiting. Indeed, if we replace the function G by qG, A will be replaced by qA, A^{-1} will be replaced by $q^{-1}A^{-1}$, $\|A^{-1}\|$ will be replaced by $q^{-1}\|A^{-1}\|$. and the constant T in (1.3) will be replaced by $q^{-1}A^{-1}$. Therefore for fairly large q all the conditions of Theorem 1.1 hold for $G_1 = qG$ with any form of analytic function G.

This chapter also proves the global version of Theorem 1.1. It is proved for operators of the form

$$Fu = -\sum_{i=1}^{m} \partial_i G_i(\nabla u) + G_0(u), \tag{1.5}$$

where $G_i(y)$ and $G_0(y)$ are functions that are analytic for real values of the arguments, and which satisfy the conditions

$$\sum G_{ij}(y)\xi_i\xi_j \geqq C_2|\xi|^2,$$
$$C_2 > 0, (G_{ij} = \partial G_i/dy_j), \quad G_0(u) \geqq C_0 > 0.$$

Theorem 1.2. Suppose the operator F, which is determined by (1.5), satisfies the conditions of Theorem 1.1 with $E = W_p^1(T^m)$, $p > m$, and with

$$E = W_q^l(T^m), \quad l - m/q > 2 - m/p, \qquad q > 1.$$

Suppose, in addition, $C_2 > 1$. Then the operators $\mathcal{E}$ and $\mathcal{E}^{-1}$, which are analytic in $\operatorname{Re} W_q^l(T^m)$, exist, such that Formula (1.4) holds, where A is an operator determined by the formula

$$Av = -\sum G_{ij}(0)\, \partial_i\, \partial_j v + G_0'(0)v. \tag{1.6}$$

Theorems 1.1 and 1.2 are used to construct eigen functionals.

The nonlinear complex-valued functional h is called eigen for F^* with the eigenvalue λ if

$$h(Fv) = \lambda h(v), \qquad \forall\, v \in \mathscr{D}(F). \tag{1.7}$$

Theorem 1.3. Suppose the operator F, determined by (1.5), satisfies the conditions of Theorem 1.2. Suppose e_i, $i \in \mathbb{N}$ is a system of eigen functions of the operator A that is complete in

$\mathrm{Re}L_2(T^m)$, where A is determined by (1.6), with eigenvalues λ_i. Then the functionals h_i, $i \in \mathbb{N}$, which are determined by the formula

$$h_i(u)=(e_i, \mathscr{E}^{-1}(u)), \tag{1.8}$$

where $\mathcal{E}$ is the same as in (1.4), are eigen for F^* with the eigenvalues λ_i and form a system of functionals that is complete in $\mathrm{Re}\, W_p^2(T^m)$ $p>m$.

Remark 1.2. It is obvious that if F is a linear operator, and h is a linear functional, definition (1.7) agrees with the definition of the operator F^*, which is adjoint to F. As in the linear case, we can use the functionals h_i to obtain information on the solution of the equation $Fu=h$. Indeed, since $h_i(Fu) = h_i(f)$, by virtue of (1.7)

$$h_i(u)= \lambda_i^{-1}h_i(f).$$

Thus the value $h_i(u)$ of the functional h_i on the solution u of the equation is expressed in terms of its value on the known right-hand side of f. Since the system $\{h_i\}$ is complete, the values $h_i(u)$ uniquely determine the solution u of the nonlinear equation.

Remark 1.3. The greatest successes in the analysis of the detailed geometric and analytic structure of nonlinear differential operators with partial derivatives were achieved in the theory of evolution equations of the Korteweg-de Fries type. The coordinates I_j, φ_j were obtained for them, in which these equations acquire the form

$$\dot{I}_j=0, \quad \dot{\varphi}_j=\omega(I), \qquad j\in\mathbb{N}$$

(see Zakharov V. E. and Faddeev L. D. [1]). The above form is the simplest form of infinite-dimensional Hamiltonian systems.

We can consider Theorem 1.3 as a theorem on the reduction of elliptic operators to the simplest form. Indeed, taking the functionals $\zeta_i = h_i(u)$ which were constructed in Theorem 1.3 as the coordinates ζ_j in $\operatorname{Re} W_2{}^P$, we obtain that the action of the operator is written in the form

$$F:\zeta_j \to \lambda_j \zeta_j, \quad \zeta_j \in \mathbb{R}, \qquad j \in \mathbb{N}.$$

This form is the simplest form of infinite-dimensional mappings.

Thus, Theorem 1.3 is, on one hand, a nonlinear analog of the theorem on the completeness of the eigenvectors of linear operators and, on the other, a stationary analog of the theorem of full integrability. Even in the linear case, obtaining eigenvectors of a differential operator with variable coefficients is an extremely difficult problem. Therefore we pose the problem of expressing $h(u) = h(F^{-1}f)$ in terms of the value h in known functions when h is not an eigen functional F^*. At the same time it is assumed that we know not only $h(f)$, but also all the terms of the sequence

$$h(f),\, h(Ff), \ldots, h(F^k f), \ldots \tag{1.9}$$

consisting of the values h on the iterations F, which are applied to the specified function f. We will assume

$$\Phi(k) = h(A^k f), \qquad k \in \mathbb{Z}. \tag{1.10}$$

It is obvious that when $k \in \mathbb{N}$ the values $\Phi(k)$ are known, and when $k = -j$ equal the value h on the solution u of the equation $A^j u = f$ they are also the required quantities. The problem of obtaining $\Phi(-j)$ and, in particular $\Phi(-1)$, reduces to extrapolating the function Φ from integer positives to negatives. Obviously, this extrapolation is only correct if Φ belongs to a fairly narrow class of functions.

When F is a linear self-adjoint positive operator, and h is a linear functional, the complex powers F^z of the operator F are determined. Therefore for smooth f the function $\Phi(k) = h(F^k z)$ extends to the entire function

$$\Phi(z) = h(F^z f), \qquad z \in \mathbb{C}. \tag{1.11}$$

Since the operator $A^{i\tau}$, $\tau \in \mathbb{R}$ is uniformly bounded with respect to τ, $\Phi(z)$ is bounded on the straight lines that are parallel to the imaginary axis.

A similar extension is also feasible when F is a nonlinear operator, and h is a nonlinear functional. This extension is based on the formula

$$F^z f = \mathscr{E} \circ A^z \circ \mathscr{E}^{-1} f.$$

With corresponding conditions on the function Φ, determined by (1.11), we were able to obtain an expression $h(F^{-j} f)$ in terms of the value (1.9). These expressions in the linear case agree with those which were obtained in Chapter 4 and Chapter 6.

The results concerning differential operators of the form (1.1) and (1.5) were obtained as a corollary of the theorems on nonlinear operators in a Banach space which are proved in this chapter. Theorem 1.1 is proved in §4 (see Theorem 4.1), Theorem 1.2 is proved in §6 (see Theorem 6.2), and Theorem 1.3 is proved in §7 (Theorem 7.2).

§2. Basic definitions.

In this chapter we shall use E to denote a complex Banach space. We shall use S to denote a linear closed operator in E, which has a domain of definition $\mathcal{D}(S)$ that is everywhere dense in E.

It will be assumed everywhere below that the operator S is invertible, whilst

$$\|S^{-1}\| \leqq 1, \quad S^{-1}: E \to E. \tag{2.1}$$

We shall denote the domain of definition $\mathcal{D}(S^k)$ of the operator S^k by E_k, and shall determine the norm $\| \ \|_k$ on E_k by

$$\|u\|_k = \|S^k u\| \quad (k = 1, 2, \ldots), \ \|u\|_0 = \|u\|, \tag{2.2}$$

where $\| \ \|$ is the norm in E. We shall denote the product j of samples of the space E_k by $E_k^j (j \in \mathbb{N})$. We shall determine the norm of the vector $\bar{u} = (u_1, \ldots, u_j) \in E^j$ using

$$\|\bar{u}\|_k = \max_i \|u_i\|_k. \tag{2.3}$$

We shall now introduce an important class of nonlinear operators in E.

Suppose $B^j : E^j \to E$ is a j-linear operator that is continuous on E^j. The norm of the j-linear operator is determined using the formula

$$\|B_j\| = \sup_{\|\bar{u}\| \leqq 1} \|B_j(\bar{u})\|.$$

We shall use $B_j \circ S^{-k}(\bar{u})$ to denote the operator determined by the equation

$$B_j \circ S^{-k}(\bar{u}) = B_j(S^{-k} u_1, \ldots, S^{-k} u_j). \tag{2.4}$$

We shall use $B_j(u)$ to denote the j-homogeneous operator specified by the correspondence $u \to B_j(u, \ldots, u)$.

When $j > 2$ we shall use $B_j(u, v)$ to denote the operator which is $(j - 1)$-homogeneous with respect to v, and is specified by the correspondence

$$(u, v) \to B_j(u, v, \ldots, v).$$

Suppose B_j changes E_k^j into E_k, $k = 0, 1, \ldots$. We shall determine the norm $|||\ |||$ of the operator B_j using the equality

$$|||B_j||| = \sup_{k \in \mathbb{Z}_+} \sup_{\|\bar{u}\| \leqq 1} \|S^k B_j \circ S^{-k}(\bar{u})\|. \tag{2.5}$$

If this norm is finite, we shall call B_j an operator that is quasi-interchangeable with S. The norm $||| \cdot |||$, more accurately, a family of norms (these norms are determined on different operators for different j), besides the usual properties of the norm has the following important properties:

$$|||S^k B_j \circ S^{-k}||| \leqq |||B_j|||, \qquad \forall k \in \mathbb{Z}_+, \tag{2.6}$$

$$|||B_j(B_{i1}, \ldots, B_{ij})||| \leqq |||B_j||| \prod_{l=1}^{j} |||B_{il}|||. \tag{2.7}$$

Inequality (2.6) directly follows from (2.5), since

$$|||S^l B_j \circ S^{-l}||| = \sup_{k \geqq l} \sup_{\|\bar{u}\| \leqq 1} \|S^k B_j \circ S^{-k}(\bar{u})\| \leqq |||B_j|||.$$

We shall now prove (2.7). We will assume $t_l = |||B_{il}|||^{-1}$. Suppose

$$\bar{u} = (\bar{u}_1, \ldots, \bar{u}_l), \ \|\bar{u}\| \leqq 1.$$

Obviously,

$$\begin{aligned}&\|S^kB_j(B^i_{i1}\circ S^{-k}(\bar u_1),\dots,B_{ij}\circ S^{-k}(\bar u_j))\|\\&=\left(\prod_{l=1}^{j}t_l^{-1}\right)\|S^kB_j(t_1S^kB_{i1}\circ S^{-k}(\bar u_1),\dots,t_jS^kB_{ij}\circ S^{-k}(\bar u_j))\|.\end{aligned}\tag{2.8}$$

It is also obvious that

$$\begin{aligned}&\|S^kB_j(t_1B_{i1}\circ S^{-k}(\bar u_1),\dots,t_jS^kB_{ij}\circ S^{-k}(\bar u_j))\|\\&\leqq\|S^kB_j\circ S^{-k}(t_1S^kB_{i1}\circ S^{-k}(\bar u_1),\dots,t_jS^kB_{ij}\circ S^{-k}(\bar u_j))\|.\end{aligned}\tag{2.9}$$

Since by virtue of (2.3)

$$\|u_l\|\leqq\|(\bar u_1,\dots,\bar u_j)\|\leqq 1$$

then by virtue of (2.5), applied to B_{il} , we have:

$$\|t_lS^kB_{il}\circ S^{-k}(\bar u_l)\|\leqq t_l|||B_{il}|||=1.$$

Therefore by virtue of (2.3)

$$\|(t_1S^kB_{i1}\circ S^{-k}(\bar u_1),\dots,t_jS^kB_{ij}\circ S^{-k}(\bar u_j))\|\leqq 1.$$

Using this estimate, we immediately derive from (2.5) that the right-hand side of (2.9) does not exceed $|||B_j|||$. Therefore from (2.9) and (2.8) follows the estimate

$$\|S^kB_j(B_{i1}\circ S^{-k}(\bar u_1),\dots,B_{ij}\circ S^{-k}(\bar u_j)\|\leqq|||B_j|||\prod_{l=1}^{j}t_l^{-1}.$$

Since $t_l^{-1}=|||B_{il}|||$, hence follows (2.7).

Suppose the operator $B_j(\bar u)$ has a strengthened property of continuity with respect to variables with numbers larger than 1, namely $B_j(u, Sv)$ maps $E_k \times E_k^{j-1}$ into E_k, $k = 0, 1,,,$. We shall introduce the norm $||| \; |||_1$ for these operators,

$$|||B_j||| = \sup_{k \geq 0} \sup_{||\bar{u}|| \leq 1} ||S^k B_j(S^{-k}u_1, S^{-k+1}u_2, \ldots, S^{-k+1}u_j)||. \tag{2.10}$$

We shall call the operator B_j, for which the norm $||| \; |||_1$ is finite, a quasi-linear operator that is quasi-interchangeable with S. Note that in the quasi-linear j-linear operators B_j the first variable is marked, and the remaining variables have a somewhat subordinate character. Since $||S^{-1}|| \leqslant 1$, by comparing (2.5) and (2.10) we obtain:

$$|||B_j||| \leqq |||B_j|||_1 ||S^{-1}||^{j-1} \leqq |||B_j|||_1. \tag{2.11}$$

The norm $||| \cdot |||_1$ has properties that are analogous to those of the norm $||| \cdot |||$:

$$|||S^k B_j \circ S^{-k}|||_1 \leqq |||B_j|||_1, \tag{2.12}$$

$$|||(B_j(B_{i1}, \ldots, B_{ij})||| \leqq |||B_j|||_1 \prod_{l=1}^{j} |||B_{il}|||_1. \tag{2.13}$$

The proof of these inequalities is analogous to that of inequalities (2.6) and (2.7).

Suppose the sequence of positive numbers a_j, $j = 2, 3, \ldots$, is such that

$$a_j \leqq C\rho^{-j}. \tag{2.14}$$

If the sequence of j-linear operators B_j satisfies the estimate

$$|||B_j||| \leqq a_j, \tag{2.15}$$

then when $v \in \mathcal{O}_\rho = \{v \in E: \|v\| < \rho\}$ the operator B is determined,

$$B(u) = u + \sum_{j=2}^{\infty} B_j(u). \tag{2.16}$$

We shall call the operator $B(u)$, which is determined by Formula (2.16), where the j-homogeneous operators $B_j(u)$ satisfy (2.15), (2.14), an operator that is quasi-interchangeable with S at zero.

If a stronger condition holds instead of (2.15)

$$|||B_j||| \leqq a_j. \tag{2.17}$$

we shall call the operator B, determined by (2.16), a quasi-linear operator that is quasi-interchangeable with S at zero.

We shall call the operator $G: \Omega \to E$, determined on the set $\Omega \subset E$, analytic in Ω, if for any point $z_0 \subset \Omega$

$$G(z_0 + u) = G(z_0) + \sum G_j(u),$$

where G_j are j-linear operators, whilst this series converges in the neighbourhood of the point z_0. If this series converges for all $u \in E$, we shall call G an entire operator on E.

Proposition 2.1. Suppose B is an operator that is quasi-interchangeable with S at zero. Then the operator B is analytic in the domain

$$\mathcal{O}_\rho^m = \{u \in E_m: \|u\|_m < \rho\}, \quad \forall\, m \in \mathbb{Z}_+,$$

and the number ρ is the same as in (2.14).

Proof. It is obvious that

$$\forall\, m \in \mathbb{Z}_+,\ \|B_j\|_m \leqq |||B_j|||,$$

where $\|B_j\|_m$ is the norm of the j-linear operator from E_m^j in E_m. Obviously,

$$\|B_j(\bar{u})\|_m \leqq \|B_j\|_m \prod_{i=1}^{j} \|u_i\|_m.$$

Therefore it follows from estimates (2.14), (2.15) that series (2.16) converges when $\|u\|_m < \rho$.

We shall now prove that $B(u + v)$ expands in a series with respect to v when $\|\bar{u}\|_m + \|\bar{v}\|_m < \rho$. We shall fix u, $\|u\| = \rho_1$, $\rho_1 < \rho$. Expanding $B_j(\bar{u} + \bar{v})$ in the sum of terms which are homogeneous with respect to u and v. we obtain

$$B_j(\bar{u}+\bar{v}) = \sum_{i=1}^{m} B_{ji}(\bar{u}, \bar{v}),$$

where

$$\|B_{ji}(\bar{u}, \bar{v})\|_m \leqq j!/((j-i)!\,i!)\,\|\bar{u}\|_m^{j-i}\,\|\bar{v}\|_m^{i}\,\|B_j\|.$$

The operator $B_i(u, \bar{v})$, which is i-linear with respect to v, has the following form in the expansion $B(u + v)$:

$$B_i(u, \bar{v}) = \sum_{j \geqq i} B_{ji}(u, \bar{v}).$$

Therefore when $\|\bar{v}\| < \varepsilon$, $\varepsilon < \rho - \rho_1$,

$$\|B_i(u, \bar{v})\|_m \leqq C \sum_{j \geqq i} \rho^{-j} j!/((j-i)!\,i!)\rho_1^{j-i}\varepsilon^i$$
$$\leqq C \sum_{j \geqq i} \rho^{-j}(\rho_1+\varepsilon)^j = C\left(\frac{\rho_1+\varepsilon}{\rho}\right)^i \frac{\rho}{\rho-\rho_1-\varepsilon}.$$

Hence it is obvious that when $\|u\| + \|v\| < \rho$ the series $\sum B_i(u, v)$ converges, and the proposition is proved.

Proposition 2.2. If B is a quasi-linear operator that is quasi-interchangeable with S at zero, when $\|u\| < \rho$

$$B(u)=P(u)u, \tag{2.18}$$

where $P(v): E \to E$ is a linear operator for each $v \in \mathcal{O}\rho$. When $\|v\|_k < \rho$ the operators $P(v)$ continuously act from E_k to E_k and from E_{k+1} to E_{k+1}, $k \in \mathbb{N}$. The mappings $v \to P(v)$, which operate from

$$\mathcal{O}^k_\rho=\{v:\|v\|_k<\rho\},$$

where ρ is fairly small, to the spaces of linear bounded operators $\mathcal{L}(E_k, E_k)$ and $\mathcal{L}(E_{k+1}, E_{k+1})$, are analytic on $\mathcal{O}^k_\rho$, $k \in \mathbb{N}$.

Proof. We shall determine $P(v)u$ using the formula

$$P(v)u=u+\sum_{j=2}^{\infty} B_j(u, v). \tag{2.19}$$

Note that the operator $B_j(u, v)$ is linear with respect to u for each v. We shall estimate the norm of this linear operator. By virtue of (2.10) when $k \in \mathbb{N}$

$$\begin{aligned}\|B_j(u, v)\|_k&=\|S^k B_j(u, v)\|=\|S^k B_j \circ S^{-k}(S^k u, S^k v)\|\\&=\|S^k u\|\,\|S^{k-1}v\|^{j-1}\|S^k B_j\\&\circ S^{-k}(\|S^k u\|^{-1}S^k u, \|S^{k-1}v\|^{-1}SS^{k-1}v)\|\\&\leqq\|u\|_k\|v\|^j_{k-1}\,|||S^k B_j\circ S^{-k}|||_1\\&\leqq\|u\|_k\|v\|^j_{k-1}\,|||B_j|||_1.\end{aligned}$$

Hence we immediately obtain that

$$\|B_j(\cdot, v)\|_{\mathcal{L}(E_k, E_k)}\leqq|||B_j|||_1\,\|v\|^{j-1}_{k-1}. \tag{2.20}$$

From this inequality and estimates (2.14), (2.17) we obtain that when $v \in \mathcal{O}_\rho^{k+1}$ in $\mathcal{L}(E_k, E_k)$ the following series of operators converges:

$$I + \sum_{j=2}^{\infty} B_j(\cdot, v) = P(v). \tag{2.21}$$

It is obvious that $P(v) \in \mathcal{L}(E_k, E_k)$ analytically depends on $v \in \mathcal{O}_\rho^{k-1}$. Since $\|v\|_{k-1} \leqslant \|v\|_k$, it follows from (2.20) that $P(v)$ analytically depends on $v \in \mathcal{O}_\rho^k$.

The proposition is thus proved for $k \in \mathbb{N}$, $k \geqslant 1$. The case $k=0$ is considered in a fully analogous way.

§3. Local linearisation.

In this paragraph we consider the following problem. The nonlinear operator F, which has a fixed point u_0, is active in the space E. It is required to construct a change of variables in the neighbourhood of this point, such that the operator F becomes the linear operator A after this change. I.e. it is required to obtain $\mathcal{E}$ and A, such that the following diagram is commutative in the neighbourhood of the point $u_0 \in E$:

$$\begin{array}{ccc} E & \xrightarrow{F} & E \\ \mathscr{E}\uparrow\downarrow\mathscr{E}^{-1} & & \mathscr{E}\uparrow\downarrow\mathscr{E}^{-1}. \\ E & \xrightarrow{A} & E \end{array} \tag{3.1}$$

We can write the requirement of commutativity of (3.1) in the form of the equations

$$\mathscr{E}^{-1} \circ F \circ \mathscr{E} = A, \tag{3.2}$$

or

$$F = \mathscr{E} \circ A \circ \mathscr{E}^{-1}. \tag{3.3}$$

Note that Eq.(3.2) has been considered by Schroder in the one-dimensional case $E = \mathbb{C}$ [1]. Let us consider the simplest example of a representation of the form (3.3). Suppose $E = \mathbb{C}$ and $F: \mathbb{C} \to \mathbb{C}$ has the form

$$F: z \to z^2. \tag{3.4}$$

We shall make the change $z = e^w$. Then (3.4) will take the form:

$$F: e^w \to e^{2w}.$$

Obviously, in this case

$$\mathscr{E}(w) = e^w,\ \mathscr{E}^{-1}(z) = \ln z,$$

and the change effected by the function $\mathcal{E}^{-1}$ is analytic in the neighbourhood of the fixed point $z = 1$ of the mapping (3.4). Eq.(3.3) takes the form:

$$z^2 = e^{2 \ln z}$$

in the neighbourhood of the point $z = 1$, i.e. the operator A is identical with the operator of multiplication by 2.

Note that the problem of analytic linearisation in the case $E = \mathbb{C}^N$ is solved using Poincaré's theorem (see the formulation and proof in Arnol'd's book [1]). This problem has been examined by several authors in the nonanalytic case for $E = \mathbb{R}^n$ (see Hartman's books [1], and its references). The problem has been examined by Smajdor [1] and Ware [1] in the infinite-dimensional case, for analytic F, which is everywhere determined in the neighbourhood of a fixed point. Here we shall consider the case when the operator

F is determined not in an open neighbourhood, but only in an everywhere dense set (in applications F will be a differential operator).

We shall now move on to a detailed discussion.

We shall use the following formula to determine the operator F in the set $\{v: \|v\|_\nu < \rho\}$, where ρ is fairly small:

$$F(u) = B \circ Au, \quad A = MS^\nu, \tag{3.5}$$

where B is an operator that is quasi-interchangeable with S (see §2), M is a number, $M > 1$, and S is the same as in §2.

Henceforth we will consider the operators F, which permit a representation of the form (3.5) in the neighbourhood of zero. The fixed point of the operator F is identical with zero. We shall seek the operator $\mathcal{E}$, which is quasi-interchangeable with S in the neighbourhood of zero, satisfies the equation

$$F \circ \mathscr{E} \circ A^{-1} = \mathscr{E}, \tag{3.6}$$

This equation is a corollary of (3.2) and (3.3). We recall that by virtue of (2.16) it follows from the fact that $\mathcal{E}$ is an operator that is quasi-interchangeable with S that

$$\mathscr{E}(0) = 0, \quad \mathscr{E}'(0) = I. \tag{3.7}$$

Conditions (3.7) and Eq.(3.6) uniquely determine the analytic operator $\mathcal{E}$ (see the proof of Theorem 3.1 below). Without the satisfaction of (3.7), the choice of $\mathcal{E}$ and A, such that (3.3) holds, is nonunique. Indeed, if we assume

$$\mathscr{E}_1 = \mathscr{E} \circ L_0,$$

where L_0 is a linear invertible operator, we have the formula

$$F=\mathscr{E}_1\circ A_1\circ\mathscr{E}_1^{-1},\quad A_1=L_0^{-1}AL_0.$$

which is similar to (3.3)

Remark 3.1. Difficulties connected with the presence of "small denominators" often arise when linearising the operators F, which are generated by operators of translation along the trajectories of dynamic systems (see Arnol'd [1], Moser [1], Nikolenko [1,2]). Fundamental difficulties arise in the case we are considering due to the unboundedness of the operator A.

Henceforth we shall need a lemma on the existence of a solution of the subsidiary equation in $\mathbb{C}$, which is close to (3.6) in form.

Lemma 3.1. Suppose the function $a(z)$ is analytic in the neighbourhood of zero and its Taylor coefficients are nonnegative, $a(0) = 0$, $a'(0) = 1$. Then when $M > 1$ there exists the solution $g(z)$ of the equation

$$a(Mg(M^{-1}z))=g(z), \tag{3.8}$$

which is analytic in the neighbourhood of zero and satisfies the conditions $g(0) = 0$, $g'(0) = 1$. If the function $a(z)$ is entire, $g(z)$ is also entire.

Proof. We shall seek the solution of (3.8) in the form of series

$$g(z)=\sum_{j=1}^{\infty} g_j z^j. \tag{3.9}$$

Expanding $a(z)$ in the neighbourhood of zero in the series

$$a(z)=z+\sum_{j=2}^{\infty} a_j z^j \tag{3.10}$$

and substituting (3.9) into (3.8), we obtain that $g_1 = 1$, and the coefficients g_n, $n \geq 2$, satisfy the recurrence formulae:

$$g_n = M^{1-n} g_n + M^{-n} \sum_{j=2}^{\infty} a_j M^j \sum_{i_1+\cdots+i_j=n} \prod_{l=1}^{j} g_{i_l} \qquad (n=2,3,\ldots). \tag{3.11}$$

From (3.11) follow the estimates

$$0 \leqq g_n \leqq M^{-n}/(1-M^{1-n}) \sum_{j=2}^{\infty} a_j M^j \sum_{i_1+\cdots+i_j=n} \prod_{l=1}^{j} g_{i_l}. \tag{3.11'}$$

Since $M > 1$, then $M^{-n}/(1 - M^{1-n}) < C$, where C does not depend on $n \geq 2$.

We shall determine successively the numbers f_n, $n = 1, 2, \ldots$ using the formulae

$$f_1 = 1, f_n = C \sum_{j=2}^{\infty} a_j M^j \sum_{i_1+\cdots+i_j=n} \sum_{l=1}^{j} f_{i_l}. \tag{3.12}$$

Comparing (3.11') and (3.12) we have:

$$0 \leqq g_n \leqq f_n, \qquad n=1,2,\ldots. \tag{3.13}$$

Consider the equation

$$f = z + C[a(Mf) - Mf], \tag{3.14}$$

where $a(z)$ is determined by (3.10), and the constant C is the same as in (3.12). It is obvious that $f = 0$, $z = 0$ are solutions of this equation. The partial derivative with respect to f from the left-hand side equals 1, and from the right-hand side equals 0 when $z = 0$, $f = 0$. By virtue of the theorem on the implicit function, Eq.(3.14) has the solution $f = f(z)$, which is analytic in the neighbourhood of zero. Therefore the function $f(z)$ is represented in the form of the converging series

$$f(z)=\sum_{k=1}^{\infty} f_k z^k. \tag{3.15}$$

Substituting $f(z)$ into (3.14), we see that the numbers f_k - the coefficients of series (3.15) - satisfy the recurrence formulae (3.12). Since series (3.15) converges when $|z| < \rho'$, $\rho' > 0$, by virtue of (3.13) series (3.9) also converges when $|z| < \rho'$. The function $g(z)$, which is defined by (3.9), satisfies (3.8) by virtue of (3.11).

Let us now consider the case when the function $a(z)$ is entire. Suppose $g(z)$ is analytic when $|z| < \rho''$, $\rho'' > 0$. Then the function $g(M^{-1}z)$ is defined when $|z| \leqslant M\rho''$. Consequently, when $|z| < M\rho''$ the function $a(Mg(M^{-1}z))$ - the left-hand side of Eq.(3.8) - is defined and analytic. Since $M > 1$, we thereby obtain an analytic extension of the function $g(z)$ - the right-hand side of Eq.(3.8) - from the neighbourhood $\{|z| < \rho''\}$ to the larger neighbourhood $\{|z| < M\rho''\}$. Repeating this argument, we obtain that $g(z)$ is extended to a function that is analytic for all z, i.e. $g(z)$ is an entire function. The lemma is thus completely proved.

We shall now formulate and prove the basic result of this paragraph.

Theorem 3.1. Suppose B is an operator that is quasi-interchangeable with S at zero, and F is determined by Formula (3.5). Then the operator $\mathcal{E}$ exists, which is quasi-interchangeable with S at zero and is the solution of Eq.(3.6). The operator $\mathcal{E}$ is expanded in the series

$$\mathscr{E}(u)=u+\sum_{j=2}^{\infty} \mathscr{E}_j(u), \tag{3.16}$$

which converges in the neighbourhood $\mathcal{O}_r = \{u : \|u\| \leqslant r,\ r > 0\}$ together with the series

$$S^k\mathscr{E}\circ S^{-k}(u)=u+\sum_{j=2}^{\infty} S^k\mathscr{E}_j\circ S^{-k}(u), \qquad k=1,2,\dots. \tag{3.17}$$

The operators $S^k \mathcal{E} S^{-k}$ are also quasi-interchangeable with S at zero and satisfy the equation

$$S^k F \circ \mathscr{E} A^{-1} S^{-k} = S^k \mathscr{E} S^{-k}. \tag{3.18}$$

If the operator B is a quasi-linear operator that is quasi-interchangeable with S at zero, $\mathcal{E}$ is also quasi-linear and quasi-interchangeable with S at zero.

If the function $a(z)$, determined by Formula (3.10), where the numbers a_j are the same as in (2.15), is entire, then series (3.16) and (3.17) converge in E.

Proof. We shall seek the solution $\mathcal{E}$ of Eq.(3.6) in the form (3.16). Equating the terms with an identical degree of homogeneity, we obtain that $\mathcal{E}_1 = I$ and that the operators $\mathcal{E}_j (j \geqslant 2)$ satisfy the recurrence formulae

$$\mathscr{E}_n - A\mathscr{E}_n \circ A^{-1} = R_n,$$

$$R_n = \sum_{j=2}^{\infty} \sum_{i_1 + \cdots + i_j = n} B_j(A\mathscr{E}_{i_1} \circ A^{-1}, \ldots, A\mathscr{E}_{i_j} \circ A^{-1}), \tag{3.19}$$

where $n \geqslant 2$. (Note that there are only n-1 nonzero terms in the sum over j). We shall construct $\mathcal{E}$ using induction. Suppose $\mathcal{E}_n$ when $n < n_0$ are already constructed. It is obvious that

$$|||R_n||| \leqq \sum_{j=2}^{\infty} \sum_{i_1 + \cdots + i_j = n} |||B_j(A\mathscr{E}_{i_1} \circ A^{-1}, \ldots, A\mathscr{E}_{ij} \circ A^{-1})|||.$$

Bearing in mind that

$$A = MS^{\nu},\ A^{-1} = M^{-1} S^{-\nu},$$

and using estimate (2.7), we obtain:

$$|||R_n||| \leqq \sum_{j=2}^{\infty} \sum_{i_1 + \cdots + i_j = n} |||B_j||| \prod_{l=1}^{j} |||M^{1-i_l} S^{\nu} \mathscr{E}_{i_l} \circ S^{-\nu}|||.$$

Using estimate (2.15), we obtain hence:

$$|||R_n||| \leq \sum_{j=2}^{\infty} a_j \sum_{i_1+\cdots+i_j=n} \prod_{l=1}^{j} |||\mathscr{E}_{i_l}||| M^{j-n}. \tag{3.20}$$

We shall seek the solution of Eq.(3.19) in the form of series:

$$\mathscr{E}_n = \sum_{\gamma=0}^{\infty} A^{\gamma} R_n \circ A^{-\gamma}. \tag{3.21}$$

Since

$$|||S^k A^{\gamma} R_n \circ A^{-\gamma} S^{-k}||| = |||S^{k+\gamma\nu} R_n \circ S^{-\nu\gamma-k}||| M^{\gamma-n\gamma} \leq M^{\gamma-n\gamma} |||R_n|||, \tag{3.21'}$$

series (3.21) converges, whilst

$$|||S^k \mathscr{E}_n S^{-k}||| \leq |||R_n||| \sum_{\gamma=0}^{\infty} M^{\gamma-n\gamma} = \frac{|||R_n|||}{1-M^{1-n}}. \tag{3.22}$$

$\mathcal{E}_n$ is thus constructed when $n = n_0$. It follows from (3.20) and (3.22) that

$$(1-M^{1-n})|||S^k \mathscr{E}_n S^{-k}||| \leq M^{-n} \sum_{j=2}^{\infty} a_j M^j \sum_{i_1+\cdots+i_j=n} \prod_{l=1}^{j} |||\mathscr{E}_{i_l}|||. \tag{3.23}$$

Suppose the function $a(z)$ is defined using Formula (3.10), in which a_j are the same as in (3.23). Comparing (3.11) and (3.23), and bearing in mind that $|||\mathcal{E}_1||| = 1$, we see that

$$|||\mathcal{E}_j||| \leqslant g_j \text{ when } j = 1, 2, \ldots . \tag{3.24}$$

Since series (3.9) converges when $|z| < \rho'$ where $\rho' > 0$ by virtue of Lemma 3.1, the following series also converges when $|z| < \rho'$:

$$\sum |||\mathscr{E}_j||| z^j. \tag{3.25}$$

It follows hence and from the definition of the norm of the j-linear operator that series (3.17) converge when $\|u\| < \rho'$ for all $k \in \mathbb{Z}_+$. In particular, series (3.16) converges. The operator $\mathcal{E}$ satisfies (3.6) and (3.18) by virtue of (3.19).

Let us now consider the case when the operator B is quasi-linear and quasi-interchangeable with S at zero. Estimates in which the norm $|||\cdot|||_1$ is taken instead of the norm $|||\cdot|||$ are made in a similar way, and the convergence of series (3.25) is established, where we have $|||\cdot|||_1$ instead of $|||\cdot|||$. The theorem is proved.

Theorem 3.2. Suppose $\mathcal{E}$ is an operator that is quasi-interchangeable with S at zero. Then the operator $\mathcal{E}^{-1}$ exists, which is quasi-interchangeable with S at zero and is inverse to the operator $\mathcal{E}$ in the neighbourhood of zero.

Proof. The existence and analyticity of $\mathcal{E}^{-1}$ in the neighbourhood of zero follow from the theorem on the implicit function. It remains to verify that $\mathcal{E}^{-1}$ is quasi-interchangeable with S at zero. Since $\mathcal{E}^{-1}$ is analytic, then

$$\mathcal{E}^{-1}=\sum_{j=1}^{\infty}\mathcal{E}_j^1,$$

where $\mathcal{E}_j^1$ are j-homogeneous operators. From the identity

$$\mathcal{E}\circ\mathcal{E}^{-1}=I$$

follow the obvious recurrence formulae for $\mathcal{E}_j^1 : \mathcal{E}_1^1 = I$,

$$\mathcal{E}_n^1=\sum_{j=2}^{\infty}\sum_{i_1+\cdots+i_j=n}\mathcal{E}_j(\mathcal{E}_{i_1}^1,\ldots,\mathcal{E}_{i_j}^1) \qquad (n=2,3,\ldots).$$

From these relations we obtain the estimates

$$|||\mathcal{E}_n^1|||\leq\sum_{j=2}^{\infty}\sum_{i_1+\cdots+i_j=n}|||\mathcal{E}_j|||\prod_{l=1}^{j}|||\mathcal{E}_{i_l}^1|||.$$

Setting $|||\mathcal{E}_j||| = a_j$ and comparing it with (3.12), we obtain that $|||\mathcal{E}_n^1||| < f_n$, where f_n are numbers from (3.12), where C=1, M=1. Since the series $\sum f_n z^n$ converges, it follows that $\mathcal{E}_n^{-1}$ is quasi-interchangeable with S at zero, and the theorem is proved.

Theorem 3.3. Suppose the operator F is determined by Formulae (3.5), where $M > 1$. Then for fairly small r the representation F in the form (3.3) holds in the neighbourhood $\{\|u\|_\nu < r\} = \mathcal{O}_r^\nu$, where $\mathcal{E}$ and $\mathcal{E}^{-1}$ are quasi-interchangeable with S at zero.

Proof. The operators $\mathcal{E}$ and $\mathcal{E}^{-1}$ are determined in the neighbourhood of zero, whilst Eq.(3.6) holds. By virtue of (3.6) the following equation holds on $\mathcal{O}_r^\nu$ for small r:

$$F \circ \mathcal{E} = \mathcal{E} A. \tag{3.26}$$

The operator $\mathcal{O}_{r1}^\nu$ is determined on $\mathcal{E}^{-1}$. The operator $\mathcal{E}$ is quasi-interchangeable with S at zero, and therefore $\|\mathcal{E}^{-1} w\|_s < C\|w\|_s$ for small $\|w\|_s$. We can therefore multiply (3.26) over

$$\mathcal{O}_{r_2}^\nu (Cr_2 < r)$$

from the left by the operator $\mathcal{E}^{-1}$ and we obtain (3.3).

§4. Local linearisation of nonlinear differential operators on a torus.

In this paragraph we shall prove the theorem that nonlinear analytic elliptic operators F on a torus $T^m = \mathbb{R}^m/(2\pi\mathbb{Z})^m$ permit analytic linearisation in the neighbourhood of zero:

$$F = \mathcal{E} \circ A \circ \mathcal{E}^{-1},$$

where A is a linear elliptic operator.

We shall consider the spaces of functions on the torus T^m, i.e. the spaces of complex-valued functions that are periodic with respect to each variable with the peiod 2π and are defined on $\mathbb{R}^m$. Henceforth the integral $m > 1$ is fixed. We shall use the Sobolev spaces $W_p^l(T^m) = W_p^l$, $p > 1$. For integer $l > 0$ the norm in W_p^l will be specified by the equation

$$\begin{aligned} \|u\|_{W_p^l}^p &= \|u\|_{l,p}^p = \sum_{|\alpha| \leq l} \|\partial^\alpha u\|_{0,p}^p, \\ \|u\|_{0,p}^p &= \|u\|_{L_p}^p = \int |u(x)|^p \, dx, \end{aligned} \tag{4.1}$$

where the integral is taken over the cube $|x_i| < \pi$, $i = 1, \ldots, m$. For noninteger l W_p^l is determined using Fourier's transform, see below (4.18). For all l

$$\|\partial^\alpha u\|_{l,p} \leq C \|u\|_{l+|\alpha|,p}. \tag{4.2}$$

It is well known that when $l > m/p$ the spaces $W_p^l(T^m) = W_p^l$ are rings with reference to the pointwise multiplication of functions, and for all u, $v \in W_p^l$ the followng inequality holds:

$$\|u \cdot v\|_{l,p} \leq C_{l,p} \|u\|_{l,p} \|v\|_{l,p}. \tag{4.3}$$

(See Remark 4.3 at the end of the paragraph for the proof for integral l.)

We shall use $\partial^{(\nu)} u$ to denote the set of all partial derivatives from $u(x)$ up to the order ν inclusively (including a zero-order derivative, i.e. u). We can obviously consider $\partial^{(\nu)} u$ like a vector of some dimension r, $r = r(m)$

$$\partial^{(\nu)} u = (\partial^{\alpha^1} u, \ldots, \partial^{\alpha^r} u).$$

For convenience, we shall number the components of the vector $\partial^{(\nu)}u$ in such a way that all the highest-order derivatives ν are components with numbers from 1 to r_1, and the derivatives of an order that is less than ν have numbers greater than r_1.

Consider a nonlinear differential operator of order

$$F(u) = G(\partial^{(\nu)}u), \tag{4.4}$$

$G(z)$ is a function that is analytic in the neighbourhood of zero in $\mathbb{C}^r$, which expands in a series with respect to $z=(z_1,\dots,z_r)$,

$$G(z) = \sum_\alpha g_\alpha z^\alpha, \quad \alpha=(\alpha_1,\dots,\alpha_r)\in \mathbb{Z}^r_+. \tag{4.5}$$

It is assumed that $G(0) = 0$, i.e. in the sum (4.5) $|\alpha| > 1$. It is obvious that we can rewrite the function $G(z)$ in the form:

$$G(z) = G_1(z) + G_{(2)}(z), \tag{4.6}$$

where

$$G_1(z) = \sum_{|\alpha|=1} g_\alpha z^\alpha, \quad G_{(2)}(z) = \sum_{|\alpha|\geqq 2} g_\alpha z^\alpha. \tag{4.7}$$

The function G_1 is obviously linear.

We shall rewrite the operator F, determined by (4.4), in the form

$$F(u) = \sum_{j=1}^{\infty} F_j(u), \quad F_j(u) = \sum_{|\alpha|=j} g_\alpha(\partial^{(\nu)}u)^\alpha,$$
$$g_\alpha = \frac{1}{\alpha!}\partial G/\partial z^\alpha|_{z=0}. \tag{4.8}$$

Proposition 4.1. Suppose

$$z(x)=(z_1(x),\dots,z_N(x))\in(W_p^l(T^m))^N,\quad l>m/p.$$

Suppose K_0 is the set of values of this function:

$$K_0=\{z\in\mathbb{C}^N : z=z(x), x\in T^m\}.$$

Suppose the function $G(z) : \mathbb{C}^N \to \mathbb{C}$ is analytic in some neighbourhood $\mathcal{O}_{2\delta}(K_0)$ of the set K_0. Then the mapping $v(x) \to G(z(x) + v(x))$ is analytic with respect to v in the neighbourhood of zero in $(W_p^l)^N$.

Proof. At each point $z \in K_0$ the function $G(z + v)$ expands in the Taylor series

$$G(z+v)=\sum\frac{1}{\alpha!}\,\partial_z^\alpha G(z)v^\alpha. \tag{4.8'}$$

The set K_0 is compact, since $W_p^l(T^m) \subset C(T^m)$ when $l > m/p$. Lemma 1.1.3 is therefore applicable, from which follows estimate (1.1.1), in which $x = z$, $m = N$, $f = G$, $K = \mathcal{O}_\delta(K_0)$. From estimate (1.1.1), by virtue of the convergence of series (1.1.5) for small y, we have the following estimate for some R:

$$\left|\partial_z^\beta G(z)\right|/\beta!\leqq CR^{|\beta|},\quad z\in\mathcal{O}_\delta(K_0)=K. \tag{4.9}$$

We shall substitute $z = z(x)$, $v = v(x)$ into (4.8'). According to (4.3)

$$\left\|\partial_z^\alpha G(z)v^\alpha\right\|_{l,p}\leqq C_{l,p}^{|\alpha|}\left\|\partial_z^\alpha G(z(x))\right\|_{l,p}\left\|v\right\|_{l,p}^{|\alpha|} \tag{4.10}$$

($\|v\|_{l,p} = \max\|v_i\|_{l,p}$). According to Proposition 4.5, which is formulated and proved at the end of this paragraph,

$$\left\|\partial_z^\alpha G(z(x))\right\|_{l,p}\leqq\left\|\partial_z^\alpha G(z)\right\|_{C^l(K)}C(\left\|z(x)\right\|_{l,p}). \tag{4.11}$$

It follows from estimate (4.9) that

$$\|\partial_z^\alpha G(z)\|_{C^l(K)} \leqq C_1(2R)^{|\alpha|}\alpha!.$$

It follows from this estimate and from (4.11) and (4.10) that when

$$2R\|v\|_{l,p}C_{l,p}<1$$

series (4.8′) converges and Proposition 4.1 is proved.

Proposition 4.2. Suppose the function $G(z)$ is defined in the neighbourhood of zero using Formula (4.5), and the operator $F(u)$ is determined using Formula (4.4) and operates from $W_p^{l+\nu}$ to W_p^l, $l > m/p$. Then F is analytic in the neighbourhood of zero.

Proof. The mapping $L: u \to \partial^{(\nu)}u$ is linear and bounded from $W_p^{l+\nu}$ to $(W_p^l)^r$. This mapping maps zero into zero. Since the mapping $G: z(x) \to G(z(x))$ is analytic in the neighbourhood of zero in $(W_p^l)^r$ by virtue of Proposition 4.1, the mapping $F = GL$ is analytic in the neighbourhood of zero in W_p^l, and Proposition 4.2 is proved.

Consider the differential operator A with the constant coefficients

$$Av = \sum_{|\beta| \leqq \nu} a_\beta \partial^\beta v. \tag{4.12}$$

A is obviously a ν-th order linear operator.

Henceforth we shall consider the linear operators $A = A(\partial)$, which satisfy the following condition.

Condition 4.1. The polynomial $a_{(\nu)}(\xi)$ (the principal symbol of the operator $A(\partial)$), which is determined using the equation

$$a_{(\nu)}(\xi) = \sum_{|\beta| = \nu} a_\beta (i\xi)^\beta, \tag{4.13}$$

is analytic, i.e.

$$C|\xi|^{\nu} \geqq |a_{(\nu)}(\xi)| \geqq C^{-1}|\xi|^{\nu}. \tag{4.14}$$

In addition, the complete symbol of the operator $A(\partial)$

$$a(\xi) = \sum_{|\beta| \leqq \nu} a_{\beta}(i\xi)^{\beta}, \qquad \xi \in \mathbb{R}^m, \tag{4.15}$$

does not vanish when $\xi \in \mathbb{R}^m$.

Condition 4.2. There is a sector

$$\{\omega_1 \leqq \arg z \leqq \omega_2\},\ \omega_1 < \omega_2,$$

in the complex plane $\mathbb{C}$, such that the function $z = a(\xi)$, where $a(\xi)$ is determined using Formula (4.15), does not take values in this sector.

If the operator $A = a(\partial)$ satisfies condition 4.2, the complex powers $(a(\xi))^{\lambda}$, $\lambda \in \mathbb{C}$ of its complete symbol are determined. Consequently the powers of the operator A are determined using the formula

$$A^{\lambda} = \mathcal{F}^{-1}(a(\xi))^{\lambda}\mathcal{F} \tag{4.16}$$

(here $\mathcal{F}$ is a discrete Fourier transform (see (5.2.13), (5.2.14)), which sets the Fourier coefficients $\tilde{u}(\xi)$ and $\xi \in \mathbb{Z}^m$ in correspondence with the function $u(x)$ on the torus), and $\alpha(\xi)$ is the operator of multiplication by the function $a(\xi)$.

Proposition 4.3. If $E = W_p^{\,l}(T^m)$, the ν-th order operator A satisfies condition 4.1, S=A, and the space E_k, which is determined by (2.2), is identical with the space $W_p^{\,l+\nu}(T^m)$. If, in addition, condition 4.2 holds, and $S = A^{1/\nu}$, then E_k is identical with $W_p^{l+k}(T^m)$.

This proposition is the corollary of the following statement.

Proposition 4.4. Suppose A satisfies conditions 4.1 and 4.2, $S = A^{1/\nu}$. Then the constant $C = C(l, p)$ exists, such that

$$C\|u\|_{l+1,p} \leq \|Su\|_{l,p} \leq C^{-1}\|u\|_{l+1,p} \tag{4.17}$$

for any $l \geqslant 0$.

Proof. As is known (see Nikol'skii [1]) we can specify the norm in $W_p^l(T^m)$ (which is equivalent for integer l to the norm (4.1)) using the equation

$$\|u\|_{W_p^l} = \|(1-\Delta)^{l/2}u\|_{L_p}, \tag{4.18}$$

where Δ is the Laplace operator, and $(1 - \Delta)^{\lambda}$ is determined using Fourier's transform by means of Formula (4.16). Thus, to prove (4.17) it is sufficient to prove the boundedness in L_p of the operators Λ and Λ^{-1},

$$\Lambda = \Lambda(\partial) = (1-\Delta)^{l/2}((a(\partial))^{\delta}(1-\Delta)^{-(l+1)/2} = (a(\partial))^{\delta}(1-\Delta)^{-1/2} \tag{4.19}$$

($\delta = 1/\nu$). The symbol of the operator Λ has the form

$$\Lambda(\xi) = (a(\xi))^{\delta}(1+|\xi|^2)^{-1/2} = \frac{a_{(\nu)}^{\delta}}{|\xi|}\left[1+\left(\frac{a-a_{(\nu)}}{a_{(\nu)}}\right)^{\delta}\left(1+\frac{1}{|\xi|^2}\right)^{-1/2}\right].$$

The factor $a^{\delta}_{(\nu)}(\xi)/|\xi|$, which is composed of the principal parts of the symbols $a(\xi)$ and $(1 + |\xi|^2)^{1/2}$, is homogeneous of the order of zero. Expanding in the neighbourhood of infinity the fractional powers in the square brackets in series, and using Euler's lemma on homogeneous functions, we obtain the well-known estimate

$$|\partial_\xi^\alpha \Lambda(\xi)| \leqq C_\alpha |\xi|^{-|\alpha|} \tag{4.20}$$

for fairly large $|\xi|$. A similar inequality also holds for $\Lambda^{-1}(\xi)$. It follows from the satisfaction of (4.20) for $\Lambda(\xi)$ and $\Lambda^{-1}(\xi)$ (see Nikol'skii [1]), that $\Lambda(\xi)$ and $\Lambda^{-1}(\xi)$ are multipliers in L_p, $1 < p < \infty$, i.e. the operators $S(1 - \Delta)^{-1/2}$ and $S^{-1}(1 - \Delta)^{1/2}$ are bounded in $L_p(T^m)$.

Remark 4.1. Proposition 4.3 follows directly from Proposition 4.4 when condition 4.2 holds. If condition 4.2 does not hold, $a(\partial)(1 - \Delta)^{\nu/2}$ is considered instead of $(a(\partial))^{\delta}(1 - \Delta)^{1/2}$, and the boundedness in L_p of this operator is derived from the theorem on multipliers, whence follows the estimate

$$C\|u\|_{l+\nu,p} \leqq \|Au\|_{l,p} \leqq C^{-1}\|u\|_{l+\nu,p}. \tag{4.21}$$

Proposition 4.2 follows from this estimate when $S = A$.

The action of differential operators with constant coefficients on a torus on the product of functions has an important property which is described by the following lemma.

Lemma 4.1. If the operator A, which is determined by Formula (4.12), satisfies condition 4.1, $l > m/p$, the constants C_0 and C_1 exist, such that for any integer $k > 0$ and any u, $v \in W_p^{\,l}(T^m)$ the following inequality holds:

$$\|A^k(A^{-k}u \cdot A^{-k}v)\|_{l,p} \leqq C_0 C_1^k \|u\|_{l,p} \|v\|_{l,p}. \tag{4.22}$$

Proof. The operator A^k represents the sum of N^k differential monomials of an order no greater than νk (N is the number of terms on the right-hand side of Formula (4.12)) It follows from the Leibnitz formula that a differential monomial of an order no higher than νk, applied to the product $A^{-k}u \cdot A^{-k}v$, equals the sum of not more than $2^{\nu k}$ terms of the form

$$\partial^{\gamma_1} A^{-k}u \cdot \partial^{\gamma_2} A^{-k}v,$$

where $|\gamma_1| + |\gamma_2| \leqslant \nu k$. Therefore $A^k(A^{-k}u \cdot A^{-k}v)$ equals the sum of not more than $2^{\nu k}N^k$ expressions of the form

$$a[\partial^{\gamma_1}(A^{-k}u)\partial^{\gamma_2}(A^{-k}v)], \tag{4.23}$$

where

$$|\gamma_1|+|\gamma_2|\leqq \nu k, \quad |a|\leqq\left(1+\max_{\alpha}|a_\alpha|\right)^k. \tag{4.24}$$

Each of these terms, in accordance with (4.3), does not exceed the following number with respect to the norm in W_p^l:

$$|a|C_{l,p}\|\partial^{\gamma_1}A^{-k}u\|_{l,p}\|\partial^{\gamma_2}A^{-k}v\|_{l,p}, \tag{4.25}$$

where $C_{l,p}$ is a constant from (4.3). Since the operators ∂^γ and A^{-k} commute, the expression

$$\partial^{\gamma_1}A^{-k}u, \quad |\gamma|\leqq k\nu,$$

can be rewritten in the form

$$\partial^{\gamma_1}A^{-k}u=\prod_{i=1}^{k}[\partial^{\alpha_i}A^{-1}]u, \qquad 0\leqq|\alpha_i|\leqq \nu. \tag{4.26}$$

Bearing in mind that the operators ∂^α, which operate from $W_p^{l+\nu}$ to W_p^l when $|\alpha| \leqslant m$, are uniformly bounded, and using (4.21), we obtain:

$$\|\partial^{\alpha_i} A^{-1} v\|_{l,p} \leqq C_2 \|v\|_{l,p}, \qquad \forall \alpha_i, |\alpha_i| \leqq \nu. \tag{4.27}$$

From (4.26) and (4.27) follows the estimate

$$\|\partial^{\gamma_1} A^{-k} u\|_{l,p} \leqq C_2^k \|u\|_{l,p}, \tag{4.28}$$

where C_2 does not depend on k. We also have a similar estimate for $\partial^{\gamma_2} A^{-k} u$. By virtue of these estimates we make an upper estimate of (4.25) by

$$|a| C_3 C_2^{2k} \|u\|_{l,p} \|v\|_{l,p}.$$

Bearing in mind estimate (4.24) and the fact that the number of terms of the form (4.23) and

$$A^k(A^{-k}u \cdot A^{-k}v)$$

is not larger than $(N2^\nu)^k$, we obtain that estimate (4.22) holds, where C_0 and C_1 are sufficiently large and do not depend on k.

Lemma 4.2. Suppose the operator A satisfies conditions 4.1 and 4.2, $l > m/p$. Suppose $S_1 = A^{1/\nu}$. Then the constants C_0 and C_{10} exist, such that

$$\|S_1^j(S_1^{-j}u \cdot S_1^{-j}v)\|_{l,p} \leqq C_0 C_{10}^j \|u\|_{l,p} \|v\|_{l,p} \tag{4.29}$$

when $j \in \mathbb{N}$, whilst we can take $C_{10} = C_1^{1/\nu}$, C_1 as C_{10}, and C_1 is the same as in (4.22).

Proof. Obviously, $j = k\nu + i$, $0 \leqslant i < \nu$. Therefore, by virtue of inequality (4.17) and (4.22)

$$\begin{aligned}\|S_1^j(S_1^{-j}u\cdot S^{-j}v)\|_{l,p}&=\|S_1^iA^k(A^{-k}S_1^{-i}u\cdot A^{-k}S_1^{-i}v)\|_{l,p}\\&\pm C_2\|A^k(A^{-k}(S_1^{-i}u)\cdot A^{-k}(S_1^{-i}v)\|_{l+i,p}\\&\leqq C_3C_1^k\|S_1^{-i}u\|_{l+i,p}\|S_1^{-i}v\|_{l+i,p}\\&\leqq C_4C_1^k\|u\|_{l,p}\|v\|_{l,p}\\&\leqq C_5C_1^{k+i/v}\|u\|_{l,p}\|v\|_{l,p}.\end{aligned}$$

Hence directly follows (4.29).

Let us now consider the operator F, which is determined by Formula (4.4). We shall use A to denote its linear part,

$$Au=G_1(\partial^{(v)}u),$$

where G_1 is determined by Formula (4.7). The linear operator A has the form

$$Av=\sum_{|\alpha|=1}\sum_{|\beta|\leqq v}g_{\alpha(\beta)}\partial^\beta v=G_1(\partial^{(v)}v). \tag{4.30}$$

Henceforth we will assume that A satisfies condition 4.1.

Theorem 4.1. Suppose the operator F is determined by Formula (4.4), the linear operator A is determined by Formula (4.30), and the space

$$E=W_p^l(T^m),\ l>m/p.$$

Suppose the operator A satisfies condition 4.1. Suppose, in addition, the norm of the operator A^{-1} from E to E is less than unity:

$$\|A^{-1}u\|_{l,p}<T_0\|u\|_{l,p},\qquad T_0<1, \tag{4.33}$$

and the constant C_1 in inequality (4.22) is also less than unity. Suppose the operator B is determined on the neighbourhood of zero in E using the formula

$$Bv = F(A^{-1}v). \tag{4.34}$$

Then:

1) $M > 1$ will be obtained, such that $A = MS_0$, the operator $S_0 = M^{-1}A$ satisfies (2.1) and the operator B, which is determined by (4.34), is quasi-interchangeable with S_0 at zero.

2) The operator F when

$$\|u\|_{l+\nu,p} \leqq \delta,$$

where $\delta > 0$ is also fairly small, is represented in the form

$$F = \mathscr{E} \circ L \circ \mathscr{E}^{-1},$$

where $\mathscr{E}$ and $\mathscr{E}^{-1}$ are operators that are quasi-interchangeable with S_0 at zero.

3) If the operator A also satisfies condition 4.2, and the operator $A^{-1/\nu}$ has a norm that is less than 1, M will be obtained, such that $M^{-1}A = S^{\nu}$, S satisifes (2.1) and the operators B, $\mathscr{E}$ and $\mathscr{E}^{-1}$ are quasi-interchangeable with S at zero, i.e. sect.1) and sect.2) hold, where S_0 is replaced by S.

Proof. We shall consider the case when condition 4.2 holds and $M^{-1}A = S^{\nu}$, $\nu > 1$. (The case when condition 4.2 does not hold is considered in a similar way, as if ν equalled 1). We will assume

$$S = (M^{-1}A)^{1/\nu} \tag{4.35}$$

($A^{1/\nu}$ is determined by Formula (4.16)). Since $\|A^{-1/\nu}\| < 1$, then when $M = 1 + \varepsilon$, ε is fairly small,

$$\|S^{-1}\| = M^{1/\nu}\|A^{-1/\nu}\| \leqq 1.$$

Since (4.22) holds with the constant $C_1 < 1$, then according to Lemma 4.2 estimate (4.29) holds with the constant $C_{10} < 1$ for $S_1 = A^{1/\nu}$. Therefore the constant C_{10} will be multiplied by $M^{1/\nu}$ for $S = M^{-1/\nu}S_1$ in estimate (4.29), and for fairly small ε, $M = 1+\varepsilon$, the constant C_{10} will not be larger than 1, i.e.

$$\|S^j(S^{-j}u\cdot S^{-j}v\| \leqq C_0\|u\|\,\|v\|. \tag{4.36}$$

Here and below $\|\ \| = \|\ \|_{l,p}$. The following estimate is derived from estimate (4.36) using induction:

$$\left\|S^j\left(\prod_{k=1}^{n} S^{-j}v_k\right)\right\| \leqq C_0^{n-1}\prod_{k=1}^{n}\|v_k\|. \tag{4.37}$$

Indeed, according to (4.36)

$$\left\|S^j\left(\prod_{k=1}^{n-1} S^{-j}v_k\cdot S^{-j}v_n\right)\right\| = \left\|S^j\left(S^{-j}\left(S^j\prod_{k=1}^{n-1} S^{-j}v_k\right)\cdot S^{-j}v_n\right)\right\|$$
$$\leqq C_0\left\|S^j\prod_{k=1}^{n-1} S^{-j}v_k\right\|\,\|v_n\|,$$

whence it is easy to derive (4.37).

We shall now prove that the operator B is quasi-interchangeable with S at zero. According to (4.8) and (4.34)

$$Bu = u + \sum_{j=2}^{\infty} F_j(A^{-1}u), \tag{4.38}$$

where F_j are the same as in (4.8). The operator B thus has the form (2.16), where

$$B_j(u) = \sum_{|\alpha|=j} g_\alpha\cdot(\partial^{(\nu)}A^{-1}u)^\alpha. \tag{4.39}$$

The term $B_{j\alpha}$ of the sum on the right-hand side of (4.39) has the form of the product j of the factors

$$B_{j\alpha}(u)=g_\alpha \partial^{\gamma_1}A^{-1}u\ldots\partial^{\gamma_j}A^{-1}u. \tag{4.40}$$

Passing from j-homogeneous to j-linear operators, we note that (4.40) can be written in the form

$$B_{j\alpha}(\bar{u})=g_\alpha \partial^{\gamma_1}A^{-1}u_1\ldots\partial^{\gamma_j}A^{-1}u_j. \tag{4.41}$$

We shall estimate the norm $|||B_{j\alpha}|||$. Obviously, since S^{-1} commutes with ∂^{γ_j} and with A^{-1}, then

$$\|S^k B_{j\alpha}\circ S^{-k}\bar{u}\|=|g_\alpha|\left\|S^k\prod_{i=1}^{j} S^{-k}(\partial^{\gamma_i}A^{-1}u_i)\right\|. \tag{4.42}$$

Using (4.37), we obtain hence the estimate

$$\|S^k B_{j\alpha}\circ S^{-k}\bar{u}\|\leqq|g_\alpha|C_0^{j-1}\prod_{i=1}^{j}\|\partial^{\gamma_i}A^{-1}u_i\|. \tag{4.43}$$

Using the boundedness in E of the operators $\partial^{\gamma_i}A^{-1}$ $(|\gamma_j| \leqslant \nu)$ with respect to the norm using constant C, from (4.43) and (4.39) we obtain:

$$\|S^k B_j\circ S^{-k}\bar{u}\|\leqq\sum_{|\alpha|=j}|g_\alpha|C_{01}^j\prod_{i=1}^{j}\|u_i\|. \tag{4.44}$$

By virtue of (4.9) $|g_\alpha| \leqslant CR^{|\alpha|}$. Therefore from (4.44) and from (2.5) we obtain:

$$|||B_j|||\leqq C'C_{01}^j(2R)^j. \tag{4.45}$$

Hence it follows that conditions (2.15), (2.14) hold, i.e. B is an operator that is quasi-interchangeable with S at zero, and sect.1) (where $S_0 = S$) is proved.

Theorem 3.3 is therefore applicable, using which we obtain that $\mathcal{E}$ and $\mathcal{E}^{-1}$, which are quasi-interchangeable with S at zero, exist, whilst Formula (3.3) holds in the domain $\{\|v\|_\nu \leq r\}$. Using Proposition 4.2, we obtain that $E_\nu = W_{l+\nu,p}$ and therefore, taking $\delta < r$, we obtain the statement of sect. 2) for $S_0 = S$. Theorem 4.1 is thus completely proved.

In the following paragraphs we shall construct a global linearisation of the operators F, which satisfy additional conditions of global character.

Remark 4.2. The conditions on the operator A - the linear part of the operator F at zero - imopsed in Theorem 4.1, hold for a wide class of operators F. Indeed, according to Lemmas 4.1 and 4.2 estimates (4.22) and (4.29) hold for arbitrary operators A, which satisfy conditions 4.1 and 4.2 respectively, with certain constants C_1 and C_{10}. It is obvious that when the operator A is multiplied by the numerical multiplier q the constants C_1 and C_{10} are multiplied by q^{-1} and $q^{-1/\nu}$ respectively. Therefore, if the operator F of the form (4.4), for which the operator A, determined by (4.30), satisfies condition 4.1, then for fairly large q the operator $F_1 = qF$ is analytically linearised in the neighbourhood of zero:

$$F_1 = \mathscr{E}_1 \circ A_1 \circ \mathscr{E}_1^{-1}, (A_1 = qA).$$

In conclusion we shall prove the statement on the superposition of functions from C^l and $W_p^{\,l}$.

Proposition 4.5. Suppose the domain

$$\Omega_1 \subset \mathbb{R}^N, \Omega_2 = T^m,$$

$$f \in C^l(\Omega_1), \quad \varphi : T^m \to \Omega_1, \quad \varphi \in (W_p^l(T^m))^N, \qquad l > m/p.$$

Then the constant $C = C(l, m, N, \Omega)$ exists, such that

$$\|f(\varphi(x))\|_{l,p} \leqq C\|f\|_{C_l}(\|\varphi\|_{l,p}^l + 1), \tag{4.46}$$

where $\| \ \|_{l,p}$ is the norm in $W_p^l(T^m)$.

Proof. Supopse φ is a smooth function. We shall use Formula (5.6.29), where $\Omega_2 = T^m$, to estimate $\partial^\alpha f(u)$ in $L_p = L_p(T^m)$; obviously, it also holds when the dimension N of the space $\mathbb{R}^N$, containing Ω_1, does not equal the dimension m of the torus $T^m = \mathbb{R}^m/(2\pi\mathbb{Z})^m$. When $|\alpha| < l$ we obtain from (5.6.29):

$$\|\partial^\alpha f(\varphi(x))\|_{0,p} \leqq C_0\|f(\varphi)\|_{C^l(\Omega_1)} \max_{\gamma_{ij}} \|\prod_i \partial^{\gamma_{ij}}\varphi_i\|_{0,p}, \tag{4.47}$$

where

$$|\gamma_{ij}| \geqq 1, \sum |\gamma_{ij}| = l_\gamma \leqq l.$$

We will assume $q_{ij} = l_\gamma/\gamma_{ij}$, and obviously $\sum 1/q_{ij} = 1$. We shall estimate the product on the right-hand side of (4.47) using Holder's inequality:

$$\|\prod \partial^{\gamma_{ij}}\varphi_i\|_{0,p} \leqq \prod \|\partial^{\gamma_{ij}}\varphi_i\|_{0,pq_{ij}} \leqq \prod \|\varphi_i\|_{|\gamma_{ij}|,pq_{ij}}. \tag{4.48}$$

According to Sobolev's imbedding theorem,

$$\|\varphi\|_{\gamma,p1} \leqq C\|\varphi\|_{l,p}. \tag{4.49}$$

if

$$\gamma \leqq l, \quad \gamma - m/(pq) \leqq l - m/p. \tag{4.50}$$

We shall prove that this condition holds when $q = q_{ij}$, $\gamma = |\gamma_{ij}|$. By virtue of the definition of the numbers q_{ij}

$$m/(pq_j) = m|\gamma_{ij}|/(l_\gamma p) \geqq m|\gamma_{ij}|/(lp).$$

Therefore to prove (4.50) it is sufficient to prove the inequality

$$\gamma + m/p \leqq l + m\gamma/(lp), \tag{4.51}$$

where $\gamma = |\gamma_{ij}| \leqslant l$, $m/p < l$. Obviously, if $l = \gamma$ or $l = m/p$, inequality (4.51) becomes an equality. Note that the right-hand side of (4.51) which is considered as a function of l, increases monotonically on the half-line $l > m/p$, $l > \gamma$. Indeed, the minimum point

$$l_0 = \sqrt{m\gamma/p}$$

cannot belong to this half-line, since the system of inequalities

$$\sqrt{m\gamma/p} > m/p, \quad \sqrt{m\gamma/p} \geqq \gamma$$

does not have solutions. Therefore (4.51), and consequently (4.50) also holds when $\gamma = |\gamma_{ij}|$, $q = q_{ij}$, i.e. estimate (4.49), where $\gamma = |\gamma_{ij}|$, $q = q_{ij}$, holds. We obtain estimate (4.46) from (4.49), (4.48) and (4.47).

Suppose φ is an arbitrary function from $W_p^{\,l}$ with values in Ω_1. Since $W_p^{\,l} \subset C(T^m)$, the set of values φ is compact in Ω_1. Suppose φ_i is a sequence of smooth functions which approximate φ in $W_p^{\,l}$ and consequently in $C(T^m)$, Their values are also contained in Ω_1 and estimate (4.46) is applicable to them. Since φ_i converge to φ in C, $f(\varphi_j)$ converge in C to $f(\varphi)$. Note that $f(\varphi_j)$ are bounded in $W_p^{\,l}$ by virtue of (4.46). Since $W_p^{\,l}$ is reflexive, according to the theorem on weak compactness some subsequence $f(\varphi_j)$ weakly converges to some function $f_0 \in W_p^l$. But since $f(\varphi_j) \to f(\varphi)$, $f_0 = f(\varphi)$, and estimate (4.46) therefore holds.

Remark 4.3. Estimate (4.3) follows from Proposition 4.5. Indeed, we will assume $f(u, v) = uv$. It follows from (4.46) that $\|uv\|_{l,p} \leqslant C$ when

$$\|u\|_{l,p} \leqq 1, \ \|v\|_{l,p} \leqq 1.$$

Using the linearity with respect to u and with respect to v, we obtain inequality (4.3).

§5. Analytic continuation

In this paragraph we shall continue our examination of nonlinear operators in a Banach space. Under the additional constraints imposed on the operator F, we will prove that the representation of F in the form (3.3) holds not only locally, but also in the neighbourhood of some real subspace V of the complex space E.

Suppose V is a linear real closed subspace of the complex space E. Henceforth we will assume that the subspaces $V_k = V \cap E_k$ are closed in E_k and are everywhere dense in V, and that the operator S maps V_{k+1} into V_k, and $S^{-1} : V_k$ into V_{k+1}.

Suppose B is a nonlinear operator which is analytic in the open neighbourhood of the subspace V and which maps V_k into V_k, $k \in \mathbb{Z}_+$. Suppose, in addition, B is quasi-interchangeable with S at zero, whilst B_j change V^j into V. We shall then call the operator B quasi-interchangeable with S on V.

We recall that a linear continuous operator A from the Banach space E to the Banach space H is called a Fredholm operator if the kernel of this operator is finite-dimensional:

$$\dim \ker A < \infty,$$

and the form $A(E)$ is a closed linear subspace in H of finite co-dimensions:

$$\dim \operatorname{coker} A < \infty.$$

The number

$$\operatorname{ind} A = \dim \ker A - \dim \operatorname{coker} A.$$

is the index of the Fredholm operator (see, for example, Palais [1] for the properties of Fredholm operators).

We shall call the operator B, which is quasi-interchangeable with S on V, an operator that is Fredholm with the index 0 and quasi-interchangeable with S on V if, firstly, it is an operator that is quasi-interchangeable with S at zero, and, secondly, B is represented on V in quasilinear form (2.18): $B(u) = P(u)u$, whilst the linear operator $P(v)$ has the following properties:

1) for all $v \in V_k$ the operator $P(v)$ maps V_k into V_k and V_{k+1} into V_{k+1} and is a Fredholm operator with the index 0 in these spaces;

2) the operator $P(v)$ analytically depends on $v \in V_k$ like an element of the space of the linear operators $\mathcal{L}(V_k, V_k)$ and $\mathcal{L}(V_{k+1}, V_{k+1})$.

Theorem 5.1. Suppose B is an operator that is quasi-interchangeable with S on V, the operator F is determined by Formula (3.5), and $\mathcal{E} = \mathcal{E}_0$ is the local solution of Eq.(3.6), given by Formula (3.16) when $\|u\| < r$. Then the analytic continuation of $\mathcal{E}$ to V is given by the formula

$$\mathscr{E}(u) = F^j \circ \mathscr{E}_0 \circ A^j u, \tag{5.1}$$

where the number j is so great that $M^{-j}\|u\| < r$. This continuation determines in V the operator $\mathcal{E}$ which is quasi-interchangeable with S on V, and satisfies Eq.(3.6).

Proof. First of all, we note that the j-linear operators $\mathcal{E}_j$, constructed in the proof of Theorem 3.1, map V_k^j into V_k $\forall\, k \in \mathbb{Z}_+$. We shall carry out the proof using induction with respect to j. When $j=1$ the statement is obvious, since $\mathcal{E}_1 = I$. Suppose

$$\mathcal{E}_j(V_k^j) \subset V_k$$

when $j < n$, $k \in \mathbb{Z}_+$. Then the operator R_n in Formula (3.19) changes V_k^n into V_k.

Indeed, according to (3.19) only $\mathcal{E}_{il}$ when $i_l < n-1$, the operators B_j, A and A^{-1} occur in the definition R_n. The operator $A^{-1} = M^{-1}S^{-\nu}$ changes V_k into $V_{k+\nu}$, since S^{-1} changes V_k into V_{k+1}. The operator $\mathcal{E}_{il}$ maps $V_{k+\nu}$ into $V_{k+\nu}$ on the assumption of induction, and the operator A maps $V_{k+\nu}$ into V_k, since $S\colon V_{k+1} \to V_k$. Therefore

$$A\mathcal{E}_{i_l} \circ A^{-1}$$

maps V_k^{il} into V_k. Since B_j maps V^j into V, the operator R_n, determined by (3.19), maps V^n into V.

We shall now prove that the operator $\mathcal{E}_n$, determined by (3.21), maps V_k^n into V_k. Each term

$$A^{\gamma}R_n \circ A^{-\gamma}$$

maps V_k into V_k, since

$$A^{-\gamma}\colon V_k \to V_{k+\nu\gamma}, \quad R_n\colon V_{k+\nu\gamma}^n \to V_{k+\nu\gamma}, \quad A^{\gamma}\colon V_{k+\nu\gamma} \to V_k.$$

Since series (3.21) converges with respect to the norm of the operators $||| \; |||$, it converges on each $\bar{v} \in V_k^n$ in the norm $\| \; \|_k$ by virtue of (3.21′)

$$\|A^\gamma R_n \circ A^{-\gamma}\bar{v}\|_k = \|S^k A^\gamma R_n \circ A^{-\gamma} S^{-k}(S_k \bar{v})\|$$
$$\leq |||R_n||| M^{-\gamma(n-1)} \|\bar{v}\|^n.$$

Since V_k is closed in E_k, the sum of series (3.21) also belongs to V_k. Thus, the induction step is completed, and it is proved that

$$\mathscr{E}_j V_k^j \subset V_k, \qquad \forall k \in \mathbb{Z}_+, \; j \in \mathbb{Z}_+. \tag{5.2}$$

Since series (3.17) converges in E when $\|u\| < r$, series (3.16) converges in E_k when $\|u\|_k < r$. By virtue of (5.2) each of its terms belongs to V_k when $v \in V_k$, and therefore

$$\mathcal{E}_0(v) \in V_k \text{ when } \|v\|_k < r \tag{5.3}$$

Suppose $\|u\| = R$ then for large j

$$\|A^{-j}u\|_j \leq M^{-j}R < r.$$

Therefore, by virtue of Theorem 3.1 the operator

$$\mathscr{E} \circ A^{-j}$$

is analytic from $\mathcal{O}_r \cap V$ to V_J. Since F^j is analytic from V_j to V, the operator $\mathcal{E}$, determined by (5.1), is analytic on $\mathcal{O}_R$. Substituting the expression $\mathcal{E}$ using Formula (4.1) into the left-hand side of (3.6), we obtain the expression

$$FF^j \circ \mathscr{E}_0 \circ A^{-j}A^{-1}(u) = F^j \circ (F \circ \mathscr{E}_0 \circ A^{-1}) \circ A^{-j}(u) \tag{5.4}$$

when $\|u\| < R$. Since $\mathcal{E}_0$ satisfies Eq.(3.6) on $\mathcal{O}_r$, then

$$F^j \circ (F \circ \mathscr{E}_0 \circ A^{-1})A^{-j}(u) = F^j \circ \mathscr{E}_0 \circ A^{-j}(u)$$

when $\|u\| < R$ also. It follows hence and from (5.4) that the operator $\mathcal{E}$ satisfies Eq.(3.6) on $\|u\| < r$. The operator $\mathcal{E}$ is identical with $\mathcal{E}_0$ when $\|u\| < r$. Indeed, the solution of Eq.(3.6) which is quasi-interchangeable with S at zero is unique, since the terms of series (3.16) are uniquely determined if $\mathcal{E}(0)=0$ and $\mathcal{E}'(0)=1$. We immediately see from (5.1) that the requirements of (3.7) hold for $\mathcal{E}$, if they hold for $\mathcal{E}_0$. By virtue of Formula (3.6) for $\mathcal{E}_0$ the operator $\mathcal{E}$, determined by Formula (5.1), does not depend on j for large j.

We shall now verify that $\mathcal{E}$ continuously maps V_k into V_k. If

$$\Omega_R^k = \{v \in V : \|v\|_k < R\}$$

the linear operator A^{-j} continuously maps Ω_R^k into $\Omega_r^{k+j\nu}$ ($r > M^{-j}R$). By virtue of Theorem 3.1, Proposition 2.1 and (5.3), $\mathcal{E}_0$ analytically maps $\Omega_r^{k+j\nu}$ into $V_{k+j\nu}$, and the operator F^j analytically maps $V_{k+j\nu}$ into V_k. Therefore the operator $\mathcal{E}$ is analytic from Ω_R^k to V_k for any R, and is consequently analytic from V_k to V_k. Thus, all the requirements of the property $\mathcal{E}$ are established, and the theorem is proved.

Remark 5.1. When $V=E$, and the operator B is entire, Theorem 5.1 is a simple corollary of Theorem 3.1.

We shall now consider the question on the domain of definition of the transformation $\mathcal{E}^{-1}$, which is inverse to $\mathcal{E}$. Note that this question is much more complex than a question about the domain of definition of $\mathcal{E}$. This is obvious from the simplest one-dimensional example

$$F(z)=(1+z)^3-1, \quad \mathcal{E}(z)=e^z-1, \quad \mathcal{E}^{-1}(z)=\ln(1+z).$$

In this example we are not able to extend $\mathcal{E}^{-1}$ to the whole real axis. However we shall indicate below the conditions on F that are sufficient for $\mathcal{E}^{-1}$ to continue analytically to the real subspace V of the complex space E.

We shall introduce the definitions we shall henceforth need.

We shall say that the operator B maps the set $\Omega \subset V$ into the set $\Omega_1 \subset V$ with preservation of smoothness, if

$$B(\Omega \cap V_k) \supset \Omega_1 \cap V_k, \qquad \forall k \in \mathbb{Z}_+. \tag{5.5}$$

The nonlinear operator F, which is determined on V_ν and which maps V_ν into V, will be called stretching to V if

$$\|F(v)\| \geq T\|v\|, \qquad \forall v \in V_\nu, \tag{5.6}$$

where $T > 1$, and constant T does not depend on v.

Theorem 5.2. Suppose B is an operator that is quasi-interchangeable with S on V. Suppose $\varkappa > 0$, $\varkappa$ is an integer and the operator B maps $V_\varkappa$ on to all $V_\varkappa$ with preservation of smoothness.

Let $F = B \circ A$ be a stretching operator on V ($A = MS^\nu$). Suppose the number $r > 0$ exists, such that $\mathcal{E}$ maps $V_\varkappa$ on to $\mathcal{O}_r \cap V_\varkappa$ with preservation of smoothness. Then $\mathcal{E}$ maps V_k on to all V_k, $k = \varkappa, \varkappa + 1, \ldots$.

Proof. We shall prove that $\mathcal{E}$ maps V_k into $\mathcal{O}_{Tr} \cap V_k$ with preservation of smoothness. For this we shall use Formula (3.6):

$$F \circ \mathcal{E} \circ A^{-1} = \mathcal{E},$$

which holds on V by virtue of Theorem 5.1. To prove that $\mathcal{E}$ has the property indicated, it is sufficient to prove that

$$F\circ\mathcal{E}\circ A^{-1}.$$

has this property. The operator $A^{-1} = M^{-1}S^{-\nu}$, which occurs on the left-hand side of Eq.(3.6), maps V_k on to $V_{k+\nu}$. Since $\mathcal{E}$ maps $V_\varkappa$ into $\mathcal{O}_r \cap V_\varkappa$ with preservation of smoothness, then

$$\mathcal{E}(V_k)\supset\mathcal{O}_r\cap V_k,\qquad \forall\, k\in\mathbb{Z}_+,\quad k\geqq\kappa,$$

and, consequently, since $A^{-1}V_k = V_{k+\nu}$, then

$$\mathcal{E}\circ A^{-1}(V_k)\supset\mathcal{O}_r\cap V_{k+\nu},\qquad \forall\, k\in\mathbb{Z}_+,\quad k\geqq\kappa-\nu. \tag{5.7}$$

We shall now prove that

$$F(\mathcal{O}_r\cap V_{k+\nu})\supset\mathcal{O}_{Tr}\cap V_k,\qquad \forall\, k\geqq\kappa, \tag{5.8}$$

where T is a constant from inequality (5.6) for the stretching operator F. Indeed, by virtue of (5.5), where $\Omega = V_\varkappa$, $\Omega_1 = V_\varkappa$, we have: $B(V_k) = V_k$ when $k > \varkappa$. Since $A(V_{k+\nu}) = V_k$, it follows that

$$F(V_{k+\nu})=B\circ A(V_{k+\nu})=V_k$$

and

$$F(V_{k+\nu})\supset\mathcal{O}_{Tr}\cap V_k,\qquad \forall\, k\geqq\kappa. \tag{5.9}$$

To derive (5.8) from (5.9), consider the equation $F(u) = g$, $g \in V$,

$\|g\| < Tr$ and $u \in V_\nu$. Since the operator F is a stretching operator, from (5.6) we obtain the estimate

$$T\|u\| \leqq \|F(u)\| = \|g\| < Tr$$

and, consequently, $\|u\| < r$. Therefore

$$F^{-1}(\mathcal{O}_{Tr} \cap V) \cap V_\nu \subset \mathcal{O}_r.$$

(5.8) follows directly from this inclusion and from (5.9). It follows from (5.7) and (5.8) that

$$F \circ \mathscr{E} \circ A^{-1}(V_k) \supset F(\mathcal{O}_r \cap V_{k+\nu}) \supset \mathcal{O}_{Tr} \cap V_k, \qquad \forall k \geqq \kappa.$$

Bearing in mind Eq.(3.6), it follows hence that

$$\mathscr{E}(V_k) \supset \mathcal{O}_{Tr} \cap V_k, \qquad \forall k \geqq \kappa,$$

i.e. $\mathcal{E}$ maps V_k on to $\mathcal{O}_{r_1} \cap V_\varkappa$ with preservation of smoothness,

$$r_1 = Tr, \ T > 1.$$

It is obvious that we can repeat all the arguments, having replaced r by r_1. Repeating this argument by a factor of n (n is such that $rT^n > R$), we obtain that $\mathcal{E}$ maps $V_\varkappa$ into $\mathcal{O}_R \cap V_\varkappa$ with preservation of smoothness for any R. Hence directly follows the statement of the lemma.

Remark 5.2. In Theorem 5.2 the most difficult condition to verify is that on $\mathcal{E}$, which consists of the fact that $\mathcal{E}$ maps $V_\varkappa$ into $\mathcal{O}_r \cap V_\varkappa$ with preservation of smoothness. We shall formulate a theorem which gives the sufficient condition for the above condition to be satisfied.

Theorem 5.3. Suppose $\mathcal{E}$ is a quasilinear operator that is Fredholm with the index 0 and quasi-interchangeable with S on V. Suppose

$$k \in \mathbb{Z}_+, \quad u \in V, \ \mathcal{E}(u) = h \in V_k.$$

Then $u \in V_k$, and the operator $\mathcal{E}$ maps $V_\varkappa$ into $V_\varkappa \cap \mathcal{O}_r$ with preservation of smoothness for any $\varkappa \in \mathbb{Z}_+$.

The proof of Theorem 5.3 is based on the following lemma.

Lemma 5.1. Suppose E_1, E_2, H_1, H_2 are real Banach spaces, and the operator $P_2: E_2 \to H_2$ is Fredholm and has index $\varkappa_0$. Suppose $E_2 \subset E_1$, $H_2 \subset H_1$ and the embeddings are continuous and everywhere dense. Suppose the operator P_2 continues to the operator $P_1 : E_1 \to H_1$, whilst the operator P_1 is Fredholm and has the same index $\varkappa_0$. Then, if $P_1 u = h$, where $h \in H_2$, then $u \in E_2$.

Proof. Since there are no more solutions of the equation $P_1 u = 0$, belonging to E_2, than of all the solutions from E_1, the dimension k_1 of the kernel of the operator P_1 is not less than the dimension k_2 of the kernel of the operator P_2, $k_1 > k_2$. Since P_1 is Fredholm with the index $\varkappa$, the image of P_1 in H_1 is specified by the equations

$$(f_1, h) = 0, \ldots, (f_{k_1 - \kappa_0}, h) = 0 \tag{5.10}$$

where $f_j \in H_1$, f_j are linearly independent on H_1. Obviously, f_j are also continuous on H_2. Since H_2 is everywhere dense in H_1, then f_j are also independent on H_2. Since $P_1(H_1) \supset P_2(H_2)$, then $h = P_2(v)$ when $v \in E_2$ needs to satisfy (5.10). Since f_j are linearly independent on H_2, the subspace $P_2(E_2)$ has a codimension in H_2 that is not less than the number of conditions (5.10), i.e.

$$\dim \operatorname{coker} P_2 \geqq k_1 - \kappa_0.$$

Consequently, $k_2 - \varkappa_0 > k_1 - \varkappa_0$, i.e. $k_2 > k_1$. Bearing in mind that $k_2 < k_1$, we obtain that $k_2 = k_1$. Therefore the subspace $P_2(E_2)$ is specified in H_2 by the same conditions (5.10). If the equation $P_1 u = h$ has a solution in E_1, conditions (5.10) hold. If, in addition, $h \in H_1$, then owing to the satisfaction of (5.10) $h \in P_2(E_2)$, and therefore the equation $P_2 u = h$ has the solution $u_2 \in E_2$. Since $k_1 = k_2$, the dimensions of the kernels P_1 and P_2 are equal, and since $\ker P_2 \subset \ker P_1$, then $\ker P_1 = \ker P_2$. Consequently $\ker P_1 \subset E_2$. Since the general solution of the equation $P_1 u = h$ has the form

$$u = u_2 + v, \; v \in \ker P_1$$

and $u_2 \in E_2$, then $u \in E_2$ also; *Q.E.D.*

Proof of Theorem 5.3. Since $\mathcal{E}$ is quasilinear, then $\mathcal{E}(u) = Q(u)u$, where Q is a linear Fredholm operator. Suppose the equation $Q(u)u = h$, where $h \in V$, has a solution in $V = V_0$. We shall use induction to prove that if $h \in V_k$, then $u \in V_k$. Suppose this is proved when $k < k_0$. We shall prove it when $k = k_0$. Suppose $h \in V_k$. Then, obviously, $h \in V_{k-1}$. On the assumption of induction $u \in V_{k-1}$. By virtue of the required properties of the operator $\mathcal{E}$ the linear operators

$$Q(u): V_{k-1} \to V_{k-1}$$

and

$$Q(u): V_k \to V_k$$

are Fredholm with a zero index. Since $h \in V_k$, then using Lemma 5.1 we obtain that $u \in V_k$.

Since $\mathcal{E}'(0) = I$, then $\mathcal{E}(V) \supset \mathcal{O}_r$ for fairly small r. Using the already proved properties of the equation $\mathcal{E}(u) = h$, we obtain that the inclusion (5.5) holds, where $\Omega = V$, $\Omega_1 = \mathcal{O}_r$, and Theorem 5.3 is proved.

The following lemma indicates the condition on the operator B which is sufficient for the conditions of Theorem 5.3 to be satisfied.

Lemma 5.2. Suppose B is a quasilinear operator that is quasi-interchangeable with S on V. Then the operator $\mathcal{E}$ (constructed in Theorem 5.1) is a quasilinear operator that is quasi-interchangeable with S on V, and consequently,

$$\mathscr{E}(u) = Q(u)u, \tag{5.11}$$

where $Q(v)$ when $v \in V_k$ is a linear operator, and $Q(v)$ continuously maps V_k into V_k and V_{k+1} into V_{k+1}. If B is a quasilinear Fredholm operator with the index 0 which is quasi-interchangeable with S on V, $\mathcal{E}$ is also a quasilinear Fredholm operator with the index 0 which is quasi-interchangeable with S on V.

Proof. According to Theorem 3.1, the operator $\mathcal{E}$ is quasilinear and quasi-interchangeable with S at zero, i.e. Formula (5.11) occurs when $\|u\| < r$, where r is fairly small.

We shall use a formula that is similar to Formula (5.1) to extend $Q(v)$ to all $v \in V$. Before calculating this formula, we shall define the law of composition of the two quasilinear operators $\Lambda(u, v)$ and $G(u, v)$ using the formula

$$\Lambda \circ G(u, v) = \Lambda(G(u, v), G(v, v)). \tag{5.12}$$

Note that Formula (3.6), which occurs for the operator $\mathcal{E}(u)$, determined by (3.16), also occurs when F is a quasilinear operator and the law of composition is understood in the sense of (5.12), i.e.

$$F \circ \mathscr{E}(A^{-1}u, A^{-1}v) = \mathscr{E}(u, v). \tag{5.13}$$

To verify the validity of (5.13) for small $\|v\|$, it is sufficient to verify that the relations (3.19) agree with (5.12). We shall write (3.19), having pointed out the arguments explicitly:

$$\begin{aligned}(\mathscr{E}_n - A\mathscr{E}_n \circ A^{-1})(u_1, u_2, \ldots, u_n) \\ = \sum_{j=2}^{\infty} \sum_{i_1 + \cdots + i_j = n} B_j(A\mathscr{E}_{i_1} \circ A^{-1}(u_1, \ldots, u_{i_j}), \\ \ldots, A\mathscr{E}_{ij} \circ A^{-1}(u_{n-i_j+1}, \ldots, u_n))\end{aligned} \tag{5.14}$$

Since $\mathcal{E}_j(u, v) = \mathcal{E}_j(u, v, \ldots, v)$, hence we obtain:

$$\begin{aligned}(\mathscr{E}_n - A\mathscr{E}_n \circ A^{-1})(u, v, \ldots, v) \\ = \sum_{j=2}^{\infty} \sum_{i_1 + \cdots + i_j = n} B_j(A\mathscr{E}_{i_1} A^{-1}(u, v, \ldots, v), \ldots, A\mathscr{E}_{i_j} A^{-1}(v, \ldots, v)),\end{aligned}$$

i.e. variable u is present only in the first place under the operator sign $\mathcal{E}_{j1}$, which stands in the first place under the sign B_j. Thus in each term of the sum on the right-hand side of (5.14) the composition of the operators B and $\mathcal{E}$ matches Formula (5.12), and Formula (5.13) consequently holds when $\|v\| < r$.

We shall now determine the operators $Q(v)$ for large $\|v\|$ using Formula (5.1), where j is fairly great, and the law of composition is understood in the sense of (5.12):

$$\mathscr{E}(u, v) = F^j \circ \mathscr{E}_0(A^{-j}u, A^{-j}v). \tag{5.15}$$

We shall write the operator F^j in quasilinear form, explicitly marking the unbounded part - the operator A^j (we recall that $F(u) = B(Au) = P(Au)Au$)

$$F^j(u, v) = P_j(A^j v)A^j u, \tag{5.16}$$

where $P_j(w)$ is a linear operator for each fixed w. Bearing in mind (5.12), we obtain recurrence formulae for the operators P_j from the identity

$$F^j = F \circ F^{j-1},$$

when $j = 1$ $P_1 = P$, where P is the same as in (2.18), and when $j = 2, 3, \ldots$

$$P_j(A^j v)A^j u = P(AF^{j-1}(v))AP_{j-1}(A^{j-1}v)A^{j-1}u.$$

Replacing $A^j v$ by v, and $A^j u$ by u, we shall rewrite this equation in the form

$$P_j(v)u = P(AF^{j-1}(A^{-j}v))AP_{j-1}(A^{-1}v)A^{-1}u. \tag{5.17}$$

We shall use induction to prove that when $v \in V_k$, $P_j(v)$ maps V_k into V_k, and V_{k+1} into V_{k+1}. Suppose this holds when $j < j_0$, $k \in \mathbb{Z}_+$. Consider (5.17), where $j = j_0$, $v \in V_k$. Then

$$AF^{j-1}(A^{-j}v) \in V_k, \quad A^{-1}v \in V_{k+\nu}.$$

The operator A^{-1} maps V_k into V_k, and V_{k+1} into $V_{k+1+\nu}$. Since $A^{-1}v \in V_{k+\nu}$, then $P_{j-1}(A^{-1}v)$ maps (on the assumption of induction) $V_{k+\nu}$ into $V_{k+\nu}$ and $V_{k+\nu+1}$ into $V_{k+\nu+1}$. Therefore the operator $AP_{j-1}(A^{-1}v)A^{-1}$ maps V_k into V_k, and V_{k+1} into V_{k+1}. Further, since

$$AF^{j-1}(A^{-j}v) \in V_k,$$

then

$$P(AF^{j-1}(A^{-j}v))$$

maps V_k into V_k and V_{k+1} into V_{k+1}. Thus $P_j(v)$ also has the property required in the definition of quasilinearity when $j = j_0$, and consequently also for all j.

We shall now use Formula (5.15), where we shall express F^j using Formula (5.16):

$$\mathscr{E}(u, v) = P_j(A^j\mathscr{E}_0(A^{-j}v, A^{-j}v))A^j\mathscr{E}_0(A^{-j}u, A^{-j}v),$$

and by virtue of (5.11)

$$Q(v) = P_j(A^j\mathscr{E}_0(A^{-j}v))A^jQ(A^{-j}v)A^{-j}. \tag{5.18}$$

Suppose $\|v\|_k < R$, j is so great that $RM^{-j} < r$, where $M > 1$ is a number from Formula (3.5). Then

$$\|A^{-j}v\|_{k-1j} < r$$

and Formula (5.18) determines the operator $Q(v)$. (Note that $Q(v)$ does not depend on j by virtue of Formula (5.13)). Since by virtue of Proposition 2.2 the operator $A^jQ(A^{-j}v)A^{-j}$ maps V_k into V_k and V_{k+1} into V_{k+1}, and P_j, as was proved above, has the same property, the linear operator Q also maps V_k into V_k and V_{k+1} into V_{k+1}. It is obvious from (5.18) that $Q(v)$ analytically depends on $v \in V_k$ by virtue of the analyticity of P_j and $Q|_{\mathscr{O}_r^k}$. Thus Q is a quasilinear operator that is quasi-interchangeable with S on V.

Let us now consider the case when B is Fredholm. Since the product of linear Fredholm operators is a Fredholm operator, and the index of the product equals the sum of the indices of the factors, the operators $P_j(v)$ (see (5.17)) are Fredholm with the index 0. Using Formula (5.18) and the fact that $Q(v)$ is invertible for small v, we obtain that $Q(v)$ is Fredholm for all v (in the spaces V_k, $k \in \mathbb{Z}_+$) with the index 0. The lemma is proved.

Proposition 5.2. If B is a one-to-one operator that is quasi-interchangeable with S on V, the operator $\mathcal{E}$ - the solution of (3.6) - is also one-to-one on V. If, in addition, B has a non-degenerate first differential at each point V (from V_k to V_k $\forall k \in \mathbb{Z}_+$), $\mathcal{E}$ has a nondegenerate first differential everywhere on V.

Proof. The statement is obvious in the small neighbourhood of zero. For large $\|v\|$ we shall use the formula

$$\mathscr{E}(v)=F^j\circ\mathscr{E}\circ A^{-j}(v),$$

the one-to-one character and nondegeneracy of the differential of the operator F^j on $V_{k+\nu j}$ and the nondegeneracy of the differential $\mathcal{E}$ in the neighbourhood of zero.

We shall now formulate the basic result of the paragraph.

Theorem 5.4. Suppose B is a quasilinear operator that is Fredholm with the index 0 and is quasi-interchangeable with S on V. Suppose the operator F is determined by Formula (3.5), maps $V_{k+\kappa+\nu}$ into $V_{k+\kappa}$, $\forall k\in\mathbb{Z}_+$ and is an operator that stretches V. Suppose the operator B is one-to-one on V, and this differential is nondegenerate on V_k, $k\in\mathbb{Z}_+$. Then the representation F in the form (3.3) holds on $V_{k+\kappa+\nu}$, $k\in\mathbb{Z}_+$,

$$F=\mathscr{E}\circ A\circ\mathscr{E}^{-1},$$

where $\mathcal{E}$ and $\mathcal{E}^{-1}$ are analytic on V_k when $k > \varkappa$.

Proof. According to Lemma 5.2 the conditions of Theorem 5.3 hold. According to Theorem 5.3 $\mathcal{E}$ maps $V_\varkappa$ into $V_\varkappa \cap \mathcal{O}_r$ with preservation of smoothness. Therefore all the conditions of Theorem 5.2 hold and $\mathcal{E}(V_\varkappa) = V_\varkappa$, whilst $\mathcal{E}(V_k) = V_k \ \forall k > \varkappa$. The operator $\mathcal{E}$ is one-to-one by virtue of Proposition 5.1, and its differential is nondegenerate. Therefore the operator $\mathcal{E}^{-1}$, which is analytic by virtue of the theorem on the implicit function, is determined on $V_\varkappa$. Formula (3.3) follows from (3.6).

§6. Global linearisation of nonlinear differential operators on a torus.

Consider on the torus T^m the nonlinear differential operator F, which is determined by Formula (4.4). We will assume that the function $G(z)$, which occurs in the definition of this operator, has the following properties: the function $G(z)$ is determined everywhere in the real subspace $R^r \subset \mathbb{C}^r$ of vectors with real components, and is real and analytic in this subspace. It is also assumed that the operator F has a quasilinear structure, namely the nonlinear part $G_{(2)}$ (see (4.6), (4.7)) has the form

$$G_{(2)}(\partial^{(\nu)}u) = \sum_{|\beta| \leqq \nu} G^0_\beta(\partial^{(\nu-1)}u)\, \partial^\beta u. \tag{6.1}$$

Obviously, the leading coefficients occur linearly in $G(\partial^{(\nu)}u)$. (In accordance with the arrangement on the numbering of variables z_i (see §4) the function $G(z)$, $z \in \mathbb{R}^N$, whose nonlinear part $G_{(2)}(z)$ is the same as in (6.1), will be linear with respect to the first N_1 variables z_j, corresponding to $\partial^\beta u$ with $|\beta| = \nu$).

We shall call the differential operators

$$F(u) = G(\partial^{(\nu)}u) = Au + G_{(2)}(\partial^{(\nu)}u),$$

- A is linear and real - of the above form quasilinear really analytic differential operators.

The operator F is called elliptic if a linear operator with constant coefficients

$$Au + \sum_{|\beta| \leqq \nu} G_\beta^0(z)\, \partial^\beta u \tag{6.2}$$

is elliptic (see condition 4.1) for any $z \in \mathbb{R}^{N-N_1}$.

Below we will consider the operators F in the functional spaces $W_p^{\,l}(T^m)$, which we shall, for brevity, denote by $W_p^{\,l}$.

Theorem 6.1. Suppose F is an elliptic quasilinear really analytic differential operator, and ν is even. Suppose the linear operator $A = F_1$, determined by (4.30), satisfies condition 4.1. Suppose

$$E = W_p^l, \qquad l > m/p,\ V = \operatorname{Re} W_p^l$$

is the subspace E, consisting of real functions. Suppose the operator $A^{1/\nu}$ satisfies the conditions of sect. 3) of Theorem 4.1 ($A^{1/\nu}$ exists, since the satisfaction of condition 4.2 follows from the realness of A and the satisfaction of condition 4.1). Suppose

$$B = F \circ A^{-1}.$$

Then B is a quasilinear operator that is quasi-interchangeable (with $S = M^{-1/\nu} A^{1/\nu}$ for some $M > 1$) and Fredholm with the index 0. If G_α^0 are integral functions, B is an integral operator from V_k to V_k, $k \in \mathbb{Z}_+$.

Proof. First of all we shall prove that the operator $B(u)$ is quasi-interchangeable with S on V. According to sect. 1) of the statement of Theorem 4.1, the operator B is quasi-interchangeable with S at zero. It therefore remained to verify that B is analytic on V and maps V_k into V_k. Note, first of all, that by virtue of inequality (4.17)

$$V_k = \operatorname{Re} W_p^{l+k}(T^m).$$

It follows from Proposition 4.5 that if

$$u_j \in W_p^l, \quad l > m/p, \quad j = 1, \ldots, N, \quad G \in C^l(\mathbb{R}^N),$$

then

$$G(u_1, \ldots, u_N) \in W_p^l(T^m).$$

Therefore, bearing in mind that A^{-1} changes $\operatorname{Re} W_p^{l+k}$ into $\operatorname{Re} W_p^{l+\nu+k}$ we obtain that B maps $\operatorname{Re} W_p^{l+k}$ into $\operatorname{Re} W_p^{l+k}$ i.e. maps V_k into V_k. The analyticity of B on V_k follows from the analyticity of the function G (it is verified in the same way as in Proposition 4.2). Therefore all the requirements from the definition of the operator B that is quasi-interchangeable with S on V hold.

It is obvious from the proof of Propositions 4.1 and 4.2 that if the function $G(z)$ is entire, i.e. the number R in (4.9) is as small as desired, series (4.8), which determines $F(u)$, converges in any W_p^{l+k}, $l + k > m/p$. Consequently the series for

$$B = F \circ A^{-1}$$

also converges, i.e. B is an entire operator on $V_k = \operatorname{Re} W_p^{l+k}$.

We shall now prove that the operator B is quasilinear and quasi-interchangeable with S at zero. The proof is carried out in the same way as that of sect. 1) of Theorem 4.1. The difference is that instead of

$$\|S^k B_{j\alpha} \circ S^{-k} \bar{u}\|$$

an expression of the following form is estimated in Formula (4.43):

$$\|S^k B_{j\alpha}(S^{-k}u_1, S^{-k+1}u_2, \ldots, S^{-k+1}u_j\|,$$

i.e. u_2, ..., u_j are taken insted of Su_2, ..., Su_j. The fact that by virtue of (6.1) the leading derivatives occur linearly in G, and consequently also in $G_{j\alpha}$, plays the main role in deriving the required estimate. Therefore only ∂^{γ_1} can have the order ν on the right-hand side of Formula (4.43), and all the remainder ∂^{γ_1} have the smaller order $|\gamma_i| < \nu-1$ when $i > 2$. Therefore the operators $\partial^{\gamma_1}A^{-1}S$ which we have instead of $\partial^{\gamma_1}A^{-1}$ in Formula (4.43), are also bounded in W_p^l, and we obtain, similarly to estimate (4.44), the estimate

$$\|S^k B_j(S^{-k}uS^{-k+1}\bar{v})\| \leqq \hat{g}_j C_{02}^{j-1}\|u_1\| \prod_{i=2}^{j} \|v_i\|.$$

It follows from this inequality that an estimate similar to (4.45) holds:

$$|||B_j|||_1 \leqq C'' C_{03}^j.$$

Hence it follows that B is a quasilinear operator that is quasi-interchangeable with S at zero.

We shall now prove that B is a quasilinear operator that is quasi-interchangeable with S on V and Fredholm with the index 0. For this it is required to obtain the representation $B(u)$ in the form (2.18): $B(u) = P(u)u$. We obtain this representation from Formula (6.1):

$$B(u) = u + \sum_{|\alpha| \leqq \nu} G_\alpha^0(\partial^{(\nu-1)}A^{-1}u)\,\partial^\alpha A^{-1}u. \tag{6.3}$$

We shall take as $P(v)$ a linear operator that is determined by the formula

$$P(v)u=u+\sum_{|\alpha|\leqq\nu} G^0_\alpha(\partial^{(\nu-1)}A^{-1}v)\,\partial^\alpha A^{-1}u. \tag{6.4}$$

If $v \in W_p^l$, then $A^{-1}v$ by virtue of the ellipticity of A belongs to $W_p^{l+\nu}$ (see (4.21)) and therefore all the components of the vector $\partial^{(\nu-1)}A^{-1}v$ belong to W_p^{l+1}. Therefore, since $l > m/p$, all the coefficients of the operator (6.4) belong to W_p^{l+1}. Multiplication by these coefficients is a bounded operator in both W_p^l and in W_p^{l+1} by virtue of (4.3). Therefore the linear operator $P(v)$, determined by (6.4), is continuous from W_p^l to W_p^l and from W_p^{l+1} to W_p^{l+1}. Since the coefficients $G^0_\alpha(\partial^{(\nu-1)}A^{-1}$ analytically depend on v in W_p^{l+1} (Proposition 4.1), the operator $P(v)$ analytically depends on $v \in W_p^l$ in $\mathcal{L}(W_p^l, W_p^l)$ and in $\mathcal{L}(W_p^{l+1}, W_p^{l+1})$.

It remained to verify that the operator P, which is determined by (6.4), is Fredholm with the index 0. Note that the operator

$$P(v)A: W_p^{l+k+\nu} \to W_p^{l+k}$$

has the form

$$P(v)Au=Au+\sum_{|\alpha|\leqq\nu} G^0_\alpha(\mathscr{D}^{(\nu-1)}A^{-1}v)\,\partial^\alpha u. \tag{6.4'}$$

If we fix $x \in T^m$ and freeze $v = v(x)$, and consequently also the coefficients at this point, we obtain an operator with constant coefficients of the form (6.2), which is elliptic by virtue of the conditions of the theorem. This means that an operator with variable coefficients $P(v)A$ is also elliptic, and consequently also Fredholm (see, for example, Bers, John, Schechter [1], Hormander [1]). Note that by virtue of the continuous dependence of $P(v)A$ on v it is sufficient to establish the Fredholm character of $P(v)A$ when $v \in C^\infty$. Since the index of the operator $P(v)A$ does not change with homotopy (see Palais [1]), then ind $(P(tv)A)$ = const when $t \in [0, 1]$. Since $P(0)A = A$, and the operator A is invertible, the index $P(v)A$ equals zero for all v. Since $P(v) = (P(v)A)A^{-1}$, and the index of composition of the operators equals the sum of the indices, $P(v)$ also has the index 0.

Theorem 6.1 is thus completely proved.

To obtain the representation F on v = Re W_p^l in the form

$$F=\mathscr{E}\circ A\circ\mathscr{E}^{-1}$$

using Theorem 5.4, it is required to impose additional requirements on F, sufficient for F to map $V_{\nu+k}$ on to V_k, for it to stretch V, and for it to be one-to-one. Below we give examples of the operators F, which have all these properties.

Consider a second-order operator of divergent form:

$$F(u)=-\sum_{i=1}^{m}\partial_i G_i(\nabla u)+G_0(u). \tag{6.5}$$

Here G_i, G_0 are real analytic functions,

$$G_i(0)=0, \quad G_0(0)=0, \quad \nabla u=(\partial_1 u,\dots,\partial_m u).$$

We shall use

$$G_{ij}(i,j=1,\dots,m)$$

to denote the derivative from $G_0(\nabla uC)$ with respect to $\partial_j u$. We impose the following requirements on G_i:

$$\sum_{i,j=1}^{m} G_{ij}(z)\xi_i\xi_j \geqq C(z)|\xi|^2, \quad C(z)>0, \qquad \forall z\in\mathbb{R}^m, \tag{6.6}$$

$$G_0'(u)\geqq K_0>0. \tag{6.7}$$

Lemma 6.1. Suppose the operator F is determined by (6.5) and satisfies conditions (6.6), (6.7). Suppose $p > m$, $p > 2$. Then

$$\forall u \in \operatorname{Re} W_p^3(T^m) \quad \|F(u)\|_{1,p} \geqq K_0 \|u\|_{1,p},$$

and the operator F will be a stretching operator $\operatorname{Re} W_p^1(T^m)$ when $K_0 > 1$.

Proof. We shall use $\varphi(v)$ to denote the function $\varphi(v) = |v|^{p-2}v$. Obviously,

$$\partial_i \varphi(v) = (p-1)|v|^{p-2}\partial_i v, \qquad i = 1, \ldots, m. \tag{6.8}$$

We shall take

$$u \in \operatorname{Re} W_p^3(T^m)$$

(this space, by virtue of Sobolev's imbedding theorem, is contained in $C^2(T^m)$). We shall multiply $F(u)$ by $\varphi(u)$ and shall integrate it over T^m. Integrating by parts, we obtain:

$$\begin{aligned}\int F(u)\varphi(u)\,dx = \int \sum_{i=1}^{m} G_i(\nabla u)\,\partial_i \varphi(u)\,dx \\ + \int G_0(u)\varphi(u)\,dx.\end{aligned} \tag{6.9}$$

Here and below $\int$ indicates integration over $T^m = [-\pi, \pi]^m$. Obviously,

$$G_i(y) = \int_0^1 \frac{d}{d\theta} G_i(\theta y)\,d\theta = \int_0^1 \sum_{j=1}^{m} G_{ij}(\theta y) y_j\,d\theta. \tag{6.10}$$

Bearing in mind this equation and (6.8), from (6.9) we obtain

$$\begin{aligned}\int F(u)\varphi(u)\,dx = (p-1)\int_0^1 \sum_{i,j=1}^{m} G_{ij}(\theta \nabla u)\,\partial_j u\,\partial_i u\,d\theta |u|^{p-2}\,dx \\ + \int G_0'(\theta u) u \varphi(u)\,dx.\end{aligned} \tag{6.11}$$

The first term on the right-hand side is nonnegative by virtue of condition (6.6), and the second, by virtue of (6.7), is less than

$$K_0 \int |u|^p \, dx = K_0 \|u\|_{0,p}^p = K_0 \|u\|^p.$$

(Below in the proof of the theorem we shall, for brevity, denote the norm $\| \ \|_{0,p}$ in $L_p(T^m)$ by $\| \ \|$.) Thus, from (6.11) follows the inequality

$$\int F(u)\varphi(u)\, dx \geqq K_0 \|u\|^p. \tag{6.12}$$

We shall estimate the left-hand side of (6.12) using Holder's inequality:

$$\begin{aligned} \left|\int F(u)\varphi(u)\, dx\right| &\leqq \left(\int (F(u))^p \, dx\right)^{1/p} \left(\int (\varphi(u))^{p/(p-1)} \, dx\right)^{(p-1)/p} \\ &= \|F(u)\| \, \|u\|^{p-1}. \end{aligned} \tag{6.13}$$

From (6.12) and (6.13) we obtain:

$$\|F(u)\| \geqq K_0 \|u\|. \tag{6.14}$$

We shall now multiply $F(u)$ by

$$-\partial_\kappa \varphi(\partial_\kappa u), \quad m \geqq \kappa \geqq 1,$$

and shall integrate the two sides by parts:

$$\begin{aligned} -\int F(u)\, \partial_\kappa \varphi(\partial_\kappa u)\, dx &= \int \partial_\kappa F(u) \varphi(\partial_\kappa u)\, dx \\ &= \int \left[-\sum_{i=1}^m \partial_i \partial_\kappa G_i(\nabla u) + \partial_\kappa G_0(u) \right] \varphi(\partial_\kappa u)\, dx \\ &= (p-1) \int \sum_{i,j=1}^m G_{ij}(\nabla u)\, \partial_j \partial_\kappa u\, \partial_i \partial_\kappa u |u|^{p-2}\, dx \\ &\quad + \int G_0'(u) |\partial_\kappa u|^p \, dx. \end{aligned} \tag{6.15}$$

By virtue of condition (6.6) the first term on the right-hand side of (6.15) is nonnegative, and the second is, by virtue of (6.7), not less than $\|\partial_\varkappa u\|^p$. We shall transform the left-hand side, integrating by parts:

$$-\int F(u)\,\partial_\kappa \varphi(\partial_\kappa u)\,dx = \int \partial_\kappa F(u)\varphi(\partial_\kappa u)\,dx.$$

Making an upper estimate of this expression in a similar way to (6.13), from (6.15) we obtain the inequality:

$$\|\partial_\kappa F(u)\| \geqq K_0 \|\partial_\kappa u\|. \tag{6.16}$$

Note that to derive (6.16) we can first take $u \in C^\infty(T^m)$. Since both sides of (6.16) continuously depend on $u \in W_p^3(T^m)$ $(p > m)$, then approximating $u \in W_p^3$ using smooth functions, we obtain (6.16) in the general case ($\partial_\varkappa F(u)$ continuously depends on u, since

$$\nabla u \in W_p^2 \subset C^1(T^m)$$

and $G_i(\nabla u)$ are analytic in W_p^2 with respect to $u \in W_p^3$). Raising (6.14) and (6.16) in powers of p and summing over $\varkappa$, we obtain:

$$\|F(u)\|_{1,p}^p \geqq K_0^p \|u\|_{1,p}^p. \tag{6.17}$$

Hence follows the statement of Lemma 6.1.

We shall use A to denote the operator - the linear part in zero of the operator F, determined by (6.5):

$$Au = -\sum G_{ij}(0)\,\partial_i \partial_j u + G_0'(0)u. \tag{6.18}$$

Lemma 6.2. Suppose F is determined by (6.5), whilst conditions (6.6), (6.7) hold, $p > m$. Then F maps Re W_p^{l+3} on to Re W_p^{l+1} when $l \geq 1$. The operator

$$B = F \circ A^{-1},$$

where A is determined by (6.18), maps Re W_p^{l+1} into Re W_p^{l+1}.

Proof. it is required to prove the solvability in Re W_p^{l+3} of the equation $F(u) = h$ when

$$h \in \mathrm{Re}\, W_p^{1+l} \subset C^1(T^m)$$

We shall first derive a priori estimates $|u|$ and $|\nabla u|$. For this we shall use estimates (6.14) and (6.16), where $\| \ \| = \| \ \|_{Lp}$. As is known,

$$\|v\|_{L_\infty} = \lim_{p \to \infty} \|v\|_{L_p}. \tag{6.19}$$

From (6.14), (6.16) and (6.19) we obtain that

$$\|u\|_{L_\infty} \leqq K_0^{-1} \|h\|_{L_\infty}, \quad \|\partial_\kappa u\|_{L_\infty} \leqq K_0^{-1} \|\partial_\kappa h\|_{L_\infty}, \qquad \kappa = 1, \dots, m.$$

Since u belongs to $C^2(T^m)$, the norms in $L_\infty(T^m)$ are identical with those in $C(T^m)$, and therefore

$$\|u\|_{C^1} \leqq K_0 \|h\|_{C^1}. \tag{6.20}$$

Differentiating (6.5) with respect to $x_\varkappa$, we obtain an equation that is linear with respect to $\partial_\varkappa u$:

$$-\sum_{i,j=1}^{m} \partial_i (G_{ij}(\nabla u)\, \partial_j \partial_\kappa u) + G_0'(u)\, \partial_\kappa u = \partial_\kappa h. \tag{6.21}$$

By virtue of (6.20) $|\nabla u|$ and $|u|$ are bounded by a constant that depends only on $\|h\|_{C^1}$. By virtue of (6.6), where z is bounded, we obtain that Eq.(6.21) is uniformly elliptic. It follows from Theorem 2.2 of Chapter IX, Ladyzhenskaya and Ural'tseva [1], (to which we refer below as [L. U.]), that

$$\|\partial_\kappa u\|_{C^\alpha} \leqq C_1, \qquad \alpha>0, \quad \kappa=1,\dots,m, \tag{6.22}$$

where α and C_1 depend only on $\|h\|_{C^1}$.

We shall now rewrite the equation $F(u) = h$ in nondivergent form:

$$F(u) = -\sum_{i,j=1}^{m} G_{ij}(\nabla u)\,\partial_i\partial_j u + G_0(u) = h. \tag{6.23}$$

By virtue of (6.22) the coefficients $G_{ij}(\nabla u)$ of this equation are bounded in C^α. Therefore, using Schauder's estimates of the solutions of these equations (see [L. U.], Ch. III), we obtain:

$$\|u\|^{C^{2+\infty}} \leqq C_2, \tag{6.24}$$

and C_2 depends only on $\|h\|_{C^1}$. Using the usual method (see L. U.]), from the presence of an a priori estimate for an equation of the form (6.23) we derive the existence of its solution in $C^{2+\alpha}(T^m)$.

It remained to prove that if $h \in W_p^{l+1}$, then $u \in W_p^{l+3}$. By virtue of (6.24) the coefficients (6.23) are bounded in $C^{1+\alpha}(T^m)$. According to Sobolev's imbedding theorem

$$W_p^2(T^m) \subset C^{1+\beta}(T^m)$$

when $\beta < 1 - m/p$. Without loss of generality we can assume that $\alpha < \beta$. Therefore $h \in C^{1+\alpha}(T^m)$. Using Schauder's estimates once more (now in $C^{3+\alpha}$), we obtain:

$$\|u\|_{C^{3+\alpha}} \leqq C_3,$$

where C_3 depends only on $\|u\|_{2,p}$. Therefore the coefficients (6.23) are bounded in $C^{2+\alpha}$. Using the estimates of the solutions of elliptic equations in W_p^l (see [L. U.]), we obtain the estimate

$$\|u\|_{4,p} \leqq C_4 \cdot \|h\|_{2,p}. \tag{6.25}$$

Since the estimate of the coefficients in

$$W_{k-1,p} \subset C^{k-2+\beta},$$

follows from the estimate $\|u\|_{k,p}$, then using estimates of the solutions of equations of the form (6.23) in W_p^l we obtain by induction that when

$$h \in W_p^{l+1} \; u \in W_p^{l+3}, \quad \forall l \geqq 1.$$

Therefore

$$F(\operatorname{Re} W_p^{l+3}) = \operatorname{Re} W_p^{l+1}.$$

Since A^{-1} is an isomorphism between $\operatorname{Re} W_p^{l+1}$ and $\operatorname{Re} W_p^{l+3}$, then

$$FA^{-1}(\operatorname{Re} W_p^{l+1}) = \operatorname{Re} W_p^{l+1}$$

and Lemma 6.2 is completely proved.

Lemma 6.3. Suppose F is determined by (6.5) and the conditions (6.6) and (6.7) hold,

$$F: \mathrm{Re}\, W_p^{l+2} \to \mathrm{Re}\, W_p^l,\ l \geqq 1,\ p > m.$$

Then the operator F is Fréchet differentiable, and its differential $F'(v)$ is determined by the following formula when $v \in \mathrm{Re}\, W_p^{l+2}$:

$$F'(v)u = -\sum_{i,j=1}^{m} \partial_i (G_{ij}(\nabla v)\, \partial_j u) + G_0'(v)u. \tag{6.26}$$

At the same time, for any $v \in \mathrm{Re}\, W_p^{l+2}$ and for any $u \in \mathrm{Re}\, W_2^l$ the following inequality holds:

$$(F'(v)u, u) \geqq K_0(u, u), \tag{6.27}$$

where $(\ ,\ )$ is a scalar product in $L_2(T^m)$.

Proof. The Fréchet-differentiabiilty of F follows from the analyticity of F, which is proved in the same way as in Proposition 4.2. We directly obtain Formula (6.26), deriving the increment $F(v + u)$ which is linear with respect to u at the point v. To obtain (6.27), we multiply (6.26) by u and integrate by parts:

$$(F'(v)u, u) = \int \sum G_{ij}(\nabla v)\, \partial_j u\, \partial_i u\, dx + \int G_0'(v) |u|^2\, dx.$$

Using conditions (6.6) and (6.7), we obtain (6.27).

Lemma 6.4. Suppose the conditions of Lemma 6.3 hold. Then the operator F is one-to-one.

Proof. Consider the difference $F(u_1) - F(u_2)$. Obviously, by virtue of the differentiability of F,

$$F(u_2) - F(u_1) = F'(u_1 + \theta(u_2 - u_1))(u_2 - u_1)$$

Multiplying this expression in L_2 by $(u_2 - u_1)$, by virtue of (6.27) we obtain:

$$K_0\|u_2-u_1\|^2 \leqq |(F(u_2)-F(u_1), u_2-u_1)| \leqq \|F(u_2)-F(u_1)\| \, \|u_2-u_1\|$$

where $\| \ \|$ is the norm in L_2. Therefore

$$\|u_1-u_2\| \leqq K_0^{-1}\|F(u_2)-F(u_1)\|,$$

whence follows the one-to-one character of F.

Theorem 6.2. Suppose the operator F is determined by Formula (6.5), whilst conditions (6.6) and (6.7) hold, where $K_0 >$ 1. Suppose

$$E = W^1_{p_0}(T^m), \quad p_0 > m.$$

Suppose the operator A is determined by Formula (6.18) and the operator A is such that the constant C_1 in (4.22) is smaller than unity, $C_1 < 1$, and the operator $A^{1/2}$ satisfies the condition $\|A^{-1/2}\| < 1$, where the norm is in the space of the operators in E.

Then

1) the operator F on Re $W_p^4(T^m)$ is represented in the form

$$F = \mathscr{E} \circ A \circ \mathscr{E}^{-1},$$

where the operators $\mathcal{E}^{-1}$ and $\mathcal{E}$ are analytic from Re W_p^{l+1} to Re W_p^{l+1} $\forall l \in \mathbb{N}$.

2) If the functions G_i are entire, the operator $\mathcal{E}$ is entire in these spaces.

3) If

$$s \geqq 2, \quad s - m/p > 2 - m/p_0,$$

and the operator A satisfies, additionally, the conditions $C_1 < 1$ and $\|A^{-1/2}\| < 1$ for the space $E' = W_p{}^s$, then $\mathcal{E}$ and $\mathcal{E}^{-1}$ are analytic from Re $W_p{}^{s+l}$ to Re $W_p{}^{s+l}$ then $l \in \mathbb{Z}_+$.

Proof. We shall use Theorem 5.4, where $B = F \circ A^{-1}$. We will assume V = Re $W_p{}^1$. Then, according to Proposition 4.3 V_k = Re W_p^{k+1}. According to Lemma 6.1 F stretches V, and according to Lemma 6.2 F maps $V_{k+\varkappa+\nu}$ on to $V_{k+\varkappa}$, where $\varkappa = 1$, $\nu = 2$. According to Lemma 6.4 the operator B is one-to-one on V, and according to Lemma 6.3 its differential is nondegenerate (since $F'(v)$ is Fredholm, it is sufficient that there should be no kernel for F' to be invertible, and the kernel equals zero by virtue of (6.27)). According to Theorem 6.1 B is a quasilinear operator that is quasi-interchangeable with $S = M^{-1/2}A^{1/2}$ for some $M > 1$. Thus, all the conditions of Theorem 5.4 hold. The statement of sect. 1) of Theorem 6.2 is a corollary of the statements of Theorem 5.4.

When G_l are entire functions, then according to Theorem 6.1 B is an entire operator, and by virtue of Theorem 3.1 the operator $\mathcal{E}$ is entire, which proves sect. 2).

We shall prove sect. 3), Note that if

$$s \quad m/p > 2 - m/p_0, \; s > 2,$$

then $s > m/p + 1$. Therefore we can take

$$\operatorname{Re} W_p^s = V', \; V' \subset V.$$

as the new space V. The operator B will, as previously, be a quasilinear operator that is quasi-interchangeable with S on V', and according to Theorem 5.1, where $V = V'$, the operator $\mathcal{E}$ is also determined on V' and is quasilinear, Fredholm with the index 0 and quasi-interchangeable with S on V'. We shall now prove that $\mathcal{E}$ maps Re $W_p{}^s$ on to Re $W_p{}^s$. According to Theorem 5.3 $\mathcal{E}$ maps V' on to $\mathcal{O}_r \cap V'$ with preservation of smoothness. Note that the continuation of $\mathcal{E}^{-1}$ to V', constructed in Theorem 5.2, is determined using the formula

$$\mathscr{E}^{-1}(u)=A^k\mathscr{E}^{-1}\circ F^{-k}u,$$

where k is fairly great. If l is fairly great, then, since F stretches to V,

$$\|F^{-l}u\|_V \leqq T^{-l}\|u\|_V,\ T>1.$$

At the same time, by virtue of Lemmas 6.2 and 6.4, if $u \in V$, then $F^{l}u \in V_l$, and if $u \in V'$, then $F^{l}u \in V'_l$. Note that since the operator F is differentiable, $F(0) = 0$ and $F'(0)$ is invertible from $V_{k+\nu}$ to V_k, then for fairly small f

$$\|F^{-1}(f)\|_{V_{k+\nu}} \leqq C\|f\|_{V_k}, \qquad k\in\mathbb{N}.$$

Therefore, for fixed i and large l

$$\|F^{-i-l}u\|_{V_{i\nu}} \leqq C_i\|F^{-l}u\|_V \leqq T^{-l}C_i\|f\|_V. \tag{6.28}$$

Since $V_i \subset V'$ for fairly large i by virtue of Sobolev's imbedding theorem it follows from (6.28) that for fairly large k

$$\|F^{-k}u\|_{V'}$$

will be as small as desired. Note that according to Theorem 5.3, used in the case $V=V'$, the operator $\mathcal{E}$ maps V' into $\mathcal{O}'_r \cap V'$

with preservation of smoothness ($\mathcal{O}'_r$ is the neighbourhood of zero in E' with respect to the norm V'). Obviously, $F^{-k}(u)$ for fairly large k enters $\mathcal{O}'_r$. At the same time, by virtue of Lemmas 6.2 and 6.4

$$F^{-k}u \in V'_{\nu k}.$$

Since $\mathcal{E}$ maps V' into $\mathcal{O}'_r \cap V'$ with preservation of smoothness, then

$$\mathcal{E}^{-1}(F^{-k}u) \in V'_{\nu k}.$$

The operator A^k is continuous from $V'_{\nu k}$ to V'. Thus the restriction of the operator $\mathcal{E}^{-1}$ over V' maps V' into V', and the operator $\mathcal{E}$ maps V' on to all V'. The analyticity of $\mathcal{E}^{-1}$ from V' to V' follows from the invertibility of F' from V'_ν to V' by virtue of Proposition 5.2, where $V = V'$.

Remark 6.1. Operators of more general form can be considered in a completely similar way to an operator of the form (6.5):

$$F(u) = -\sum_{i=1}^{m} \partial_i G_0(\partial^{(1)}u) + G_0(\partial^{(1)}u),$$

which are subject to some conditions (see Babin [7], Formulae (5.21) and (5.22)). We also have a representation of the form

$$F = \mathcal{E} \circ A \circ \mathcal{E}^{-1}$$

for higher-order monotonic operators on a torus, for example an operator on the torus of the form

$$F(u) = (-1)^{n/2} \sum_{i=1}^{m} a_i \partial_i^n u + a_0 u + \sum_{|\alpha| < n/2} (-1)^{|\alpha|} \partial^\alpha \alpha_\alpha(\partial^\alpha u),$$

where the real analytic functions a_α satisfy some conditions on increase (see Babin [7], Example 5.1), which guarantee the smoothness of the solutions of the equations $F(u) = h$ for smooth h (see Skrypnik [1]). See Babin [7], Example 5.1, for details.

Matrix operators, and also operators that are active on functions defined on $\mathbb{R}^m$, can be considered besides scalar operators.

Remark 6.2. If F is an arbitrary operator of the form (6.5), satisfying (6.6) and (6.7), the operator $F_1 = qF$, where q is fairly great, satisfies the conditions of Theorem 6.2. This follows from Remark 4.2, and also from the obvious fact that, when F is multiplied by q, the constant K_0 in (6.7) is also multiplied by q and $K_0 q > 1$ for fairly large q.

§7. Eigen functionals of the operator F^*, which is adjoint to the nonlinear operator F.

Suppose F is a nonlinear differential operator, which acts on functions defined on the torus T^m. The adjoint (or linear) operator F^*, which act on functionals that are determined on these functions, is attached in a natural way, to this operator. If the operator and the functional are linear, the adjoint operator is also a linear differential operator. In the linear case the classical results guarantee, on natural assumptions, the existence of a complete system of eigen functionals (i.e. functions, in the linear case) of the operator F^*. Here we shall consider the nonlinear case, and shall show that the linear operators F^*, which are adjoint to the nonlinear differential operators F, considered in the previous paragraph, also have a complete set of eigen functionals.

Suppose V_s, $s \in \mathbb{Z}_+$ is a scale of real Banach spaces, $V_{s1} \subset V_{s2}$, and the imbedding is continuous when $s_1 > s_2$. In this paragraph we will consider the operator F, which acts in this scale, whilst the operator

$$F: V_{s+\nu} \to V_s \tag{7.1}$$

is analytic for each $s \in \mathbb{Z}_+$.

Differential operators of the form (6.5) are an example of these spaces and operators, and we shall take the spaces

$$V_s = W_p^{l+s}(T^m),$$

where $l > 2$, $l - m/p > 2$, as the spaces V_s. Suppose $\mathcal{A}(V_s)$ is a set of functionals that is analytic on V_s. Suppose $F: V_{s+\nu} \to V_s$ is an analytic operator. Then we shall attach the functional

$$g(F(\cdot)) \in \mathcal{A}(V_{s+\nu})$$

to any functional $g(\cdot) \in \mathcal{A}(V_s)$. Thus the mapping

$$F^*: \mathcal{A}(V_s) \to \mathcal{A}(V_{s+\nu}),$$

is determined,

$$F^* g(\cdot) = g(F(\cdot)). \tag{7.2}$$

It is obvious that since $V_{s+\nu} \subset V_s$, then $\mathcal{A}(V_{s+\nu}) \supset \mathcal{A}(V_s)$. It is easy to see that the mapping F^* is linear. We shall call the functional g eigen for F^*, if the following equation holds:

$$F^* g = \lambda g, \qquad \lambda \in \mathbb{C}, \tag{7.3}$$

i.e.

$$g(F(u)) = \lambda g(u), \qquad \forall u \in V_{s+\nu}. \tag{7.4}$$

It is obvious that if the operator F and the functional g are linear, the definition given here agrees with the usual definition of an eigen functional (vector) of an adjoint operator.

We shall say that the operator F permits analytic linearisation on $V_{s+\nu}$, if there exists the analytic on V_s operator

$$\mathscr{E}: V_s \to V_s, \quad \mathscr{E}(0)=0,\ \mathscr{E}'(0)=I,$$

which has the inverse operator $\mathscr{E}^{-1}$ which is analytic on $V_{s+\nu}$, whilst the following equality holds:

$$F(u)=\mathscr{E} \circ A \circ \mathscr{E}^{-1}(u), \qquad \forall u \in V_{s+\nu}, \tag{7.5}$$

where A is a linear operator that is continuous from $V_{s+\nu}$ to V_s. The set $\{g_l\}$ of functionals g_l, determined on V_s, is called complete if i exist for any two u, $v \in V_s$, $u \neq v$, such that $g_l(u) \neq g_l(v)$.

Theorem 7.1. Suppose the operator $F: V_{s+\nu} \to V_s$ permits the analytic linearisation

$$F=\mathscr{E} \circ A \circ \mathscr{E}^{-1},$$

on $V_{s+\nu}$, whilst the linear operator

$$A^*,\ A^* V_s^* \to V_{s+\nu}^*,$$

which is adjoint to A, has a system of eigen functionals

$$\{e_i\},\ e_i \in V_s^*.$$

which is complete on V_s. Then the operator F_0^*, which is adjoint to F, has a system of eigen analytic functionals $\{g_i\}$ that is complete on V_s,

$$g_i(F(u)) = \lambda_i g_i(u), \qquad \forall u \in V_{s+\nu}, \tag{7.6}$$

whilst the eigenvalues λ_i are identical with the eigenvalue's of the operator A^*.

Proof. We define the functionals g_i using the formula

$$g_i(u) = (e_i, \mathscr{E}^{-1}(u)), \tag{7.7}$$

where e_i is the eigen functional of A^*, i.e.

$$(e_i, Av) = (A^* e_i, v) = \lambda_i(e_i, v), \qquad \forall v \in V_{s+\nu}. \tag{7.8}$$

Using (7.5) and (7.8), we obtain:

$$\begin{aligned} g_i(F(u)) &= (e_i, \mathscr{E}^{-1}(F(u))) = (e_i, \mathscr{E}^{-1}(\mathscr{E} \circ A\mathscr{E}^{-1}(u))) \\ &= (e_i, A\mathscr{E}^{-1}(u)) = \lambda_i(e_i, \mathscr{E}^{-1}(u)) = \lambda_i g_i(u). \end{aligned}$$

Formula (7.6) therefore holds.

We shall now prove that $\{g_i\}$ forms a complete set. Indeed, if $u \neq v$, then $\mathcal{E}^{-1}u \neq \mathcal{E}^{-1}v$, since the operator $\mathcal{E}^{-1}$, like $\mathcal{E}$, is one-to-one. Since $\{e_i\}$ form a complete system, the vector e_i exist, such that

$$(e_i, \mathscr{E}^{-1}(u)) \neq (e_i, \mathscr{E}^{-1}(v))$$

and by virtue of (7.2) $g_i(u) \neq g_i(v)$, i.e. $\{g_i\}$ is a complete system of functionals. Note that if g is an eigen functional of the operator F^*, it is very easy to find the value of this functional on the solution of the equation $F(u) = h$. Indeed,

$$g(h)=g(F(u))=\lambda g(u),$$

whence

$$g(u)=\lambda^{-1}g(h). \tag{7.9}$$

In the linear case the method of solving linear differential equations with constant coefficients using Fourier's transform is based on this fact. In this case $g(u) = g_\xi(u)$ are Fourier coefficients, and the Fourier coefficient of the solution is determined by the Fourier coefficient of the right-hand side h using formula (7.9).

Theorem 7.2. Suppose F is a differential operator that is determined by Formula (6.5) and which satisfies the conditions of sect. 1) of Theorem 6.2. Then the system $\{g_l\}$ of analytic eigen functionals of the operator F which is complete on

$$\operatorname{Re} W_p^2(T^m)=V_0,\ p>m,$$

exists.

Proof. Linearisation of the operator

$$F:\operatorname{Re} W_p^4\to\operatorname{Re} W_p^2$$

exists by virtue of sect. 1) of Theorem 6.2. The operator A^*, like A, is a differential operator with constant coefficients by virtue of (6.18), and the trigonometric polynomials form a complete system of eigen vectors of the operator A^*. Therefore, using Theorem 7.1, where

$$V_0=V=\operatorname{Re} W_p^2,\ V_\nu=V_2=\operatorname{Re} W_p^4,\ (\nu=2),$$

we obtain the required statement.

Theorem 7.2 is a theorem on existence and does not provide a method of explicitly constructing the functionals $\{g_l\}$. Note, incidentially, that in the linear case in even the simplest situation of linear self-adjoint elliptic differential operators with analytic coefficients on a torus, the explicit construction of eigen functions is possible only in rare cases.

We shall now give an extremely simple example, in which linearisation of the nonlinear operator F is trivially constructed in explicit form, like the eigen functionals F^*.

Example 7.1. Suppose $\varphi(y)$, $y \in \mathbb{R}$ is a real analytic monotonic function, $\varphi'(y) > C > 0$, $\varphi(0) = 0$, $\varphi'(0) = 1$. Suppose A is a linear differential operator with real constant coefficients on the torus T^m. We will assume

$$F(u) = \varphi(A(\varphi^{-1}(u))), \tag{7.4}$$

where φ^{-1} is a function that is inverse to φ. In this case F is obviously a nonlinear differential operator on the torus. According to Proposition 4.2 the operator F is analytic from

$$\operatorname{Re} W_p^{l+\nu}(T^m)$$

to

$$\operatorname{Re} W_p^l(T^m),$$

where ν is the order of the operator A, under the condition that $l > m/p$. It is obvious that the operator that is determined on W_p^l using the correspondence

$$v \to \varphi(v). \tag{7.5}$$

is the operator $\mathcal{E}$, which gives the linearisation

$$F = \mathcal{E} \circ A \circ \mathcal{E}^{-1}$$

of the operator F. In this case the adjoint eigen functionals have the form

$$g_i(u) = \int_{T^m} e_i(x)\varphi^{-1}(u(x))\,dx, \tag{7.6}$$

where e_i is a trigonometric monomial and therefore e_i is an eigen vector of the differential operator with constant coefficients A.

The explict construction $g_i(u)$ is not possible in the general case. Therefore, as in the linear case, we shall make constructions which enable us to obtain information on the solutions of the equation $F(u) = f$, without constructing in explicit form the reduction of F to diagonal form, i.e. without requiring the explicit construction of $\mathcal{E}$ and $\mathcal{E}^{-1}$. The calculation of the functions of operators, which is constructed either using a resolvent or on the basis of the spectral expansion of self-adjoint operators, was our basic apparatus in the linear case. We shall construct complex powers of the operator F for similar purposes in the nonlinear case.

§8. Real non-integer and complex powers of nonlinear operators.

In this paragraph we shall use the definitions and notation of the previous paragraph. We will assume

$$V_\infty = \bigcap_{s=0}^{\infty} V_s. \tag{8.1}$$

Suppose the operator F permits an analytic linearisation on V_s for all fairly large s. We will assume that the corresponding operator A satisfies the following condition.

Condition 8.1. The powers A^{α}, which are continuous operators from V_{s1} to V_s when

$$s, s_1 \geqq 0, \; s_1 - \alpha \geqq s.$$

are determined in the operator A for all real $\alpha \in \mathbb{R}$. (In particular A^{α} is continuous from V_s to V_s when $\alpha < 0$). In addition, $A^{\alpha} v$ when $v \in V_{\infty}$ analytically depends on the parameter $\alpha \in \mathbb{R}$ in each V_s. We shall determine the real powers of the operator on V_{∞} using the formula

$$F^{\alpha} = \mathscr{E} \circ A^{\alpha} \circ \mathscr{E}^{-1}. \tag{8.2}$$

The following theorem describes the basic properties of the power of the operator F thus introduced.

Theorem 8.1. Suppose the operator F permits an analytic linearisation on $V_{s+\nu}$ for all $s > s_0 > 0$, whilst the operator A satisfies condition 8.1. Suppose F^{α} are determined on V_{∞} using Formula (8.2). Then:

1) for integer on $\alpha \in \mathbb{Z}$ F^{α} agrees with the usual composition of the operators F or F^{-1};

2) the operator F^{α} changes V_{∞} into V_{∞} for each α;

3) when $v \in V_{\infty}$ the vector $F^{\alpha} v$ is an analytic function of the parameter α in each V_s, $s > 0$;

4) we have the composition formula

$$F^{\alpha_1} \circ F^{\alpha_2} = F^{\alpha_1 + \alpha_2}, \qquad \forall \alpha_1, \alpha_2 \in \mathbb{R},$$

Proof. We shall first verify sect. 2). Since $\mathcal{E}^{-1}$, like $\mathcal{E}$, maps V_s into V_s for each s, then $\mathcal{E}^{-1}(V_{\infty}) = V_{\infty}$. The operator A^{α} changes V_{s1} into V_s, $s_1 > s + \alpha$, and therefore $A^{\alpha}(V_{\infty}) = V_{\infty}$. Since $\mathcal{E}^{-1}$ is determined on V_{s1} for large s, then $\mathcal{E}(V_\infty) = V_\infty$ hence we obtain a statement of sect. 2). Sect. 1) for integer positive α is a direct corollary of Formula (7.5):

$$F^\alpha = \mathcal{E} \circ A^\alpha \circ \mathcal{E}^{-1} = \mathcal{E} \circ A \circ \ldots \circ A \circ \mathcal{E}^{-1}$$
$$= \mathcal{E} \circ A \circ \mathcal{E}^{-1} \circ \mathcal{E} \circ \ldots \mathcal{E}^{-1} \circ \mathcal{E} \circ A \circ \mathcal{E}^{-1} = F \circ \ldots \circ F. \quad (8.3)$$

The operator F^{-1}, which is determined by Formula (8.2)

$$F^{-1} = \mathcal{E} \circ A^{-1} \circ \mathcal{E}^{-1},$$

is inverse for F by virtue of (7.5). Therefore the statement of sect. 1) for negative α is verifieid in a similar way to (8.3).

We shall verify sect. 3). Suppose $\alpha \in]\alpha_0, \alpha_1[$. We shall fix $s > 0$. If $v \in V_\infty$, then $w = \mathcal{E}^{-1}(v) \in V_\infty$. The vector $A^\alpha w$ analytically depends on α in V_s by virtue of condition 8.1, and analytically depends on α and

$$F^\alpha v = \mathcal{E}(A^\alpha w),$$

Q.E.D., by virtue of the analyticity of the operator $\mathcal{E}$ on V_s.

Sect. 4) directly follows from (7.5):

$$F^{\alpha_1} \circ F^{\alpha_2} = \mathcal{E} \circ A^{\alpha_1} \circ \mathcal{E}^{-1} \circ \mathcal{E} \circ A^{\alpha_2} \circ \mathcal{E}^{-1}$$
$$= \mathcal{E} \circ A^{\alpha_1 + \alpha_2} \circ \mathcal{E}^{-1} = F^{\alpha_1 + \alpha_2}.$$

The theorem is completely proved.

We shall now determine the complex powers of the operator F, which permits an analytic linearisation using Formula (7.5). We will assume that the real spaces V_s are real subspaces of the complex Banach spaces E_s, $V_s \subset E_s$, $s \in \mathbb{Z}_+$. We shall impose an additional condition on the operator A.

Condition 8.2. For any $\lambda = \alpha + i\beta \in \mathbb{C}$ the complex power A^{λ} of the operator A is determined. At the same time the operator $A^{\lambda} v$ holomorphically depends on λ when

$$v \in V_s, \quad \alpha = \operatorname{Re} \lambda < s,$$

and the operators A^{λ} are continuous from E_{s1} to E_{s2} when $s_1 - s < s_2$.

Theorem 8.2. Suppose the operator F permits an analytic linearisation on V_s for each $s > s_0$, whilst the operator

$$\mathscr{E}: V_{s+\nu} \to V_{s+\nu}$$

permits an extension to an operator that is analytic on all $E_{s+\nu}$ for each $s > s_0$. Suppose the operator A satisfies conditions 8.1 and 8.2. Then for all $\lambda \in \mathbb{C}$ the operator F^{λ} is determined on V_{∞},

$$F^{\lambda} = \mathscr{E} \circ A^{\lambda} \circ \mathscr{E}^{-1}. \tag{8.4}$$

This operator has the following properties:

1) when $\lambda = \alpha \in \mathbb{R}$ F^{λ} has all the properties formulated in Theorem 8.1;

2) when $v \in V_{\infty}$ the vector $F^{\lambda} v$ is an entire function of the parameter λ;

3) when $\alpha \in \mathbb{R}$, $\lambda \in \mathbb{C}$ the following composition formula holds on V_{∞}:

$$F^{\lambda} \circ F^{\alpha} = F^{\lambda + \alpha}; \tag{8.5}$$

4) if the operator F extends to an operator that is analytic on all E_s for each s, then

$$F^k \circ F^\lambda v = F^{k+\lambda} v, \qquad \forall k \in \mathbb{N}, \quad \lambda \in \mathbb{C}, \quad v \in V_\infty; \tag{8.6}$$

5) if the following estimate holds for the purely imaginary powers of the operator A:

$$\|A^{i\beta} v\|_s \leqq C \|v\|_s, \qquad \forall v \in V_s, \quad \beta \in \mathbb{R}, \tag{8.7}$$

where C does not depend on v and β,

$$\|A^{-1} v\|_s \leqq M^{-1} \|v\|_s, \qquad \forall v \in E_s (M > 1), \tag{8.8}$$

and the operators $\mathcal{E}$ are bounded from E_{s+1} to E_s, then when $v \in V_\infty$ for each $s > 0$ the following estimate holds:

$$\sup_{\alpha \leqq d} \sup_{\beta} \|F^{\alpha + i\beta} v\|_s \leqq C(d) < \infty. \tag{8.9}$$

At the same time, if $d < 0$, then

$$C(d) \leqq C_0 M^d. \tag{8.10}$$

Proof. Sect. 1) is obvious. We shall proceed to sect. 2). If $v \in V_\infty$, then

$$w = \mathcal{E}^{-1} v \in V_\infty$$

and by virtue of condition 8.2 $A^\lambda w$ is an analytic function of λ for all $\lambda \in \mathbb{C}$. At the same time for any N and any s

$$A^\lambda w \in E_{s+1}$$

when $\mathrm{Re}\,\lambda < N$. By virtue of the analyticity of $\mathcal{E}$ on E_{s+1} $\mathcal{E}(A^\lambda w)$

analytically depends on λ when $\operatorname{Re}\lambda < N$. Since N is arbitrary, $\mathcal{E}(A^{\lambda}w)$ is consequently an eigenfunction of λ in E_s for any s.

Sect. 3) is a direct corollary of Formula (7.5).

We shall move on to sect. 4). The right- and left-hand sides of (8.6) are entire functions of the parameter λ. (8.6) follows from (8.2) for real λ. By virtue of the uniqueness of the analytic extension Formula (8.6) holds for all λ.

We shall proceed to sect. 5). Since $v \in V_{\infty}$, then

$$v \in V_{s+r+1}\ \forall r \geqq 0.$$

Consequently

$$w = \mathscr{E}(v) \in V_{s+r+1}.$$

The vector $u(\alpha) = A^{\alpha}w$ is an analytic function of the parameter α in the space E_{s+1} when $r > \alpha > -\infty$. As $\alpha \to -\infty$ by virtue of (8.8) $u(\alpha) \to 0$. Therefore the set $\Omega = \{u(\alpha) : \alpha < r\}$ is precompact, and is consequently also a bounded set in E_{s+1}. By virtue of estimate (8.7) the sets $A^{i\beta}\Omega$ are bounded in E_{s+1} at the norm by a constant independent of β. Since the operator $\mathcal{E}$ is bounded from E_{s+1} to E_s the set

$$\{\varphi : \varphi = \mathscr{E}(A^{\alpha+i\beta}\mathscr{E}^{-1}v), \beta \in \mathbb{R}, \alpha \leqq r\}$$

is bounded in E_s by the constant $C(r)$. Thus (8.9) is compact when $dV = r > 0$. If $d < 0$, we can obviously use inequality (8.9) when $d = 0$. To prove (8.8), note that by virtue of (8.8) and (8.7)

$$\|A^{\alpha+i\beta}w\|_s \leqq C_1 M^{\alpha} \tag{8.11}$$

when $\alpha \leqslant 0$. Indeed, the set

$$\{A^{\theta}w, \theta \in [0, 1]\}$$

is bounded in E_s by the constant C_2. Therefore when $k \in \mathbb{N}$

$$\|A^{-k+\theta}w\|_s = \|A^{-k}A^{\theta}w\|_s \leqq M^{-k}\|A^{\theta}w\|$$
$$\leqq C_2 M^{-k} < C_2 M^{-k+\theta}.$$

Since when $\alpha < 0$ $\alpha = -k + \theta$, $k \in \mathbb{N}$, we thereby obtain the estimate

$$\|A^{\alpha}w\| \leqslant C_2 M^{\alpha} \text{ when } \alpha \leqslant 0.$$

Using estimate (8.7), we obtain that

$$\|A^{\alpha + i\beta}w\|_s \leqslant CC_2 M^{\alpha} \text{ when } \alpha \leqslant 0. \tag{8.12}$$

Since $\mathcal{E}(0) = 0$, and the operator $\mathcal{E}$ is differentiable at zero, it follows from (8.12) that

$$\|\mathcal{E}(A^{\alpha + i\beta}w\|_s \leqslant C_3 M^{\alpha} \text{ when } \alpha \leqslant -N,$$

where N is fairly great. Hence follows estimate (8.10) when $d \leqslant -N$, and consequently also when $d \leqslant 0$. Theorem 8.2 is thus proved.

We shall now use the existence of complex powers of the operator F to obtain $g(F^{-1}h)$, i.e. the value of g in the solution $u = A^{-1}h$ of the equation $F(u) = h$, using the known quantities

$$g(h), g(Fh), \ldots, g(F^j h), \ldots, \tag{8.13}$$

where g is some functional.

Theorem 8.3. Suppose g is a functional on E_s, which is bounded on bounded sets and is analytic on E_s, $g(0) = 0$. Suppose the operator F, which is determined on V_∞, has complex powers

$$F^\lambda,\ \lambda \in \mathbb{C},\ F^\lambda : V_\infty \to E_\infty$$

and suppose for each $v \in V_\infty$ the vector $F^\lambda v$ holomorphically depends on λ in E_s for $\lambda \in \mathbb{C}$, whilst estimates (8.9) and (8.10), where $M > 1$, hold. Then for any $h \in V_\infty$ the function $\Phi(\lambda)$, which is determined by the formula

$$\Phi(\lambda) = g(F^\lambda h), \tag{8.14}$$

is an entire function which satisfies the estimate

$$\sup_{\alpha \leqq d,\, \beta \in \mathbb{R}} \left|\Phi(\alpha + i\beta)\right| \leqq C_1(d), \tag{8.15}$$

whilst

$$C_1(d) \leqslant C_1 M^d \text{ when } d \leqslant 0. \tag{8.16}$$

Proof. Since the functional g is analytic on all E_s, the function $\Phi(\lambda)$, which is determined by (8.14), is analytic in the whole plane $\mathbb{C}$. It follows from estimate (8.9) and from the boundedness of g that estimate (8.15) holds. When $d < 0$, from estimate (8.9) and (8.10) we obtain:

$$\sup_{\alpha \leqq d}\sup_{\beta \in \mathbb{R}} \left\|F^{\alpha + i\beta} h\right\|_s \leqq C_0 M^d. \tag{8.16'}$$

Since $g(0) = 0$ and g is analytic, then

$$\|g(u)\|_s \leqslant C\|u\|_s \text{ when } \|u\|_s \leqslant C_0, \tag{8.17}$$

From estimate (8.16') and (8.17) when $d \leqslant 0$ we obtain (8.15), where

$$C_1(d) \leq CC_0 M^d,$$

i.e. inequality (8.16) holds.

As Theorem 8.3 shows, the problem of expressing $g(F^{-1}h)$ in terms of the numbers (8.13) is, on natural assumptions, equivalent to expressing $\Phi(-1)$ in terms of

$$\Phi(0), \Phi(1), \ldots, \Phi(j), \ldots, \tag{8.18}$$

where the function $\Phi(\lambda)$ is an entire function that is determined by Formula (8.14) and satisfies estimates (8.15), (8.16). As the results of Ch. 2 show, when $F=A$ is a second-order linear self-adjoint differential operator, g is a linear continuous functional, and h is an analytic function, $\Phi(-1)$ is uniquely determined by the numbers (8.18), and the polynomial representations of the solutions of the equation $Au = h$ which are constructed in Ch. 4 and 6 give the explicit expression $\Phi(-1) = (g, A^{-1}h)$ in terms of

$$\Phi(j) = (g, A^j h),$$

The following example shows that in the nonlinear case, if we do not impose additional conditions, $\Phi(-1)$ is not uniquely determined by the values $\Phi(j)$, $j = 0, 1, \ldots$.

Example 8.1. We shall use $\varphi = \varphi_\gamma$ to denote the function

$$\varphi_\gamma(u) = u + \gamma \sin(\pi u), \qquad (|\gamma| < 1/\pi). \tag{8.19}$$

It is obvious that when $|\gamma| < 1/\pi$ this function is invertible on $\mathbb{R}$. It is also obvious that the inverse function φ^{-1} is analytic on $\mathbb{R}$. Consider in the neighbourhood $T^1 = [0, 2\pi]$ a second-order linear differential operator

$$Au \equiv -\partial_x^2 u + 3u. \tag{8.19'}$$

We shall determine the nonlinear differential operator $F = F_\gamma$ using the formula

$$F(v) = \varphi(A\varphi^{-1}(v)). \tag{8.20}$$

The operator F is analytic by virtue of Proposition 4.2 from

$$W_p^{l+2}(T^1)$$

to $W_p^l(T^1)$ when $l > 1/p$. We shall take as the functional g the delta-function $\delta(x - \pi/2)$:

$$g(u) = u(\pi/2) \tag{8.21}$$

We shall denote the function $\varphi(\sin x)$ by h:

$$h(x) = \sin x + \gamma \sin(\pi \sin x). \tag{8.22}$$

It is obvious that $h(x)$ is entire. We shall determine the following complex powers for the operator F, which is determined by (8.20):

$$F^\lambda(v) = \varphi_\gamma(A^\lambda \varphi_\gamma^{-1}(v)) \tag{8.23}$$

It is obvious that the operator F, which is determined by (8.20), satisfies all the conditions of Theorem 8.3

$$(V_\infty = \mathrm{Re}\, C^\infty(T^1)).$$

We shall construct the function $\Phi(\lambda)$ using Formula (8.14), where g is determined by (8.21), and h is determined by (8.22). Obviously,

$$F^\lambda h(x) = \varphi_\gamma(A^\lambda \varphi_\gamma^{-1} \varphi_\gamma(\sin x)) = \varphi_\gamma(A^\lambda \sin x).$$

Since the function $\sin x$ is eigen for A with the eigenvalue 4, then

$$F^\lambda h(x) = \varphi_\gamma(4^\lambda \sin x).$$

By virtue of (8.21) and (8.14)

$$\Phi(\lambda) = \varphi_\gamma(4^\lambda \sin x)\big|_{x=\pi/2} = \varphi_\gamma(4^\lambda).$$

Therefore for integer nonnegative $\lambda = 0, 1, \ldots$ by virtue of (8.19) we obtain:

$$\Phi(k) = \varphi_\gamma(4^k) = 4^k + \delta \sin(\pi 4^k) = 4^k.$$

It is obvious that the values $\Phi(k)$ when $k = 0, 1, \ldots$ do not depend on γ. At the same time the number

$$\Phi(-1) = 4^{-1} + \gamma \sin \pi/4,$$

obviously depends on γ.

In the next paragraph we shall formulate additional conditions on Φ, under which $\Phi(-1)$ is uniquely determined by the numbers $\Phi(j)$, $j = 0, 1, \ldots$. Under these conditions $\Phi(-1)$ will be represented by $\Phi(j)$ in explicit form.

§9. An extrapolation problem

Consider a class of functions bounded on lines that are parallel to an imaginary axis and satisfy the inequality

$$\sup_{\xi \leqq s} \sup_{\tau \in \mathbb{R}} \left|\Phi(\xi + i\tau)\right| = \hat{\Phi}(s) < \infty, \qquad \forall s \in \mathbb{R}. \tag{9.1}$$

We shall introduce the following notation. If $\Phi(s)$ is a function that is determined on $\mathbb{Z}_+$ ($s = 0, 1, \ldots$), and $P_n(x)$ is a polynomial of the n-th degree,

$$P_n(x) = \sum_{j=0}^{n} a_j x^j, \tag{9.2}$$

we shall note the following number by $P_n * \Phi$:

$$P_n * \Phi = \sum_{j=0}^{n} a_j \Phi(j). \tag{9.3}$$

We shall use $\mathcal{M}(R_0, \rho)$, where $R_0 > 0$, $\rho > 0$, to denote the set of functions satisfying condition (9.1), where $\hat{\Phi}(s)$ satisfies the following inequalities:

$$\sum_{k=0}^{\infty} \hat{\Phi}(k) R^{2k}/(2k)! < \infty \qquad 0 < R < R_0, \tag{9.4}$$

$$\hat{\Phi}(s) \leqq C\rho^s, \qquad \forall s \leqq 0, \tag{9.5}$$

where $C = C(\Phi)$.

Theorem 9.1. Suppose $\Phi \in \mathcal{M}(R_0, \rho)$. Suppose

$$g(\lambda) = (\lambda + \rho)^{-k}, \quad 0 < R < R_0, \; r = 2R/\pi,$$

and the polynomial

$$P_n^0 = \Pi_r^{+n} g, \; P_n(\lambda) = P_n^0(\lambda - \rho).$$

Then

$$\Phi(-k)=\lim_{n\to\infty} P_n * \Phi, \tag{9.6}$$

whilst $\forall \varepsilon > 0$, C exists, such that $\forall n \in \mathbb{N}$

$$|\Phi(-k)-P_n * \Phi| \leqq Cn^{-\sigma}, \quad \sigma = 2R\sqrt{\rho}/\pi - \varepsilon. \tag{9.7}$$

We shall prove the theorem later. The proof will be based on an examination of the translation operator A, which acts on $\mathscr{M}(R_0,\rho)$ using the formula

$$A\Phi(z)=\Phi(z+1). \tag{9.8}$$

The values $\Phi(j)$, $j \in \mathbb{Z}$ are expressed using the operator A and the delta-function $\delta(z)$ (the functional $\delta(z)$ is determined, as usual, by the equation $(\Phi, \delta(z)) = \Phi(0))$ by the formula

$$\Phi(j)=(A^j\Phi, \delta(z)). \tag{9.9}$$

Thus, to obtain a representation of the form (9.6) it is sufficient to obtain a representation of the form (4.1.1) for the operator $g(A) = A^{-k}$. At the same time the limit in (4.1.1) must be understood in a fairly wide sense in order that we can proceed to the limit pointwise, i.e. in order that we can substitute a delta-function under the limit sign in (4.1.1). We shall use the results obtained in the sixth chapter to obtain Formula (4.1.1). However, to use these results we need to introduce the space E, in which the operator A is active, in order for it to possess the necessary properties.

We shall use $H = H_x$ to denote the Hilbert space $L_2(\mathbb{R}) = L_2(\mathbb{R}_x)$ with a standard scalar product. As is well known, the Fourier transform F

$$\mathcal{F}u(\xi)=\tilde{u}(\xi)=\frac{1}{2\pi}\int e^{-ix\xi}u(x)\,dx, \tag{9.10}$$

is an isomorphism between $H_x = L_2(\mathbb{R}_x)$ and $H_\xi = L_2(\mathbb{R}_\xi)$. (In the paragraph the integral sign indicates integration from $-\infty$ to $+\infty$.

Fourier's inverse transform is determined using the formula

$$u(x)=(\mathcal{F}^{-1}\tilde{u})(x)=\int e^{ix\xi}\tilde{u}(\xi)\,d\xi. \tag{9.11}$$

Parseval's equality holds:

$$(f,\varphi)=2\pi(\mathcal{F}f,\mathcal{F}\varphi). \tag{9.12}$$

This equality enables us to determine Fourier's transform on generalised functions f, which are functionals on a set of smooth rapidly decreasing functions (for a detailed discussion of the theory of generalised functions see, for example, Gel'fand and Shilov [1], Vladimirov [1]).

Consider the operators

$$p_x=(1-\partial_x^2),\quad q_x=(1+x^2). \tag{9.13}$$

These operators are obviously determined on smooth rapidly decreasing functions, and are continued to functions from H_x and to generalised functions in the standard way. We will assume

$$H_2=P_x^{-1}q_xH_x. \tag{9.14}$$

It is easy to derive from (9.10) and (9.11) that

$$p_x = \mathscr{F}^{-1} q_\xi \mathscr{F}, \quad q_x = \mathscr{F}^{-1} p_\xi \mathscr{F}. \tag{9.15}$$

Therefore, since by virtue of (9.12) $\mathscr{F} H_x = H_\xi$, then

$$\mathscr{F} H_2 = q_\xi^{-1} p_\xi H_\xi. \tag{9.15'}$$

The norm $\| \ \|$ in the space H_2 is determined by the following equation by virtue of (9.14):

$$\|f\|_2 = \|q_x^{-1} p_x f\|, \tag{9.16}$$

where $\| \ \|$ is the norm in H_x. By virtue of (9.12) the following equation holds:

$$\|f\|_2 = \sqrt{2\pi} \|p_\xi^{-1} q_\xi \mathscr{F} f\|. \tag{9.17}$$

We shall use C_b^2 (ℝ) to denote a set of functions that are doubly continuously differentiable on ℝ, and which have bounded derivatives. We shall assume for $u \in C_b^2$ (ℝ)

$$\|u\|_{C^2} = \sup_x (|u(x)| + |\partial_x u(x)| + |\partial_x^2 u(x)|). \tag{9.18}$$

Lemma 9.1. C_b^2 (ℝ) $\subset H_2$ whilst the constant $C > 0$ exists, such that

$$\|f\|_2 \leqq C \|f\|_{C^2}, \qquad \forall f \in C_b^2. \tag{9.19}$$

Proof. Obviously, by virtue of (9.13)

$$q_x^{-1} P_x f(x) = (1 + x^2)^{-1} (f(x) - \partial_x^2 f(x)).$$

Therefore, obviously,

$$|q_x^{-1}p_xf(x)|\leqq(1+x^2)^{-1}(|f(x)|+|\partial_x^2f(x)|).$$

Hence we obtain, by virtue of (9.18),

$$\|q_x^{-1}p_xf(x)\|\leqq\|(1+x^2)^{-1}\|\,\|f\|_{C^2}=C\|f\|_{C^2}.$$

Bearing in mind (9.16), hence we obtain (9.19).

Let us now consider the set $\mathcal{D}_0$ of functions which are a restriction to $\mathbb{R}_x$ of entire functions,

$$\mathcal{D}_0=\{f(x):f(x)=\Phi(ix),\,\Phi\in\mathcal{M}(R_0,\rho)\},\tag{9.20}$$

and the set $\mathcal{M}(R_0, \rho)$ is introduced at the beginning of the paragraph. By virtue of (9.1), estimating the derivative from $\Phi(z)$ using the Cauchy formula, and using the monotony of $\Phi(s)$, we obtain the estimate

$$\|\Phi(i(\cdot-is))\|_{C^2}\leqq4\hat{\Phi}(s+1),\qquad\forall s\in\mathbb{R}.\tag{9.21}$$

It follows from (9.21) and from Lemma 9.1 that $\mathcal{D}_0 \subset H_2$. We shall use E to denote the closure $\mathcal{D}_0$ in H_2.

Lemma 9.2. Suppose $u \in E$. Then $\mathcal{F}u(\xi) = 0$ when $\xi < \ln\rho$.

Proof. We shall use $C_0^\infty = C_0^\infty(\mathbb{R})$ to denote a set of infinitely smooth functions with a compact support (the support of φ is the closure of the set of those points where $\varphi \neq 0$). To prove the lemma it is sufficient to prove that if $u \in E$, then for any function such that the support of φ is contained in the domain $\xi < \ln \rho$, $(\mathcal{F}u, \varphi) = 0$. Since $(\mathcal{F}u_j, \varphi) \to (\mathcal{F}u, \varphi)$ when $\varphi \in C_0^\infty$, $u_j \to u$, it is sufficient to consider the case when $u \in \mathcal{D}_0$.

Suppose

$$\varphi \in C_0^\infty(\mathbb{R}_\xi),\ \varphi(\xi) = 0$$

when

$$\xi \geqq \ln \rho_1, \quad \rho_1 = \rho - \varepsilon,\ \varepsilon > 0,\ \rho_1 > 0.$$

According to (9.12)

$$2\pi(\mathcal{F} f, \varphi) = (f, \mathcal{F}^{-1}\varphi) = \int \Phi(ix)\overline{\mathcal{F}^{-1}}\varphi(x)\, dx. \tag{9.22}$$

According to (9.11)

$$\overline{\mathcal{F}^{-1}}\varphi(x) = \int e^{-ix\xi}\bar{\varphi}(\xi)\, d\xi = \varphi_1(x). \tag{9.23}$$

Hence it is obvious that $\varphi_0(x)$ analytically extends to complex values $z = x + iy$ using the formula

$$\varphi_1(z) = \int e^{-iz\xi}\bar{\varphi}(\xi)\, d\xi.$$

We shall estimate $|\varphi_1(z)|$ for complex z. Obviously,

$$\begin{aligned} z^k\varphi_1(z) &= \int [(i\,\partial_\xi)^k e^{-iz\xi}]\bar{\varphi}(\xi)\, d\xi \\ &= \int e^{-iz\xi}(-i\,\partial_\xi)^k\bar{\varphi}(\xi)\, d\xi \\ &= \int_{-\infty}^{\ln \rho_1} e^{y\xi} e^{-ix\xi}(-i\,\partial_\xi)^k\bar{\varphi}(\xi)\, d\xi. \end{aligned}$$

Therefore when $y \geq 0$

$$|\varphi_1(x+iy)|\,|x^2+y^2|^{k/2} \leqq e^{y \ln \rho_1} C, \qquad C=C(k,\varphi). \tag{9.24}$$

From (9.22) and (9.2) we obtain:

$$2\pi(\mathscr{F} f, \varphi)=\int \Phi(ix)\varphi_1(x)\,dx=\int \Phi(ix-y)\varphi_1(x+iy)\,dx. \tag{9.25}$$

(A shift translation of the integration contour by iy is possible since φ_1, by virtue of (9.24), rapidly decreases at infinity.) Using estimate (9.24), (9.21) and (9.5), from this equation we obtain:

$$|(\mathscr{F} f, \varphi)| \leqq C_1 \rho_1^y \hat{\Phi}(-y-1) \leqq C_2 \rho_1^y \rho^{-y}.$$

Since $0 < \rho_1 <$, and $y > 0$ is arbitrarily great, it follows hence that $(\mathscr{F} f, \varphi) = 0$, and the lemma is proved.

We shall now note the action on $\mathcal{D}_0$ of the operator A, which is determined by Formula (9.8). By virtue of (9.20) the action of A on $f \in \mathcal{D}_0$ is specified by the formula

$$Af(x)=A\Phi(ix)=\Phi(1+ix)=\Phi(i(x-i))=f(x-i). \tag{9.26}$$

Lemma 9.3. The operator A, which is determined on $\mathcal{D}_0$ by Formula (9.26), maps $\mathcal{D}_0$ into $\mathcal{D}_0$. At the same time

$$\sum \|A^k f\|_2 R^{-2k}/(2k)! < \infty, \qquad 0<R<R_0. \tag{9.27}$$

Proof. The function $\Phi_1(z) = \Phi(z + 1)$ is entire, and satisfies (9.1). Since

$$\hat{\Phi}_1(k)=\hat{\Phi}(k+1),$$

then to verify (9.4) for Φ_1 consider the following:

$$\sum \hat{\Phi}(k+1)\frac{R^{2k}}{2k!} = R^{-2}\sum \hat{\Phi}(k+1)R^{2(k+1)}\frac{(2k+1)(2k+2)}{(2k+2)!}. \tag{9.28}$$

If $R < R_0$, then this series also converges by virtue of the convergence when $R < R_1 < R_0$ of series (9.4). Inequality (9.5) also holds for Φ_1. Then $\Phi_1 \in \mathscr{M}(R_0, \rho)$. It follows from estimate (9.21) and (9.19) that

$$\sum \|A^k f\|_2 R^{2k}/(2k)! \leqq C_1 \sum \hat{\Phi}(k+1)R^{2k}/(2k)!.$$

As shown above, this series converges when $0 < R < R_0$, and (9.27) is proved.

Lemma 9.4. The operator A, determined on $\mathcal{D}_0$ by Formula (9.26), is represented in the form

$$Af = \mathscr{F}^{-1} e^{\xi} \mathscr{F} f. \tag{9.29}$$

Proof. Suppose $\varphi \in C_0^\infty(\mathbb{R}_\xi)$. Consider $(\mathscr{F}Af, \varphi)$. By virtue of (9.12) and (9.26)

$$2\pi(\mathscr{F}Af, \varphi) = (Af, \mathscr{F}^{-1}\varphi) = \int \Phi(i(x-i))\varphi_1(x)\,dx,$$

where $\varphi_1(x)$ is determined by Formula (9.23). As with 9.25), we obtain:

$$2\pi(\mathscr{F}Af, \varphi) = \int \Phi(ix)\varphi_1(x+i)\,dx. \tag{9.30}$$

According to (9.23)

$$\varphi_1(x+i) = \int e^{-ix\xi} e^{\xi}\bar{\varphi}(\xi)\,d\xi = \overline{\mathscr{F}^{-1} e^{\xi}\varphi(\xi)}.$$

Hence and from (9.30) follows:

$$2\pi(\mathscr{F}Af,\varphi)=(f,\mathscr{F}^{-1}e^{\xi}\varphi)=2\pi(\mathscr{F}f,e^{\xi}\varphi)=2\pi(e^{\xi}\mathscr{F}f,\varphi).$$

Consequently, (9.29) holds, and Lemma 9.4 is proved.

Let us now consider the operator A_1 on $\mathfrak{D}_0$,

$$A_1=A-\rho I. \tag{9.31}$$

Obviously,

$$A_1 f=\mathscr{F}^{-1}(e^{\xi}-\rho)\mathscr{F}f. \tag{9.32}$$

We shall construct the resolvent of the operator A_1. We will assume

$$(A_1-\zeta^2 I)^{-1}=\mathscr{F}^{-1}(e^{\xi}-\rho-\zeta^2)^{-1}\mathscr{F}. \tag{9.33}$$

Lemma 9.5. The operator $(A_1-\zeta^2 I)^{-1}$, determined by Formula (9.33) on E, is a bounded operator when $\operatorname{Im}\zeta\neq 0$, and all the points ζ^2 with $|\operatorname{Im}\zeta|>0$ belong to the resolvent set of the operator A_1. At the same time

$$\|(A_1-\zeta^2 I)^{-1}\|_E\leqq C \text{ when } |\operatorname{Im}\zeta|\geqq\omega>0. \tag{9.34}$$

Proof. We shall fix $\omega>0$ and the number ζ, $|\operatorname{Im}\zeta|\geqslant\omega$. We will assume

$$h(\xi)=(e^{\xi}-\rho-\zeta^2)^{-1} \text{ when } \xi\geqq\ln\rho. \tag{9.35}$$

Note that when $|\operatorname{Im}\zeta|\geqslant\omega$ the distance from the point ζ^2 to the real positive semiaxis in $\mathbb{C}$ is not less than ω^2. Therefore the function $h(\xi)$, determined by (9.35), is bounded and all its derivatives are bounded when $\xi > \ln \rho$ and rapidly aproach 0 as $\xi \to +\infty$. According to Lemma 9.2 when $u \in E$ the generalised function $\mathcal{F}u(\xi)$ has a support in the domain $\xi > \ln \rho$. Therefore the effect of the operator (9.33) on functions from E does not change if we extend the function (9.35) when $\xi < \ln \rho$ using any method. We shall extend $h(\xi)$ thus:

$$h(\xi) = \chi(\xi)(e^{\xi} - \rho - \zeta^2)^{-1},$$

$$\chi(\xi) \in C^{\infty}(\mathbb{R}_{\xi}),\ \chi(\xi) = 0 \text{ when } \xi < \ln \rho - 1,\ \chi(\xi) = 1 \text{ when } \xi > \ln \rho. \tag{9.36}$$

We shall rewrite the operator (9.33) in the form

$$Tu = (A_1 - \zeta^2 I)^{-1} u = \mathcal{F}^{-1} h(\xi) \mathcal{F} u, \qquad u \in E, \tag{9.37}$$

where $h(\xi)$ is determined by (9.36). We shall estimate the norm of the operator T in E. Obviously, according to (9.16)

$$\begin{aligned} \|Tu\|_2 &= \|q_x^{-1} p_x Tu\| = \|q_x^{-1} p_x T p_x^{-1} q_x q_x^{-1} p_x u\| \\ &\leq \|q_x^{-1} p_x T p_x^{-1} q_x\| \, \|q_x^{-1} p_x u\| = \|q_x^{-1} p_x T p_x^{-1} q_x\| \, \|u\|_2, \end{aligned} \tag{9.38}$$

where $\| \ \|$ is the norm of the vectors in H and the norm of the operators from H to H. We shall estimate the norm of the operator

$$p_x q_x^{-1} T q_x p_x^{-1}.$$

From (9.12) and (9.37) we obtain:

$$\begin{aligned} \|q_x^{-1} p_x T p_x^{-1} q_x v\|^2 &= 2\pi \|p_\xi^{-1} q_\xi h(\xi) q_\xi^{-1} p_\xi \mathcal{F} v\|^2 \\ &\leq \|p_\xi^{-1} q_\xi h(\xi) q_\xi^{-1} p_\xi\| \, \|v\|^2. \end{aligned} \tag{9.39}$$

Obviously

$$q_\xi h(\xi) q_\xi^{-1} = (1+\xi^2) h(\xi)(1+\xi^2)^{-1} = h(\xi). \tag{9.40}$$

By virtue of (9.38) and (9.39)

$$\|Tu\|_2 \leqq \|p_\xi^{-1} h(\xi) p_\xi\| \, \|u\|_2. \tag{9.41}$$

We shall prove that the operator $p_\xi^{-1} h(\xi) p_\xi$ is bounded in H_ξ. Note that

$$hp_\xi v = h(v - \partial_\xi^2 v) = hv - \partial_\xi^2(hv) + 2(\partial_\xi h)\,\partial_\xi v + (\partial_\xi^2 h)v$$

$$= p_\xi hv + 2(\partial_\xi h)\,\partial_\xi v + (\partial_\xi^2 h)v.$$

Therefore

$$p_\xi^{-1} h p_\xi = h + 2p_\xi^{-1}(\partial_\xi h)\,\partial_\xi + p_\xi^{-1}(\partial_\xi^2 h). \tag{9.42}$$

The first and last terms in (9.42) are bounded operators in L_2, since the functions h and $\partial_\xi^2 h$ are bounded on $\mathbb{R}$, and p_ξ^{-1} is also bounded. It remained to consider the operator $p_\xi^{-1}(\partial_\xi h)\,\partial_\xi$ Obviously,

$$(\partial_\xi h)\,\partial_\xi v = \partial_\xi((\partial_\xi h)v) - (\partial_\xi^2 h)v.$$

Therefore

$$p_\xi^{-1}(\partial_\xi h)\,\partial_\xi = p_\xi^{-1}\,\partial_\xi(\partial_\xi h) - p_\xi^{-1}(\partial_\xi^2 h).$$

Since the functions $\partial_\xi h(\xi)$ and $\partial_\xi^2 h(\xi)$ are bounded on $\mathbb{R}$, it is sufficient to prove the boundedness of the operator $p_\xi^{-1}\partial_\xi$. Obviously,

$$p_\xi^{-1}\partial_\xi = \mathscr{F}(-ix(1+x^2)^{-1})\mathscr{F}^{-1},$$

and since the function $x/(1 + x^2)$ is bounded, this operator is bounded in H_ξ. Thus, the operator $p_\xi^{-1}h(\xi)p_\xi$ is bounded in H_ξ, and by virtue of (9.41) the operator $T = (A_1 - \zeta^2 I)^{-1}$ is bounded.

To derive estimate (9.34), note that

$$|h(\xi)| + |\partial_\xi h(\xi)| + |\partial_\xi^2 h(\xi)|,$$

where $h(\xi)$ is determined by (9.36), are uniformly bounded on $\mathbb{R}_\xi$ with respect to ξ and ζ when $|\mathrm{Im}\,\zeta| > \omega > 0$. As is obvious from the proof of the boundedness of the operator T, the uniform boundedness of $\|T\|$ follows hence, whence follows (9.34). By virtue of (9.32) and (9.33) the operator $(A_1 - \zeta^2 I)^{-1}$ is right inverse to $A_1 - \zeta^2 I$ on E, and since $A_1\mathcal{D}_0 \subset \mathcal{D}_0 \subset E$, it is also left inverse on $\mathcal{D}_0$. Therefore $(A_1 - \zeta^2 I)^{-1}$ is inverse to $A_1 - \zeta^2 I$ on $(A_1 - \zeta^2 I)\mathcal{D}_0 \subset \mathcal{D}_0$. It follows from the definition of A (see (9.8)) and from (9.4) and (9.5), that if $\Phi \in \mathcal{A}(R_0, \rho)$, then

$$A^{-1}\Phi(z) = \Phi(z-1) \in \mathscr{A}(R_0, \rho).$$

Therefore $A(\mathcal{D}_0) = \mathcal{D}_0$ and therefore when

$$\zeta^2 = -\rho \quad (A_1 - \zeta^2 I)\mathscr{D}_0 = \mathscr{D}_0.$$

Since $\mathcal{D}_0$ is everywhere dense in E, then $\zeta^2 = -\rho$ belongs to the resolvent set of the operator A_1. By virtue of Proposition 6.1.3 all the points ζ^2, $|\mathrm{Im}\,\zeta| > 0$ then belong to the resolvent set of the operator A_1. The lemma is proved.

Theorem 9.2. The operator A_1, which is determined on $\mathcal{D}_0$ by Formula (9.31), where A is determined by Formula (9.26), is a second-order self-determined operator (see Definition 6.2.2). At the same time $R_{02}(A_1, f) > R_0$, where R_0 is a number from condition (9.4), and $\omega_0 = 0$ in (6.1.19).

Proof. Condition 6.1.2 holds by virtue of Lemma 9.5. Condition (6.2.24) holds for A by virtue of Lemma 9.3 $\forall R$, $0 < R < R_0$, and therefore $R_{02}(A, f) > R_0$, According to Lemma 6.2.5

$$R_{02}(A, f) = R_{02}(A - \rho I, f),$$

and therefore $R_{02}(A_1, f) > R_0$. Thus all the requirements of Definition 6.2.2 hold with the above values of the parameters ω_0 and R_0, and Theorem 9.2 is proved.

Theorem 9.3. Suppose $g^+(\lambda) = (\lambda + \rho)^{-k}$, A_1 is the same operator as in Theorem 9.2, $f \in \mathcal{D}_0$. Suppose $0 < R < R_0$, $r = 2R/\pi$, and the polynomial $P_n^{\,0} = \Pi_r^{+\,n} g$. Then $\forall \varepsilon > 0 \ \exists C$: $\forall n \in \mathbb{N}$

$$\|(A_1 + \rho)^{-k} f - P_n^0(A_1) f\|_E \leq C n^{-\sigma}, \quad \sigma = 2R\sqrt{\rho}/\pi - \varepsilon. \tag{9.43}$$

Proof. The function

$$(\lambda + \rho)^{-k} \in \mathcal{A}(\mathcal{J}_\beta^+)$$

when $\beta < \sqrt{\rho}$. Using (6.2.27), where $l = 0$, $\omega > \omega_0 = 0$ we obtain (9.43).

We need a lemma in order to derive Theorem 9.1 from Theorem 9.3.

Lemma 9.6. The delta-function $\delta(x)$ is a continuous functional on E.

Proof. We shall obtain fundamental solutions of the equation $p_x u = f$, i.e. the function u_0, such that $p_x u_0 = \delta(x)$. This function is determined by the formula

$$u_0(x) = \tfrac{1}{2} e^{-|x|}. \tag{9.44}$$

We shall verify that $u_0(x)$ is a fundamental solution. Obviously

$$\partial_x e^{-|x|} = -e^{-|x|} \operatorname{sign} x, \quad \partial_x^2 e^{-|x|} = e^{-|x|} + e^{-|x|} \partial_x \operatorname{sign} x. \tag{9.45}$$

(The derivatives are understood in a generalized sense, i.e. $\partial_x u$ is a functional on C_0^∞ which is determined by the formula $(\partial_x u, \varphi) = -(u, \partial_x \varphi)$.) From (9.45) we obtain:

$$(I - \partial_x^2) e^{-|x|} = e^{-|x|} \partial_x \operatorname{sign} x = 2e^{-|x|} \delta(x) = 2\delta(x),$$

i.e. (9.44) indeed determines the fundamental solution.

Consider the functional $\delta(x)$ on E. Obviously,

$$\begin{aligned}(u, \delta(x)) &= (p_x u, p_x^{-1} \delta(x)) = (p_x u, u_0(x)) \\ &= (q_x^{-1} p_x u, q_x u_0),\end{aligned}$$

where u_0 is determined by (9.44). Since

$$(1 + x^2) e^{-|x|} \in L_2(\mathbb{R}),$$

then

$$|(u, \delta(x))| \leqq \|q_x^{-1} p_x u\| \, \|q_x u_0\| = C \|u\|_2,$$

and Lemma 9.6 is proved.

Proof of Theorem 9.1. From inequality (9.43) and from Lemma 9.6 follows:

$$|(A_1 + \rho)^{-k} f, \delta(x)) - (P_n^0(A_1), \delta(x))| \leqq C_1 n^{-\sigma}.$$

Since

$$A_1 + \rho = A, \quad P_n^0(\lambda - \rho) = P_n,$$

hence we obtain

$$|A^{-k}f(0)-P_n(A)f(0)|\leq C_1 n^{-\sigma}. \tag{9.46}$$

It follows from (9.26) that $A^k f(0) = \Phi(k)$. Consequently $P_n(A)f(0) = P_n * \Phi$, where $P_n * \Phi$ is determined by (9.3). Therefore (9.7) and, consequently, (9.6), follows from (9.46). Theorem 9.1 is proved.

Remark 9.1. The class of functions $\mathcal{M}(R_0, \rho)$, in which the extrapolation problem is considered and solved in this chapter, differs from those classes usually considered in the theory of interpolation of entire functions (see Levin [1], Gel'fond [1], Leont'ev [1]).

Remark 9.2. We can also obtain a representation of the form (9.6) when the function $\hat{\Phi}$ satisfies the following weaker condition instead of (9.4):

$$\hat{\Phi}(k)\leq R^{2k}M(2k), \tag{9.47}$$

where the sequence $M(k)$ satisfies conditions (2.2.1), (2.2.8) and (2.4.16). The polynomials P_n will differ, as will the rate of convergence in (9.6).

Remark 9.3. It follows from the results obtained in the second chapter that the conditions on the increase in $M(2k)$ can not be weakened considerably. We cannot, for example, take

$$M(2k)=((2k)!)^{1+\varepsilon},\ \varepsilon>0.$$

Moreover, we cannot take in (9.47) the sequence $M(k)$, which satisfies (2.2.1) and (2.2.8), but does not satisy (2.4.16).

Indeed, in this case $C(M(k))$ is a non-quasianalytic class of functions. Consider the operator B from Theorem 2.3.1, and also the function $f \in C(M(k))$ and the function χ from the proof of this theorem. We will assume

$$\Phi(z) = (B^z f, \chi).$$

This function, as follows from the proof of Theorem 2.3.1, equals zero when $z = 0, 1, \dots$, and differs from zero when $z = -1$. At the same time, by virtue of Lemma 2.4.1 this function satisfies (9.47).

§10. Applications to differential equations.

We shall first prove theorems which prove that the results of §8 and §9 are applicable to the differential equations considered in §6. We shall then give an example of the equation $F(u) = h$, for which we able to indicate the functionals g explicitly, such that all the conditions necessary to apply the results of §9 hold.

Theorem 10.1. Suppose F is a differential operator that is determined by Formula (6.5) on $C^\infty(T^m)$. Suppose all the conditions of sects. 1) and 2), where $p=2$ and 3) of Theorem 6.2 hold (in particular, the functions G_j are entire). Then the following statements hold:

1) the complex powers of the operator F are determined on $C^\infty(T^m)$ whilst for any

$$v \in \operatorname{Re} C^\infty(T^m),$$

$F^\lambda v$ is a function that is analytic on $\mathbb{C}$ with values in $W_2^l(T^m)$ for any l;

2) if g is a functional that is determined and analytical on $W_2^l(T^m)$ for some $l > 0$ and $h \in C^\infty(T^m)$, the function $\Phi(\lambda)$, which is determined by Formula (8.14), is an entire function that satisfies (9.1), where $\hat{\Phi}(s)$ satisfies estimate (9.5).

Proof. According to Theorem 6.2 the operator F permits an analytic linearisation on

$$\operatorname{Re} W^s_p,\ s-m/p>2-m/p_0,\ p_0>m.$$

We shall take $p = 2$, $s > 2 + m/2$. At the same time the operator

$$\mathscr{E}: W_2^{2+m/2+s} \to W_2^{2+m/2+s}$$

is analtyic in the whole complex space. We will assume

$$E_s = W_2^{m/2+2+s}(T^m),\ V_s = \operatorname{Re} E_s.$$

The operators $\mathcal{E}: E_s \to E_s$ are analytic for integer $s \in \mathbb{Z}_+$. Note, now, that the symbol $a(\xi)$ of the operator A, determined by formula (6.18), is real, and therefore the operator A^λ is determined

$$A^\lambda = \mathscr{F}^{-1} a(\xi)^\lambda \mathscr{F}, \tag{10.1}$$

where $\mathscr{F}$ is a Fourier transform. The operator A satisfies conditions 8.1 and 8.2, as is immediately obvious from Formula (10.1). Therefore sect. 1) of Theorem 10.1 immediately follows from Theorem 8.2.

We shall proceed to prove sect. 2). If g is analytic on $W_2^l(T^m)$ then g is also analytic on $W_2^{l+1}(T^m)$. At the same time the set which is bounded in $W_2^{l+1}(T^m)$ is compact in $W_2^l(T^m)$ and therefore g is bounded on this set. Thus g is a functional that is bounded and analytic on all E_s for fairly large s. Note that according to Theorem 8.2 estimate (8.9) holds. Condition (8.8), where $M>1$, holds, and estimate (8.10) holds. Using Theorem 8.3, we obtain estimate (9.1) from (8.15), and estimate (9.5) from (8.10). The theorem is completely proved.

Note that we are not generally able to construct an example of the functional g, such that estimate (9.4) holds. The existence of these functionals follows, for example, from the results of §7, since eigenfunctionals obviously satisfy this condition. A functional of the form

$$g(u)=(\varphi, \mathscr{E}^{-1}(u)), \tag{10.2}$$

where φ is a functional that is linear and continuous on $W_2^l(T^m)$, is an example of a functional that is not eigen. Obviously,

$$\Phi(z)=g(F^z h)=(\varphi, A^z \mathscr{E}^{-1}(h)). \tag{10.3}$$

If the function $\mathscr{E}^{-1}(h)$ is analytic, then, as follows from the results of Chapter 2, estimate (9.4) holds for $\Phi(z)$. Thus, passing from the functions $\Phi(z)$ of the form λ^z to the functions $\Phi(z)$, which satisfy condition (9.4), enabled us to widen the class of functionals for which $g(F^{-1}h)$ is expressed in terms of $g(F^k h)$, $k \in \mathbb{Z}_+$. In the linear case (when $h(x)$ is analytic) this extension enables us to pass from eigen functionals to all the continuous linear functionals (see Chapters 4 and 6). Unfortunately, it is rare that we are able to indicate easily verifiable conditions on the functional g which guarantee the satisfaction of estimate (9.4).

We shall now give an example of equations for which we are able to indicate the functions Φ, such that condition (9.4) holds.

Consider the operator F, specified by Formula (4.4), where G is an entire function, and suppose the operator A is determined by Formula (4.30). It is assumed that the conditions of Theorem 4.1 hold for $W_2^l(T^m)$, $l > m/2$.

We shall use

$$S_\kappa W_2^l(T^m) = S_\kappa W,\ \kappa \geqq 0,$$

to denote the space of functions from $W_2^{\,l}(T^m)$, $l > m/2$, the Fourier expansion of which has the form

$$u(x) = \sum_{|k| \geqq \kappa}' C_k e^{ikx}. \qquad (10.4)$$

The sum's prime indicates that only the coefficients C_k in which all the components k_j in the index k are nonnegative are nonzero. The spaces of these (and more general) functions have been considered by Vishik and Fursikov [1]. It follows from their results, in particular, that the solution of the equation $F(u) = h$, where $h \in S_\varkappa W_2^{\,l}$, is unique in $S_\varkappa W_2^{\,l}$ when $\varkappa > 0$. The following proposition is therefore obvious.

Proposition 10.1. The operator

$$B = F \circ A^{-1}$$

is one-to-one on $S_\varkappa W_2^{\,l}$ when $\varkappa > 0$. The following function is defined on $\mathcal{E}(S_\varkappa C^\infty(T^m))$:

$$\Phi(z) = \int e^{ik\cdot x} F^z(v)\, dx,\ v \in \mathcal{E}(S_\kappa C^\infty(T^m)), \qquad (10.5)$$

where the integral will be taken over

$$T^m,\ \kappa > 0,\ k = (k_1, \ldots, k_m) \in \mathbb{Z}^m.$$

Theorem 10.2. If the operator $A = F'(0)$ is self-adjoint, the number $\sigma > 0$ exists, such that the function $\Phi(z)$, determined by

Formula (10.5), satisfies the inequality

$$|\Phi(\alpha+i\beta)|\leqq C\,e^{\sigma\alpha}, \qquad (\alpha>0) \tag{10.6}$$

See Babin [7], Theorem 6.1, for the proof.

§ 11. Another application of the extrapolation theorem.

As is obvious from the above, of the conditions of Theorem 9.1 that are imposed on $\Phi(\lambda)$, determined by (8.14), the most limiting is condition (9.4). Note that if $F(0) = 0$ and $h = 0$, then $\Phi(\lambda) \equiv 0$, and (9.4) trivially holds. However, if $h \neq 0$ and is close to zero, the increase in $\hat{\Phi}(s)$ (see (9.1)) as $s \to +\infty$ can be very rapid. We shall therefore assume when $\mu \in \mathbb{Z}_+$

$$\Phi_\mu(z)=\frac{d^\mu}{dt^\mu}\,g(F^z(th))\big|_{t=0}. \tag{11.1}$$

At the same time we assume that the function $g(F^z(th))$ is bounded and analytic with respect to z for fairly small t. As usual, the powers of F are determined by the formula

$$F^z(th)=\mathscr{E}\circ A^z\circ\mathscr{E}^{-1}(th). \tag{11.2}$$

It is assumed that the linear operator A satisfies conditions 8.1 and 8.2.

If z belongs to some bounded set $\omega\subset\mathbb{C}$, and the operator

$$B=F\circ A^{-1}$$

is quasi-interchangeable with $S = (AM^{-1})^{1/\nu}$ at zero, then by virtue of Theorem 3.3, the powers (11.2) are determined if for fairly small t. Consequently, by virtue of Theorem 4.1, the powers of the differential operator F, which is determined by (4.4) on the function th, where $h \in C^\infty(T^m)$, t is fairly small, are determined using Formula (11.2). Therefore the function (11.1) is also determined if the functional g is analytic in the neighbourhood of zero.

Let us now consider Formulae (11.1) and (11.2) in more detail. We will assume that the operators $\mathcal{E}$ and $\mathcal{E}^{-1}$, and also the functional g, expand in series that converge in the neighbourhood of zero in each of E_k, $k \in \mathbb{Z}_+$ (it is assumed that the radius of the convergence in E_k is different for different k):

$$\mathcal{E}(u)=u+\sum_{j=2}^{\infty}\mathcal{E}_j(u),\quad \mathcal{E}^{-1}(u)=u+\sum_{j=2}^{\infty}\mathcal{E}_j^{(-1)}(u),$$
$$g(u)=\sum_{j=1}^{\infty}g_j(u), \tag{11.3}$$

$\mathcal{E}_j$, $\mathcal{E}_j^{(-1)}$, g_j are j-linear mappings. The convergence of series for $\mathcal{E}$ and $\mathcal{E}^{-1}$ is assumed in the weakened sense, namely like that of the operators from E_{k-1} and E_k. The convergence of g is from E_l to $\mathbb{C}$ for some $l > 0$.

Theorem 11.1. Suppose $\mathcal{E}$, $\mathcal{E}^{-1}$ and g satisfy (11.3), the operator A satisfies conditions 8.1 and 8.2, and $h \in E_\infty$. Then the function $\Phi(z)$, which is determined by (11.1), is holomorphic for all $z \in \mathbb{C}$ and we can rewrite it in the form

$$\Phi_\mu(z)=\frac{d\mu}{dt^\mu}\bigg|_{t=0} g_{(\mu)}(\mathcal{E}_{(\mu)}(A^z\mathcal{E}_{(\mu)}^{(-1)}(th))), \tag{11.4}$$

where

$$\mathcal{E}_{(\mu)},\quad \mathcal{E}_{(\mu)}^{(-1)}$$

and $g(\mu)$ are partial sums of series (11.3) from the first μ terms.

Proof. The terms from $j > \mu$ with the substitution (11.3) into (11.1) give zero, which proves (11.4).

Theorem 11.2. Suppose the operator

$$B = F \circ A^{-1}$$

is quasi-interchangeable with S, $S = (M^{-1}A)^{1/\nu}$, $M > 1$ at zero. Suppose the linear operator A satisfies conditions 8.1 and 8.2, and also condition (5.9):

$$\|A^{i\beta}v\|_s \leqq C\|v\|_s, \qquad \forall \beta \in \mathbb{R}. \tag{11.6}$$

Suppose the functional g is analytic in the neighbourhood of zero in E_l, $l > 0$. Then

$$|\Phi_\mu(\Theta + i\tau)| < C_\mu M^k \|h\|^\mu_{\nu k + l} \text{ when } \Theta < k, \tag{11.7}$$

where C_μ does not depend on k. If $\Theta \leq 0$, then

$$|\Phi_\mu(\theta + i\tau)| \leqq C_\nu M^{\nu\theta} \|h\|^\mu_l. \tag{11.8}$$

Proof. According to Theorem 3.3 the operator F is represented by the formula

$$F = \mathscr{E} \circ A \circ \mathscr{E}^{-1},$$

where $\mathcal{E}$ and $\mathcal{E}^{-1}$ are operators that are quasi-interchangeable with S in the neighbourhood of zero. Consequently, the operators $\mathcal{E}_j$ and $\mathcal{E}_j^{(-1)}$ are quasi-interchangeable with S and by virtue of (3.5)

$$\|S^k \mathscr{E}_j S^{-k} u\| \leqq b_j \|u\|^j_0,$$

where b_j does not depend on k. Assuming $S^{-k}u = v$ and bearing in mind that

$$\|v\|_k = \|S^k v\|_0,$$

we obtain:

$$\|\mathscr{E}_j(v)\|_k \leq b_j \|v\|_k^j. \tag{11.9}$$

Similarly

$$\|\mathscr{E}_j^{(-1)}(v)\|_k \leq b_j' \|v\|_k^j. \tag{11.10}$$

Therefore, by virtue of (11.6), and bearing in mind the boundedness of S^s when $s \leq 0$, we obtain:

$$\|A^{\alpha+i\beta}\mathscr{E}_j^{(-1)}(tv)\|_l \leq C|t|^j \|v\|_{k+l}^j M^\alpha, \qquad \forall \alpha \leq k \tag{11.11}$$

when $1 < j < \mu$, C does not depend on j or k. Bearing in mind that it is sufficient to leave in Formula (11.4) the terms having an order of homoegeneity μ with respect to t, from (11.11) we obtain the estimate (11.7), using the boundedness of g_j and $\mathcal{E}_j$ in E_l . We obtain estimate (11.8), noting that

$$A^s = (MS^\nu)^s = M^s S^{s\nu},$$

and the operator $S^{-\gamma}$ bounded when $\gamma > 0$. At the same time we use the equation $\mathcal{E}(0)=0$. The theorem is proved.

Note that by virtue of Theorem 4.1 the operator F, which is determined by Formula (4.4), satisfies the conditions of Theorem 11.2, where

$$E_s = W_2^{l+s},\ l > m/2.$$

If the operator $A = F'(0)$ has a real symbol, i.e. if it is self-adjoint, then A satisfies conditions 8.1, 8.2 and (11.6).

We shall now give an example of the functions h, such that it follows from estimate (11.7) that the function $\Phi = \Phi_\mu$ satisfies estimate (9.4).

Theorem 11.3. Suppose $h(x) \in C^\infty(T^m)$ and is written in the form of the series

$$h(x) = \sum_{k \in Z^m} h_k e^{ik \cdot x}, \tag{11.12}$$

where

$$|h_k| \leqq C e^{-\psi(|k|)}, \tag{11.13}$$

and the function $\Psi(|k|)$ satisfies the condition

$$\psi(|k|) \geqq \rho(1+|k|)^\delta, \qquad \rho > 0, \quad \delta > 0. \tag{11.14}$$

Suppose P is a ν-th order differential operator with constant coefficients. Then when $s > 0$, $p \in [1. \infty[$,

$$\|P^j h\|_{s,p}^2 \leqq C_1 R_1^j \Gamma\left(\frac{\nu_j}{\delta}\right). \tag{11.15}$$

where C and R_1 do not depend on j and Γ is a gamma-function.

Proof. According to (11.12)

$$\|h\|_{l,p} \leqq C_1 \sum_k |h_k|(1+|k|)^l.$$

Bearing in mind that the following estimate holds for the symbol $P(\xi)$ of the operator P:

$$|P(\xi)| \leqq C_2(1+|\xi|)^{\nu},$$

we obtain the estimate

$$\begin{aligned} \|P^j h\|_{s,p} &\leqq C_3 C_2^j \sum_k e^{-\psi(|k|)}(1+|k|)^{s+\nu j} \\ &\leqq C_3 C_2^j \sum_k e^{-\rho(1+|k|)^{\delta}}(1+|k|)^{s+\nu j}. \end{aligned} \tag{11.16}$$

We estimate the sum on the right-hand side of (11.16) by the integral and, passing to the spherical coordinates, we will obtain:

$$\begin{aligned} \|P^j h\|_{s,p} &\leqq C_4 C_2^j \int_0^{\infty} e^{-\rho(1+t)^{\delta}}(1+t)^{s+\nu j+m-1}\,dt \\ &\leqq C_5 \rho^{-\nu j/\delta} C_3^j \int_0^{\infty} e^{-\tau}\tau^{(s+\nu j+m)/\delta-1}\,d\tau \\ &= C_5 R_2^j \Gamma\left(\frac{\nu j}{\delta}+\frac{m+s}{\delta}\right) \\ &\leqq C_6 R_3^j \Gamma\left(\frac{\nu j}{\delta}\right). \end{aligned}$$

Corollary 11.1. If in the condition (11.14) the number $\delta = \mu\nu/2$ and the operator F is determined by Formula (4.4), where $A = F'(0)$ is an operator with a real symbol, $F_1 = qF$, and the number q is fairly great, the function Φ_ν, which is determined by Formula (11.1), where $F=F_1$, belongs to the class $\mathcal{M}(R_0,\rho)$, where $R_0 > 0$, $\rho = M > 1$.

Proof. Estimate (11.7) holds by virtue of Theorem 11.2. Since the operator A has constant coefficients, Theorem 11.3 is applicable. Bearing in mind that

$$\|v\|_{E_{k+1}} = \|v\|_{l_0+(l+k)\nu,2},$$

where $l_0 > m/2$, from inequality (11.7) we obtain the following estimate:

$$|\Phi_\mu(s+i\tau)| \leqq C\|A^k h\|^{\mu}_{l_0+\nu l+\nu k} \qquad (s \leqq k). \tag{11.17}$$

From estimate (11.15) we obtain:

$$|\Phi_\mu(s+i\tau)| \leqq CR_2^k\left(\Gamma\left(\frac{\nu k}{\delta}\right)\right)^\mu \qquad \text{when } s \leqslant k.$$

Since $\delta = \mu\nu/2$, hence we obtain:

$$\Phi_\mu(s+i\tau) \leqq CR_2^k\Gamma\left(\frac{2k}{\mu}\right)^\mu \leqq CR_3^k(2k)!$$

when $s \leq k$, and inequality (9.4) holds. Inequality (9.5) follows from (11.8).

The conditions of Theorem 9.4 thus hold and Formula (9.6) enables us to express

$$\frac{d^\mu}{dt^\mu} g(F^{-1}(th))\big|_{t=0},$$

where F is the operator of (4.4), in terms of

$$\frac{d^\mu}{dt^\mu} g(F^j(th))\big|_{t=0}, \qquad j=0,1,2,\ldots,$$

if the Fourier coefficients of the function h decrease very rapidly.

REFERENCES

AKHIEZER N. I.

1 Weighted approximation of continuous functions on a whole numerical axis. - Uspekhi Matem. Nauk., vol.11, No.4, pp.3-43, 1956.

2 Lectures on the theory of approximation. - Moscow, Nauka, 1965.

ARNOL'D V. I.

1 Additional chapters of the theory of ordinary differential equations. - Moscow: Nauka, 1978.

BABIN A. V.

1 A formula expressing the solution of a differential equation with analytic coefficients on a manifold without boundary in terms of data of the problem. - Matem. Sb., vol.101(143), No.4, pp.610-638. (Engl. transl. - Math USSR Sb., 30, No.4, 539-565, 1976)

2 An expression for A^{-1} in terms of iterates of an unbounded self-adjoint operator A on analytic vectors. - Funkts. Analiz., vol.11, no.4, pp.3-5, 1977.

3 A formula expressing the solution of a differentiaal equation with analytic coefficients on a manifold without boundary in terms of local data of the problem. - Uspekhi Matem. Nauk, vol.33, No.1, pp.203-204, 1978.

4 An expression for the solution of a differential equation in terms of an iteration of differential operators. - Matem. Sb., vol.105(147), No.4, pp.467-484, 1978. (Engl. transl. - Math. USSR Sb.34, No.4, 411-424, 1978)

5 An expression for A^{-1} in terms of iterates of the operator A, which acts in a Banach space. - Funkts. Analiz, vol.12, no.4, pp.77-78, 1978.

6 Expressing the solution of the equation $Au=f$ in terms of iterations of an unbounded operator A and the weighted approximation of functions. - Uspekhi Matem. Nauk., vol.34, no.3, p.189, 1979.

7 Fractional powers of a nonlinear analytic differential operator. - Matem. Sb., No.1, pp.42-45, 1979. (Engl. transl. - Math. USSR Sb.37, No.1, 9-38, 1980)

8 Analytic linearisation and complex powers of a nonlinear differential operator. - Funkts. Analiz, vol.14, no.3, pp.61-62, 1980.

9 Nonlinear eigen functionals. - Uspekhi Matem. Nauk, vol.35, No.4, pp.145-146, 1980.

10 Representation of solutions of differential equations in polynomial form. - Uspekhi matem. Nauk, vol.38, No.2, pp.228-229, 1983.

11 An iterative method applicable directly to differential equations. - Zh. Vychisl. Matem. i Matem. Fiz., No.4, pp.771-784, 1983. (Engl. transl. USSR Comput. Math. Math. Phys. 23, No.4, 1-9. 1983)

12 Polynomial solvability of differential equations with coefficients from classes of infinitely differentiable functions. - Matem. Zametki, vol.34, No.2, pp.249-260, 1983. (English translation - Math. Notes 34, 608-614, 1984)

13 Solution of the Cauchy problem by means of weighting approximations of exponents by polynomials. - Funkts. Analiz, vol.17, no.4, pp.75-76, 1983. (Engl. transl. Funct. Anal. Appl., 17, 305-307, 1983)

14 Construction and investigation of solutions of differential equations using the methods of the theory of approximation of functions. - In: Conference on Nonlinear Problems of Mathematical Physics, IPMM AN UkrSSR, Donetsk, p.9, 1983.

15 Construction and investigation of solutions of differential equations by methods in the theory of approximation of functions. - Matem. Sb., vol.123, No.2, pp.147-173, 1984. (Engl. transl. - Math. USSR Sb.51, 141-167, 1985)

16 On the membership of solutions of differential equations in Nikol'skij spaces - Doklady AN SSSR, vol.289, No.6, 1289-1293 (1987). (Engl. transl. - Soviet Math. Dokl., vol.34 (1987), No.1, 210-214.

BAKHVALOV N. A.

1 Numerical methods. - Moscow: Nauka, 1975.

BAOUENDY M. S. and GOULAOUIC C.

1 Regularite analitique et iteres d'operateurs elliptiques degeneres. - J. of Funct. Anal., vol.9, pp.208-248, 1972.

BAOUENDY M. S. and METIVIER G.

1 Analytic vectors of hypoelliptic operators of principal type. - Amer. J. Math., vol.104, No.2, pp.287-319, 1982.

BERNSHTEIN S. N.

1 The extremal properties of polynomials. - Moscow-Leningrad: ONTI, 1937.

BERS L., JOHN F. and SCHECHTER M.

1 Partial differential equations, New York, Interscience Publ., 1969.

BESOV O. V., IL'IN V. P. and NIKOL'SKII S. M.

1 Integral representations of functions and imbedding theorems. - Moscow: Nauka, 1975.

BONY J. M. and SCHAPIRA.

1 The existence and extension of solutions of partial differential equations. - Sb. Perevodov Matematika, vol.17:1, pp.162-171, 1973. (Russian translation)

2 Solutions-hyperfunctions of Cauchy's problem. - Sb. Perev. Matematika, vol.17:2, pp.98-107, 1973. (Russian translation)

3 Existence et prolongement des solutions holomorphes des equations aux derivees partielles. - Invent. Math., Vol.17, no.2, pp.95-105, 1972.

BOLLEY P., CAMUS J., MATTERA C.

1 Analyticite microlocale et iteres d'operateurs, sem. Goulaonic-Schwartz, 1978-1979, exposé No.XIII.

CHERNYAVSKII A. G.

1 Operator quasi-analyticity for functions of several variables. - DANSSSR, vol.244, No.2, pp.296-299, 1979.

COHN J. J. and NIRENBERG L,

1 Non-coercive boundary value problems. - Comm. Pure Appl. math., vol.18, No.3, pp.443-492, 1965.

DUNFORD N. and SCHWARTZ J. T.

1 Linear operators. Part 1. General theory. - Interscience Publishers, 1958.

2 Linear operators, Part 2. Spectral theory. - Interscience Publishers, 1963.

GEL'FAND I. M. and SHILOV G. E.

1 Generalised functions, no.1-3, Fizmatgiz, 1958.

GEL'FOND A. O.

1 The calculus of finite differences. - Moscow, Nauka, 1967.

GODUNOV S. K. and RYABEN'KII V. S.

1 Difference schemes. - Moscow: Nauka, 1977.

HADAMARD J.

1 Le problème de Cauchy, et les écuations aux derivées partielles linéares hperboliques. - Paris, Hermann, 1932.

HARTMAN F.

1 Ordinary differential equations. - John Wiley & Sons, 1964.

HORMANDER L.

1 Linear partial differential operators. - Springer-Verlag, 1963.

2 Uniqueness theorems and wave front sets for solutions of linear diferential equations with analytic coefficients. - Comm. Pure Appl. Math., vol.24, pp.671-704, 1971.

KATO T.

1 Perturbation theory for linear operators. - Springer-Verlag, 1966.

KHRYPTUN V. G.

1 Classes of functions that are quasi-analytic with reference to a second-order linear hyperbolic operator. - Izv. AN SSSR, Ser. Matem., vol.34, No.5, pp.1127-1141, 1970.

KOTAKE T. and NARASIMHAN M. S.

1 Fractional powers of a linear elliptic operaotr. - Bull. Soc. Math. France, vol.90, pp.449-471, 1962.

KREIN S. G.

1 Linear differential equations in a Banach space. - Moscow: Nauka, 1967.

LADYZHENSKAYA O. A. and URAL'TSEVA N. I.

1 Linear and quasilinear equations of elliptic type. - Moscow: Nauka, 1973.

LEON'TEV A. F.

1 Entire functions. Series of exponential functions. - Moscow: Nauka, 1983.

LEVIN B. YA.

1 Distribution of the roots of entire functions. - Moscow: Gostekhizdat, 1956.

LIONS J. L. and MAGENES E.

1 Problemes aux limites non-homogenes, Vol.1-3, Paris, Dunod, 1968.

LYUBICH YU. I. and TKACHENKO V. A.

1 The criterion of quasi-analyticity for abstract operators. - DAN, vol.190, No.4, pp.772-774, 1970.

MALLIAVIN P.

1 Sur quelques procedes d'extrapolation. - Acta Math., vol.93, No.3-4, pp.179-255, 1955.

MANDELBROIJT S.

1 Séries adhérentes. Régularisation des suites. Applications. - Paris, 1952.

MARCHUK G. I.

1 Methods of computational mathematics. - Moscow: Nauka, 1980.

MARKUSHEVICH A. I.

1 Theory of analytic functions, vol.1,2. - Moscow: Nauka, 1968.

MERGELYAN S. N.

1 Weighting approximations using polynomials. - Uspekhi Matem. Nauk., vol.11, No.5, pp.107-152, 1956.

2 Approximations of functions of a complex variable. - In: Forty Years of Mathematics in the USSR. Moscow: Fizmatgiz, vol.1, pp.383-397, 1959.

MOSER J.

1 A rapidly converging iteration method and nonlinear differential equations. - Annali della Sc. Norm. Super. de Pisa ser. III, v.20, No.2, p.265-315, No.3, p.499-535, 1966).

NIKOLENKO N. V.

1 The complete integrability of Schrodinger's nonlinear equation. - DAN SSSR, vol.227, No.4, pp.235-238, 1976.

2 The reducibility of nonlinear evolution equations to linear form. - DAN SSSR, vol.268, No.3, pp.545-547, 1982.

NIKOL'SKII S. M.

1 Approximation of the functions of several variables to the theory of imbedding. - Moscow: Nauka, 1977.

OLEINIK O. A.

1 The smoothness of solutions of elliptic and parabolic equations. - DAN SSSR, vol.163, No.3, pp.577-580, 1965.

OLEINIK O. A. and RADKEVICH E. V.

1 Second-order equations with a nonnegative characteristic form. - Itogi Nauki, Matematicheskii Analiz, 1969, Moscow: 1971.

PALAIS R.

1 Seminar on the Atiyah-Singer index theorem - Princeton University Press, 1965.

SAMARSKII A. A.

1 Theory of difference schemes. - Moscow: Nauka, 1977.

SAMARSKII A. A. and NIKOLAEV E. S.

1 Methods of solving net-point equations. - Moscow: Nauka, 1978.

SCHRODER E.

1 Uber iterierte Funktionen. - Math. Ann., vol.3, pp.296-322, 1871.

SHUBIN M. A.

1 Pseudo-differential operators and spectral theory. - Moscow: Nauka, 1978.

SEDENKO V. I.

1 The normal form of nonlinear partial differential equations on a real axis. - Matem. Sb., vol.105(147), pp.121-127, 1978.

SIEGEL K. L.

1 Vorlesungen über Himmelsmechanik. Berlin, Springer, 1956.

SKRYPNIK I. V.

1 The solvability and properties of solutions of nonlinear elliptic equations. - Itogi Nauki i Tekhniki, Seriya "Sovremennie Problemy Matematiki", vol.9, Moscow: VINITI, 1976.

SMAJDOR W.

1 Analytic solutions of Schroder's equation. - Publ. Math. Debrecen, vol.26, No.3-4, pp.149-153, 1979.

VISHIK M. I. and FURSIKOV A. V.

1 Some questions of the theory of nonlinear elliptic and parabolic equations. - Matem. Sb., vol.94(136), pp.300-334, 1974.

VLADIMIROV V. S.

1 Generalised functions in mathematical physics. - Moscow: Nauka, 1984.

WARE B.

1 Infinite-dimensional versions of two theorems of Carl Siegel. - Bull. Amer. Math. Soc., vol.82, No.4, pp.613-615, 1976.

YOSIDA K.

1 Functional analysis. - Berlin, Springer, 1965.

NOTATIONS

INDEX